Nano Drug Delivery for Cancer Therapy

Firdos Alam Khan
Editor

Nano Drug Delivery for Cancer Therapy

Principles and Practices

Springer

Editor
Firdos Alam Khan
Stem Cell Research
Imam Abdulrahman Bin Faisal University
Dammam, Saudi Arabia

ISBN 978-981-99-6939-5 ISBN 978-981-99-6940-1 (eBook)
https://doi.org/10.1007/978-981-99-6940-1

This Springer imprint is published by the registered company Springer Nature Singapore Pte Ltd.
The registered company address is: 152 Beach Road, #21-01/04 Gateway East, Singapore 189721, Singapore

Preface

This book discusses the various modes and methods of nano-based drug delivery in different types of cancers such as colon, breast, cervical, ovarian, and lung cancer. The book is divided into 10 chapters. Topics such as the significance of nano drug delivery in cancer therapy, the application of nanoparticles in overcoming drug resistance, targeted therapy, and immunotherapy; synthesis of different types, shapes, and sizes of nanocarriers using synthetic and biological approaches; synthesis of nanocarriers loaded with anticancer drugs by using functionalization or conjugations, and encapsulation methods; testing of nanocarriers as delivery vehicles for chemotherapeutic agents in different types of cancers by using in vitro methods; testing of nanocarriers loaded with chemotherapeutic agents in the animal models of cancers—preclinical trials for efficacy and safety profiling of nanocarriers; testing of nanocarriers loaded with chemotherapeutic agents in cancer patients-clinical trials like phase 1 to phase 4; nanocarrier-based products in the market, FDA approval, commercialization of nanocarriers, and global market; limitations of nanocarriers such as cell and tissue toxicity, genotoxicity, and scale-up of nanomaterials; health and environmental hazards associated with nanoformulation synthesis, respiratory diseases, government regulations, and ethical issues; and finally, future trends and innovation in nano drug delivery for cancer therapy, application of siRNA (nanoparticle-based RNA) therapy, ultrasound-linked nano-cancer therapeutics, and application of exosomes-based cancer therapy are included.

First, I am grateful to the Almighty Allah, who gave me the strength to complete this edition of the book on the stipulated time. I am thankful to many people, especially to Dr. Bhavik Sawhney, Editor-Biomedicine, Springer Nature, who supported me to complete the task. I am thankful to all Springer Nature production team members for their support and cooperation. I am grateful to all the authors and especially to all corresponding authors for their immense contributions and timely completion of the work. I want to thank the entire management team of the Institute for Research and Medical Consultations (IRMC), Imam Abdulrahman Bin Faisal University, Dammam, Saudi Arabia, for their support, especially to Professor Ebtesam Al-Suhaimi, Dean, IRMC, Imam Abdulrahman Bin Faisal University, Dammam, Saudi Arabia, for her constant encouragement. I am thankful to all my teachers and mentors, especially Professor Nishikant Subhedar and Late Professor Obaid Siddiqi, FRS, for their immense contributions to shaping my research career. I am also thankful to all my friends, well-wishers, and colleagues for their support and cooperation.

I am grateful to my entire family members, especially to my father late Nayeemuddin Khan and mother Sarwari Begum, my brothers, my sisters, my wife Samina Khan, my sons (Zuhayr Ahmad Khan, Zaid Ahmad Khan, and Zahid Ahmad Khan) and my daughter (Azraa Khan), my father-in-law (Abdul Qayyum Siddiqi), and late mother-in-law (Uzma Siddiqi). All of them, in their own ways, supported me.

Enjoy reading!

Dammam, Saudi Arabia Firdos Alam Khan

About the Book

This book discusses the various modes and methods of nano-based drug delivery in different types of cancers such as colon, breast, cervical, ovarian, and lung cancer. The book is divided into 10 chapters. Topics such as the significance of nano drug delivery in cancer therapy, the application of nanoparticles in overcoming drug resistance, targeted therapy, and immunotherapy; synthesis of different types, shapes, and sizes of nanocarriers using synthetic and biological approaches; synthesis of nanocarriers loaded with anticancer drugs by using functionalization or conjugations, and encapsulation methods; testing of nanocarriers as delivery vehicles for chemotherapeutic agents in different types of cancers by using in vitro methods; testing of nanocarriers loaded with chemotherapeutic agents in the animal models of cancers—preclinical trials for efficacy and safety profiling of nanocarriers; testing of nanocarriers loaded with chemotherapeutic agents in cancer patients-clinical trials like phase 1 to phase 4; nanocarrier-based products in the market, FDA approval, commercialization of nanocarriers, and global market; limitations of nanocarriers such as cell and tissue toxicity, genotoxicity, and scale-up of nanomaterials; health and environmental hazards associated with nanoformulation synthesis, respiratory diseases, government regulations, and ethical issues; and finally, future trends and innovation in nano drug delivery for cancer therapy, application of siRNA (nanoparticle-based RNA) therapy, ultrasound-linked nano-cancer therapeutics, and application of exosomes-based cancer therapy are included.

Contents

Editor and Contributors

About the Editor

Firdos Alam Khan is the chairman of the Department of Stem Cell Biology, Institute for Research and Medical Consultations (IRMC), Imam Abdulrahman Bin Faisal University, Dammam, Saudi Arabia. In addition to his research work, he has been teaching (cell physiology and biochemistry and stem cell biology and regenerative medicine) courses to MSc students. Prof. Khan has done his Ph.D. degree in zoology with a neuroscience specialization from Nagpur University, India. Over the past 26 years, Prof. Khan has been involved in teaching various courses such as cell biology, pharmacology, business of biotechnology, biomedicine, cell and tissue engineering, and bioethics-IPR to undergraduate and postgraduate students. Prof. Khan worked with Manipal Academy of Higher Education, Dubai Campus, United Arab Emirates, as a professor and chairperson, School of Life Sciences. Prof. Khan did his first postdoctoral fellowship from the National Centre for Biological Sciences, Bangalore, India, and a second postdoctoral fellowship from Massachusetts Institute of Technology (MIT), Cambridge, USA. Professor Khan has also worked with Reliance Life Sciences as a Research Scientist in the Stem Cell Research Laboratory. Prof. Khan has been granted eight US patents and published five US patents. Prof. Khan has published more than 119 research papers in peer-reviewed journals.

Contributors

Kanwal Abbasi Pakistan Council of Scientific and Industrial Research Karachi (PCSIR), Karachi, Pakistan

Priyanca Ahlawat Translational Health Science and Technology Institute, Faridabad, Haryana, India

Zohreh Bahrami Faculty of New Sciences and Technologies, Department of Nanotechnology, Semnan University, Semnan, Iran

Saeeda Bano Pakistan Council of Scientific and Industrial Research Karachi (PCSIR), Karachi, Pakistan

Khalid Umar Fakhri Department of Biosciences, Jamia Millia Islamia, New Delhi, India

Mishaal Fareed Department of Pharmacy, Forman Christian College University, Lahore, Pakistan

Mohaddeseh Fatemi Faculty of New Sciences and Technologies, Department of Nanotechnology, Semnan University, Semnan, Iran

Piyush Kumar Gupta Department of Life Sciences, Sharda School of Basic Sciences and Research, Sharda University, Greater Noida, India
Department of Biotechnology, Graphic Era Deemed to be University, Dehradun, Uttarakhand, India
Centre for Development of Biomaterials, Sharda University, Greater Noida, India

Dilawar Hassan Tecnologico de Monterrey, School of Engineering and Sciences, Atizapan de Zaragoza, Estado de Mexico, Mexico

Zaib Un Nisa Munawar Hussain School of Life Sciences, Manipal Academy of Higher Education, Dubai Campus, Dubai, UAE

Laiba Iftikhar Department of Pharmacy, Forman Christian College University, Lahore, Pakistan

Samina Iqbal Pakistan Council of Scientific and Industrial Research Karachi (PCSIR), Karachi, Pakistan

B. Rabindran Jermy Department of Nanomedicine Research, Institute for Research and Medical Consultations, Imam Abdulrahman Bin Faisal University, Dammam, Saudi Arabia

Firdos Alam Khan Department of Stem Cell Research, Institute for Research and Medical Consultations (IRMC), Imam Abdulrahman Bin Faisal University, Dammam, Saudi Arabia

Naeem Ullah Khan Department of Pharmacy, Forman Christian College University, Lahore, Pakistan

Akhilesh Kumar Division of Medicine, ICAR Indian Veterinary Research Institute, Izatnagar, Bareilly, Uttar Pradesh, India

Rohit Kumar Department of Life Sciences, Sharda School of Basic Sciences and Research, Sharda University, Greater Noida, India
Centre for Development of Biomaterials, Sharda University, Greater Noida, India

Dhvani H. Kuntawala Department of Medical Sciences, Institute of Biomedicine (iBiMED), University of Aveiro, Aveiro, Portugal

Dora I. Medina Tecnologico de Monterrey, Institute of Advanced Materials for Sustainable Manufacturing, Monterrey, Nuevo Leon, Mexico

Vijaya Ravinayagam Deanship of Scientific Research & Department of Nanomedicine Research, Institute for Research and Medical Consultations, Imam Abdulrahman Bin Faisal University, Dammam, Saudi Arabia

Farah Rehan Department of Pharmacy, Forman Christian College University, Lahore, Pakistan

Ayesha Sani Tecnologico de Monterrey, School of Engineering and Sciences, Atizapan de Zaragoza, Estado de Mexico, Mexico

Shagufta A. Shaikh Pakistan Council of Scientific and Industrial Research Karachi (PCSIR), Karachi, Pakistan

Kauser Siddiqui Pakistan Council of Scientific and Industrial Research Karachi (PCSIR), Karachi, Pakistan

1 Significance of Nano-drug Delivery in Cancer Therapy, Application of Nanoparticles in Overcoming Drug Resistance, Targeted Therapy, and Immunotherapy

Dhvani H. Kuntawala and Zaib Un Nisa Munawar Hussain

Abstract

Cancer remains a global leading cause of death and is a complex disease that needs a versatile method to treat. Despite strategies to reduce deaths, alleviate chronic pain, and improve quality of life, existing shortfalls in cancer therapies still face limitations. The use of nanotechnology has developed as a promising avenue in cancer treatment, specifically nanoparticles (NPs) ranging from 1 to 100 nm. The design of these NPs plays an important role since physical and structural features control the pharmacokinetics, internalization, and safety of the drugs. Moreover, with the emerging multidrug resistance mechanisms studied, NPs are being subjected to more renewed research. Recently, scientists have started to explore the role of NPs in immunotherapy. Potential advantages of using nanocarriers in different biomedical applications, especially targeted therapy have been extensively studied for cancer treatment. This chapter discusses numerous types of NPs, targeting mechanisms, immunotherapy, and the importance of nanotechnology in cancer treatment. Further, we also summarize the future perspectives. Overall, further research in this area will lead to new innovative methodologies for cancer treatment that can improve patient outcomes.

Keywords

Cancer therapies · Cancer immunotherapy · Nano-drug delivery systems · Nanoparticles · Targeted cancer therapy · Nanocarriers · Drug resistance

D. H. Kuntawala (✉)
Department of Medical Sciences, Institute of Biomedicine (iBiMED), University of Aveiro, Aveiro, Portugal

Z. U. N. M. Hussain
School of Life Sciences, Manipal Academy of Higher Education, Dubai Campus, Dubai, UAE

F. A. Khan (ed.), *Nano Drug Delivery for Cancer Therapy*,
https://doi.org/10.1007/978-981-99-6940-1_1

1.1 Introduction

Globally, cancer ranks as one of the leading causes of death, accounting for almost 10 million deaths in 2020 according to the World Health Organization (WHO). An estimated 19.3 million new cancer cases and nearly 10.0 million cancer deaths occurred in 2020. Despite effective treatments and devices, cancer remains a devastating disease. Cancer develops from the alteration of normal cells into tumor cells, which can spread throughout the body. In general, cancer is considered a result of genetic mutations (Sung et al. 2021). Many therapeutic techniques have been tested against cancer diagnosis, each yielding various outcomes of effectiveness. The most common prevailing traditional treatments are chemotherapy, surgery, and radiotherapy. The application of conventional therapy has also shown improvement in cancer management. However, a leading challenge posed by this approach is the killing of cancerous and noncancerous cells, resulting in severe side effects such as bone marrow depression, hair loss, and other toxic reactions (Zitvogel et al. 2008). As a result, cancer research conducted over the past few decades has prioritized the development of creating medications that target tumor cells rather than affecting normal cells. Precision therapy is an example of targeted therapy, yet there are still several side effects, and drug resistance has remained an enduring challenge. Chemotherapy is widely used for various cancer cases and metastatic cancers as a standard treatment. Nonetheless, over 90% of chemotherapy failures are due to the emergence of resistance to the available drugs. This resistance pattern resembles the resistance seen in the treatment of infectious diseases and signifies a challenging feature of managing and preventing cancers (Goodman et al. 1946). During the last few decades, resistance pattern has appeared as a major complication and this condition enables the cancer cells to proliferate in the presence of a chemotherapeutic agent. Significant resistance leads to a repeated treatment of a single type of anticancer agent and extends to drugs with similar or different mechanisms of action. This mechanism is known as multidrug resistance (MDR), which can be inherent or acquired (Ozben 2006). To address this issue, in recent years, nanotechnology has emerged as a promising field in the treatment of cancer. By definition, nanotechnology focuses on the scientific and engineering disciplines involved in the design, synthesis, and construction of devices that exhibit functional structures at the nanoscale. Nanotechnology has many applications in medicine and plays a role in various areas such as molecular imaging, drug delivery, diagnosis of diseases, biomarker mapping, gene therapy, and biosensors. Research into precise drug therapies targeting specific sites and diagnostic approaches for different pathologies have been of interest (Sahoo and Labhasetwar 2003; Nasrollahzadeh et al. 2019). Currently, this field plays a pivotal role aimed at designing applications for safer and more efficient tumor targeting, detection, and treatment. The introduction of nanotechnology has had a significant impact on the clinical management of tumors, leading to the rapid development of targeted therapy combined with drug therapy and early tumor detection. Among them, nanoparticle-based drug delivery systems have become a research hotspot, due to their efficient loading, targeted therapy, and other roles for drugs that show promise in the biomedical field (Jain et al. 2021).

Various nanoparticle-based drug delivery systems include organic, inorganic, and hybrid NPs. Nanoparticle-based drug delivery systems are designed on their size and characteristics according to the pathophysiology of the cancerous growth. Different nanoparticle-based drug delivery systems have excellent candidate materials for therapeutic drug delivery, mainly based on their binding capacity and biocompatibility (Lu et al. 2020). Interestingly, nanocarriers appear to show a promising strategy to overcome drug resistance to cancer. Drug resistance is the most notable cause of cancer therapy failure. However, overcoming multidrug resistance displays promising new horizons for cancer therapy with upcoming advanced designs and drug delivery strategies in various nanomaterial fields (Cao et al. 2021). Given the above-mentioned facts, we outline the application of the nanocarrier system in cancer therapy and mechanisms of drug resistance and describe the directions of future research in cancer.

1.2 Overview History of Nanotechnology and Nanoparticles

In the pre-nanotechnology era, people were unknowingly coming across different nanoscale objects and utilizing nano-level procedures. For instance, in ancient primitive Egypt, black hair dying paste containing lead oxide, lime, and water was discovered. This resulted in the development of galenite NPs (Pbs and lead sulfide), allowing for consistent hair dyeing. The ancient cup exhibits optical properties; it changes depending on the location of the light source. Under natural light, the cup is green, but the illumination turns red. Analysis of this cup showed the presence of gold and silver measuring 50–100 nm Au and Ag nanoparticles, which prompt the unusual coloring via plasmon excitation of electrons. These examples show the ancient use of nanotechnology extended beyond Rome and Egypt. There is further evidence for the early use of nanotechnology processes in Ancient India, Mesopotamia, and the Maya (Joudeh and Linke 2022).

During the 1950s, the history of nanotechnology with a polymer-drug conjugate was designed by Jatzkewitz, followed by Bangham who invented the liposomes, which were used as carriers for both drugs and proteins in the 1960s. Over time, a wide range of materials is fabricated into NPs, serving as effective drug delivery systems (DDS) (Fig. 1.1). In 1972, Scheffel and colleagues first discovered albumin-bound NPs, which formed the basis of Abraxane which was approved by the US Food and Drug Administration (FDA) in 2005 for lung and breast cancer treatment. Another FDA nanoparticle-based therapy that was approved for the treatment of systemic fungal disease in cancer treatment was Abelcet, an amphotericin B lipid complex (Tereza Cerna et al. 2016). In the 1980s, Marda and colleagues discovered the accumulation of NPs in tumor sites. They located the structure of tumor blood vessels, identified by unusual and leaky design, resulting in an "enhanced permeability and retention (EPR) effect. This effect appears due to decreased lymphatic derangement and larger arrangements between endothelial cells ranging from 200 to 1200 nm compared to normal endothelium having 10–50 nm pores. As a result, the EPR effect enables increased nanoparticle buildup in the tumor site due to "passive

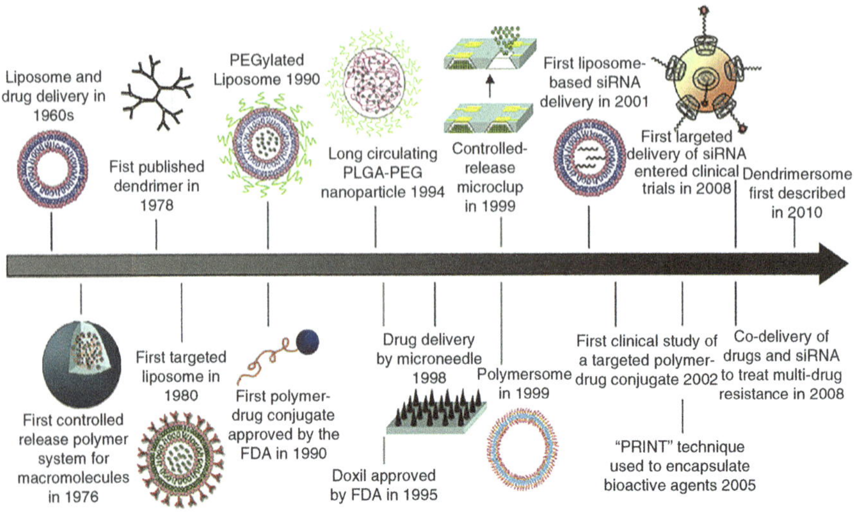

Fig. 1.1 Timeline of nanoparticle drug delivery system development. Reprinted with permission from (Shi et al. 2010). Copyright (2010) American Chemical Society (ASC)

targeting." Herewith, drug carriers display increased therapeutic advantage in tumors, as well as reduced side effects and toxicity (Yuan et al. 1995).

1.3 Significance of Nano-Based Carriers Carrying Anticancer Agents

Over the traditional methods for drug delivery NPs have many advantages which include the ability to mainly target the cancer cells by minimizing the healthy tissue damage, pharmacokinetics, and increased stability. NPs can be engineered to carry various types of therapeutics, including chemotherapeutic drugs, nucleic acids, and peptides, and can deliver these agents directly to the tumor site, where they can exert their anticancer effects. Moreover, NPs can improve the solubility and bioavailability of poorly soluble drugs, which can increase their therapeutic efficacy. Overall, the use of NPs in drug delivery has the potential to enhance the efficacy and reduce the toxicity of cancer therapies (Ertas et al. 2021).

In recent years, nano-based carriers have gained considerable attention as vehicles for drug delivery due to their unique properties. They can improve drug solubility, stability, bioavailability, and targeting to specific tissues or cells (Cheng et al. 2021). Anticancer drugs often cause significant side effects as they distribute non-specifically throughout the body, leading to toxicity in healthy tissues. Nano-based carriers can be engineered to target cancer cells specifically, reducing the damage to healthy tissues and potentially mitigating side effects (Dadwal et al. 2018). Furthermore, these carriers can increase drug circulation time, resulting in sustained drug release, reduced frequency of drug administration, improved patient compliance, and ultimately better therapeutic outcomes. Overall, using nano-based carriers for delivering anticancer drugs has the potential to revolutionize cancer treatment by enhancing drug efficacy, decreasing side effects, and improving patients' quality of life (Dong et al. 2019). Nano-based carriers have been attracting considerable attention in cancer research and therapy for delivering anticancer agents due to several reasons. Firstly, these carriers can be designed to specifically target cancer cells, delivering anticancer agents to the tumor site while minimizing harm to healthy cells (Bae et al. 2011). Secondly, nano-based carriers can increase the efficacy of anticancer agents compared to traditional delivery methods by protecting them from degradation and allowing for gradual release. Thirdly, targeted delivery by nano-based carriers can reduce the side effects associated with traditional chemotherapy by focusing only on cancer cells and sparing healthy ones. Lastly, nano-based carriers can enhance the pharmacokinetics of anticancer agents, resulting in better absorption, distribution, metabolism, and excretion of the drugs. Overall, the use of nano-based carriers for delivering anticancer agents has the potential to enhance cancer therapy by improving efficacy, minimizing side effects, and achieving a targeted delivery (Dadwal et al. 2018).

Nano-based carriers that transport anticancer agents hold significant importance in cancer treatment. They are composed of materials like lipids, polymers, and metals, which can be manipulated into NPs within the 1–100 nm range. These NPs can be utilized to carry various anticancer agents like chemotherapy drugs, peptides, or small interfering RNA (siRNA) (Mu et al. 2020). The significance of nano-based carriers carrying anticancer agents can be summarized as follows:

1. Enhanced drug delivery: Nano-based carriers can deliver anticancer agents more effectively and specifically to the tumor site while minimizing exposure to healthy tissues, reducing side effects and toxicity.
2. Improved drug efficacy: Nano-based carriers can increase the solubility, stability, and bioavailability of anticancer agents, improving their efficacy.
3. Targeted therapy: Nano-based carriers can be functionalized with ligands or antibodies to selectively target cancer cells or tumor tissues.
4. Combination therapy: Nano-based carriers can be loaded with multiple anticancer agents, enabling combination therapy to increase treatment efficacy and overcome drug resistance.
5. Personalized medicine: Nano-based carriers can be designed for specific cancer types or genetic profiles, allowing for personalized medicine approaches.

Overall, nano-based carriers carrying anticancer agents have the potential to revolutionize cancer treatment by improving drug delivery, efficacy, and specificity, enabling targeted and combination therapies, and ultimately improving patient outcomes (Hossen et al. 2019).

Nano-based carriers for anticancer agents offer several advantages over conventional chemotherapy drugs. The use of nanosized drug carriers allows for targeted drug delivery, which increases the drug's efficacy and reduces its side effects. The small size of the NPs enables them to selectively accumulate in tumor tissues due to the enhanced permeability and retention (EPR) effect, which is a characteristic of solid tumors. This selective accumulation of the NPs in tumors reduces the systemic toxicity of the drug and enhances its therapeutic efficacy (Rudzińska et al. 2021). Moreover, the nanocarriers can be engineered to release the drug in a controlled manner, increasing its bioavailability and reducing its toxicity. Additionally, the surface of the nanocarriers can be modified to increase their circulation time, stability, and targeting specificity. Overall, the use of nano-based carriers for anticancer agents offers a promising strategy for the development of effective and safe cancer treatments (Rodríguez et al. 2022).

The use of nano-based carriers for anticancer agents has the potential to revolutionize cancer treatment by improving drug delivery, efficacy, and specificity, enabling targeted and combination therapies, and ultimately improving patient outcomes. This is because nanosized drug carriers offer several advantages over conventional chemotherapy drugs. They can be targeted specifically to cancer cells, reducing toxicity to healthy tissues and increasing the drug's effectiveness. Additionally, the nanocarriers can be engineered to release the drug in a controlled manner, improving bioavailability and reducing toxicity. Moreover, the surface of the nanocarriers can be modified to increase their stability, targeting specificity, and circulation time. The integration of nanotechnology and biology provides opportunities for the development of new materials in the nanometer size range that can be applied to many potential applications in clinical medicine, including the delivery of anticancer agents to cancer stem cells (Raj et al. 2021). The use of NPs as carriers for delivering anticancer agents to cancer stem cells (CSCs) is highly beneficial. One of the studies discusses the biology of CSCs and the latest advancements in designing and synthesizing NPs as carriers to target cancer drugs specifically to CSCs. That study encompasses the development of various types of NPs, including synthetic and natural polymeric NPs, lipid NPs, inorganic NPs, self-assembling protein NPs, antibody–drug conjugates, and extracellular nanovesicles for CSC targeting. The integration of nanotechnology and biology presents opportunities to create novel materials in the nanometer size range with potential applications in the clinical medicine (Yadav et al. 2022). In the mid-1970s, liposomes were first studied for their potential as nano-drug carriers in clinical applications. The treatment of mice with tumors using liposome-entrapped actinomycin D was found to prolong survival significantly. Today, the use of nanomaterials for drug and diagnostics delivery is still a major focus of nanomedicine. Recent advancements have involved the conjugation of cell-specific ligands to the surface of NPs for greater control of drug targeting at the cellular and tissue levels. Additionally, encapsulating drugs within NPs has significantly improved drug release profiles (Yadav et al. 2022).

1.4 Various Types of Nanoparticles Are Used in Drug Delivery Systems

Over recent years, nanocarrier-assisted drug delivery systems have gained recognition for biomedical applications. Various designs of nanocarriers are actively developed to meet clinical demands that require diverse drug delivery systems. Due to their tiny size and colloidal property, nanocarriers are capable of transporting anticancer agents, like small molecular drugs, genes, or proteins. By doing so, it allows these anticancer agents to accumulate in tumors by avoiding normal tissues. In the same way that antibodies and peptides–drug conjugates do it also results in high cytotoxic concentration in tumors with reduced toxicity for the rest of the body (Pérez-Herrero and Fernández-Medarde 2015). With a size span ranging between 10 and 1000 nm with biomimetic attributes NPs are composed of particles dispersed below nanometers or solid particles. The biomimetic attributes of NPs increase as the surface-to-volume ratio becomes higher. The ability to enhance these properties for biomedical applications is expanding, with potential applications in the imaging, diagnosis, and therapy (Patra et al. 2018). All in all, the different features of NPs control their therapeutic outcome in managing cancer. NPs are placed into three forms depending on their composition materials and structure: organic, inorganic, and carbon-based. Various types of NPs for cancer therapy are depicted in Fig. 1.2.

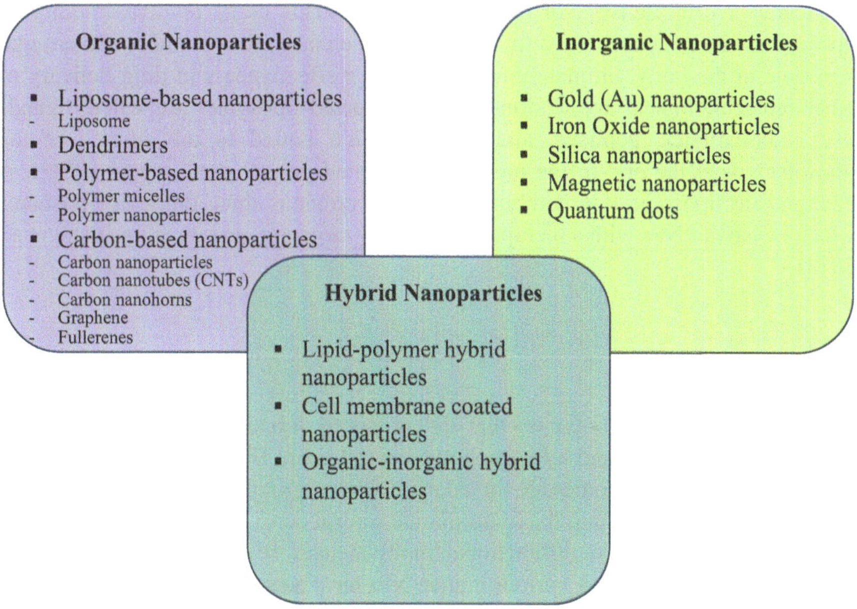

Fig. 1.2 Various types of nanoparticles for cancer therapy. Nanoparticles in drug delivery systems encompass organic, inorganic, and hybrid nanoparticles

1.5 Organic Nanoparticles

1.5.1 Liposomes

Liposomes are commonly used for the transportation of molecules in the pharmaceutical and cosmetic industry. They are small spherical form vesicles ranging from 50 to 450 nm in size composed of steroids and phospholipids which are an established formulation strategy for enhancing drug delivery. Since their membrane is similar to cell membranes and they facilitate easy incorporation into them, they are marked as better drug delivery vehicles. It has been proved that liposomes can be used with hydrophilic and hydrophobic drugs, they are biodegradable and compatible and most importantly they make therapeutic compounds stable and improve their biodistribution (Nsairat et al. 2022). Liposomes are categorized into four types: Conventional liposomes consist of a lipid bilayer surrounded by an aqueous core material and can make either cationic, anionic, or neutral cholesterol or phospholipids. Both the aqueous space and lipid bilayer can be filled with hydrophilic and hydrophobic materials, respectively. PEGlyted Polyethylene glycol (PEG) is incorporated into the surface of the liposomes to attain steric equilibrium. The main purpose of incorporating PEG into the surface of liposomes is to improve their circulation time in the body. PEGylation creates a protective layer around the liposomes, reducing their recognition and clearance by the immune system. This allows the liposomes to remain in the bloodstream for a longer period, enhancing the efficiency of drug delivery or other applications. This usage involves modifying liposomes with pegylated polyethylene glycol to enhance their stability and circulation time in the body, ultimately improving their effectiveness in drug delivery or other biotechnological applications. Ligand-targeted liposomes are where ligands like carbohydrates, peptides, and antibodies are bound to the surface of the liposomes or at the end of the previously attached polyethylene glycol (PEG). A theragnostic liposome is a combination of the previous three types of liposomes which consist of NPs with a therapeutic element, targeting, and imaging (Patra et al. 2018).

1.5.2 Dendrimers

Dendrimers are circular polymers that are composed of branched monomers, operative peripheral groups, and a hydrophobic central core. The immunotherapy drugs such as therapeutic antibodies are linked in peripheral groups, while small molecular agents are encapsulated into the hydrophobic central core. Due to the extraordinary structural properties like multifunctional, multivalences, structural transparency, and adjacent-to-monodispersed, various unique NPs have been marked and scientifically gained attention. Currently, the broadly used dendrimers are poly-propylene imine (PEI), peptide dendrimers, and poly-amidoamine (PAMAM) (Mu et al. 2020). There are two methods for the synthesis of dendrimers, the first is the passage through which the dendrimers are synthesized in the core and extending outwards and the other is concurrent which starts from the outside of the dendrimers. Based on their functionalization components, dendrimers are categorized into various types such as PPI, core-shell, glycodendrimers, liquid crystalline, PAMAMOS, chiral, peptide, and PAMAM. Due to its water-soluble nature and its ability to pass over the epithelial tissues enhancing the delivery through paracellular pathways, PAMAM is among the extensively studied type for drug delivery. New dendrimer-based NPs are currently being utilized to enable efficient chemoimmunotherapy by achieving deep penetration of the loaded drugs (Patra et al. 2018).

1.5.3 Carbon Nanoparticles and Nanotubes

As their name indicates, carbon NPs are based on the element carbon, and they are broadly known for their application in mechanical, optical, and electronic properties combined with biocompatibility in medicine. They are capable of enclosing drugs through π–π stacking because of their inherent hydrophobic nature. Carbon NPs were discovered in the late 1980s and are cylindrical tubes that are commonly referred to as graphene. Carbon NPs are categorized into various types, including carbon nanotubes, carbon nanohorns, graphene, and fullerenes. Carbon NPs are classified into two groups, single-walled carbon nanotubes and multi-walled carbon nanotubes due to their carbon-based property, by inhibiting the tumor cells they can induce an immune response by interacting with immune cells (Cheng et al. 2021). Along with the advantages carbon nanotubes possess a few drawbacks such as causing health issues and toxicity as they are insoluble in most solvents. To address this issue, chemical modification of nanotubes is done and by this approach, nanotubes can be linked to a broad range of effective molecules such as nucleic acids, therapeutic drugs, proteins, and peptides (Dadwal et al. 2018).

1.5.4 Polymeric Nanocarriers

The most common nanoparticle drug carriers are polymers. The polymers which are used in the controlled release of drugs should be biocompatible and nontoxic without any leaking impurities. They should have a physically appropriate structure with a suitable half-life. Those polymers which are used in the structure of polymer NPs can be natural or synthesized. Polymer NPs are mostly selected as biodegradable types. The advantages of polymer NPs are their high stability and mass production. Polymer NPs contain vesicular (nanocapsules) and matrix systems (nanospheres). In nanocapsules, the drug is stored in a polymer cistern. However, in nanospheres, the drug is scattered on a polymer matrix (Aghebati-Maleki et al. 2020).

1.5.5 Polymeric Micelles

Polymer micelles are a type of nanoparticle that is composed by the self-assembly of amphiphilic block copolymers. These are stable colloidal solutions with hydrophobic nature. Hydrophilic drugs could be targeted through chemical conjugation or physical interaction and hydrophobic agents could be inserted into the micelles by their hydrophobic core. In cancer treatment, Nanoxel®-loaded docetaxel (DTX) and Genexol®-loaded PTX are the approved drugs. In cancer chemoimmunotherapy, these molecules are extensively studied (Mu et al. 2020). Loading of the drugs into polymer micelles can be done through two approaches: one is through chemical covalent binding and the second is physical encapsulation. According to a study, polymeric micelles consist of multiple polymers as they are nucleated shell molecules that contain various hydrophobic N-iso-propyl acrylamide/vinyl pyrrolidone heat-resistant nano-polymer micelles that were targeted with paclitaxel and reviewed for their anticancer properties both in breast cancer cell line (MCF-7) and skin melanoma cell line(B16F10) (Marzi et al. 2022).

1.5.6 Polymeric Nanoparticles

Polymer NPs consist of more than one polymer with different structures, hydrophobic chains, and body weights. Nanospheres, nanocapsules, polymer micelles, nanosponges, and many others are the forms available in these NPs. Due to their ability to stable release, these nanoparticle systems show high therapeutic efficiency and exhibit biocompatible characteristics (Marzi et al. 2022). Among the naturally occurring nano-polymers heparin, chitosan and albumin have been utilized for the transportation of drugs along with oligonucleotide, proteins, and DNA delivery. Lately, serum albumin is used as a transporter for the preparation of paclitaxel NPs which are nanometer-sized albumin-bound paclitaxel that is being approved for clinical use in the treatment of metastatic breast cancer (Dadwal et al. 2018).

1.5.7 Inorganic Nanoparticles

Inorganic NPs are used both in passive and active agents for tumor targeting because of their ultra-small size and even circulation. But due to the reticuloendothelial system (RES) they are swiftly cleared from circulation along with that they are rapidly identified and eradicated by the scavenger receptors on Kupffer cells in the liver (Ertas et al. 2021).

1.5.8 Gold NPs

In cancer therapy, surface-modified and exposed gold (Au) nanoparticles have gained significant consideration as both active and passive drug carriers, because of their properties like non-toxicity, narrow distribution, and biocompatibility. Their ability for targeted drug delivery has improved in vivo performance by binding surface-modified gold NPs by ligand immobilization. Additionally, their physiological biocompatibility has been enhanced by the surface functionalization of gold (Au) nanoparticles along with biocompatible coatings (Ertas et al. 2021).

1.5.9 Iron Oxide NPs

Biocompatibility, non-toxicity, gene delivery to the target site, and high efficacy of drugs are particular functions of magnetic iron oxide (IO) nanoparticles, and they are broadly used in cancer diagnosis and treatment. Iron oxide nanoparticles possess some additional properties which make them strong nanocarrier agents for the delivery of drugs to the central serous chorioretinopathy (CSC) subpopulations of cancer cells. These properties include excellent drug binding, ease of surface functionalization, feasible large-scale production, and high colloidal stability in the physiological media (Zhang et al. 2022).

1.5.10 Silica NPs

Over the last ten years, the use of silica NPs in gene and drug delivery has been expanding. Among the different types, mesoporous silica (MS) nanoparticles have been used as a drug transporter and acquire particular functions such as ease of functionalization, large surface area, tunable size, and porous design for drug delivery. They efficiently transport nucleic acids and hydrophobic anticancer agents to tumors due to their composed pore size and definite shape (Marzi et al. 2022).

1.5.11 Magnetic Nanoparticles

Magnetic nanoparticles contain metal oxide or metal. They are usually coated with organic materials, polymers, and fatty acids to enhance biocompatibility and stability. Magnetic nanoparticles are known to have high efficacy in gene therapy and chemotherapy for cancer management (Mandriota et al. 2019).

1.5.12 Quantum Dots

Quantum dots are semiconductor nanocrystals that emit fluorescent colors with imaging and drug delivery applications in cancer. They possess high targeting abilities due to their surface functions, and hence, can be utilized with many targeting molecules on the surface. Some of the key aspects of quantum dots include bioconjugation, size-tunable emission, and photovoltaics (Shao et al. 2011; Yao et al. 2020).

1.5.13 Hybrid Nanoparticles

Combining organic and inorganic NPs in a single hybrid drug delivery system builds a multifunctional carrier encompassed with better biological properties that can increase treatment effectiveness and lessen drug resistance. Lipid-polymer hybrid nanoparticles have shown promising drug delivery in treating various types of cancers. This type of hybrid nanoparticle is proficient in encapsulating hydrophilic and hydrophobic drugs to achieve better therapeutic effects. Cell membrane-coated NPs are coated with a thin layer of cell membrane obtained from cells. They include advantages from both biological and synthetic components (Al-Zoubi and Al-Zoubi 2022). Cell membrane-coated nanoparticles are used in many applications such as cancer therapeutics, regenerative medicine, targeted drug delivery, enhanced drug stability, and many more. Nanoscale structures that combine organic and inorganic materials are known as organic–inorganic hybrid NPs. Some of the potential uses are controlled drug delivery, catalysis, optoelectronics, and environmental remediation. Other hybrid NPs are being developed and explored for the best drug delivery efficacy (Yao et al. 2020).

1.6 Application of Nanocarriers in Drug Resistance to Cancer

Set side by side with conventional medication for curing clinical conditions, nanodrugs possess far better efficacy. Nano-based drugs have shown better stability in severe biological habitats which in turn gives strengthened and long medicinal effects. Examples of such biological habitats are concentrated levels of proteases and various enzymes in the bloodstream and exceedingly acidic surroundings in the stomach. They have a prolonged plasma half-life. Furthermore, the delivery mechanism of nanodrugs highly increases the chances of the drug being delivered to the tumor site and these drug carriers also respond to changes in temperature or pH levels which allow the release of drugs in a controlled manner over a period of time. Nanomaterials are combinable in groups to perform combination therapy reducing multidrug resistance (Hu et al. 2018). When compared with individual molecules or bulk solids various nanomaterials such as metal, semiconductor, and polymeric particles possess innovative electronic, magnetic, optical, and structural properties. With recent advancements in research NPs that can link themselves covalently with biological molecules have been developed. These NPs have found medical applications such as using superparamagnetic iron oxide NPs as a contrast agent for lymph node prostate cancer detection and polymeric NPs for targeted gene delivery to tumor vasculatures. Researchers are carrying out extensive research to develop modern technologies employing metal and semiconductor NPs for molecular profiling research and multiplexed biological assays (Nie et al. 2007).

Cancer nanotechnology is a collaborative research field between medicine, engineering, and science with a wide range of applications for targeted therapy, molecular diagnosis, and molecular imaging. The underlying motive for the use of NPs is that, when compared with individual molecules or bulk solids, various nanomaterials such as metal, semiconductor, quantum dots, and iron oxide nanocrystals possess innovative electronic, magnetic, optical, and structural properties. These NPs target tumor antigens as well as tumor vasculatures with high affinity and precision, by conjugating with ligands such as monoclonal antibodies, peptides, or small molecules. Moreover, NPs in the size range of 5–100 nm offer large surface areas and functional groups that allow for conjugation with multiple diagnostic and therapeutic agents. Recent advancements have resulted in integrated nanodevices for early cancer detection and screening, targeted nanoparticle drugs for cancer therapy, and bio-affinity nanoparticle probes for molecular and cellular imaging. These developments have opened up exciting possibilities for personalized oncology, in which genetic and protein biomarkers are used to diagnose and treat cancer based on individual patients' molecular profiles (Nie et al. 2007).

One of the major challenges faced in cancer therapy is resistance to drugs, it can occur because of various reasons such as an increase in drug efflux, alteration in drug metabolism, and decreased drug uptake. NPs have shown effective results in overcoming resistance to drugs by enhancing drug delivery mechanisms to affected sites and also by overcoming resistance mechanisms. For instance, chemotherapeutic nanoparticle-based formulations highly enhance drug uptake by cancer cells and decrease their efflux resulting in higher curing efficacy. In addition, NPs can be designed to target specific signaling pathways or molecular targets which cause resistance to drugs, thus increasing the response to therapy (Cao et al. 2021).

1.6.1 Application of Nanoparticles in Overcoming Drug Resistance

In the past, most therapies in cancer treatment rendered unsuccessful results because of the matter of strong hindrance of drug delivery to the affected area. But that might not be the obstruction anymore. With technological advancements and studies of the numerous ways in which NPs can be used to deliver the drug, researchers are very optimistic about finding very promising techniques for delivering the drug. They are attempting to focus on those passageways and biological structures that are responsible for the hindrance of drugs not being delivered to the target area. Proposed nanoparticle-based systems have shown that they have significantly increased the effectiveness of cancer-curing medicines. These systems are designed such that the NPs prohibit efflux pumps from transporting drugs out of affected cells and overpowering the environment around the cell which is another factor of drug being resisted (Yao et al. 2020).

1.6.2 Application of Nanoparticles in Targeted Therapy

Therapies including NPs have shown that they yield more accurate results by increasing the effectiveness of the treatment, this is achieved by releasing the therapeutic agents only if the desired conditions are met. In this method, NPs are engineered to be able to attach themselves to targeted receptors on cancer cells. The

NPs are also able to react to the biological environment. An example of this technique is being able to release the deliverable in reaction to changes in pH levels or temperature conditions. They can also respond to individual enzymes and proteins that comprise cancer cells (Sajja et al. 2009).

1.6.3 Application of Nanoparticles in Immunology

Therapies including NPs have shown that they yield more accurate results by increasing the effectiveness of the treatment, this is achieved by releasing the therapeutic agents only if the desired conditions are met. In this method, NPs are engineered to be able to attach themselves to targeted receptors on cancer cells. The NPs are also able to react to the biological environment. An example of this technique is being able to release the deliverable in reaction to changes in pH levels or temperature conditions. They can also respond to individual enzymes and proteins that comprise cancer cells (Su and Kang 2020; Yang et al. 2021).

The biggest advantage of advancement in nanotech is the shrinking of the scale of drug delivery mechanisms, which results in accurately transporting the drug to the affected site more precisely than before. This improvement also has introduced new methods for transporting cancer-curing drugs such as being able to attach, enclose, or adsorb the curing and probing agents onto the NPs which can effectively address drug solubility problems. Combinatorial testing programs show that almost 40% of active substances have poor solubility. To address this problem more screening is done to identify a molecule with good solubility. The widely used attempt at enhancing solubility is to formulate salt. For non-ionizable compounds, micronization, soft-gel technology, cosolvents, surfactants, or complexing agents have been used. This process is much faster and more cost-effective to reformulate the drug than developing a new one. Thus, a broadly applicable technology that can enhance the solubility of these poorly soluble drugs could have a significant impact (Sutradhar and Amin 2014; Su and Kang 2020).

1.6.4 Mechanisms of Nanocarriers in Overcoming Drug Resistance Problems

Both science and medicine show a strong emphasis on the development of nanosized drug transport mechanisms. Various materials are used as NPs and they range on the nanoscale between 10 nm and <1000 nm, serving as a way to deliver the curing and probing agents. These materials can be liposomes, solid lipids, polymers, silicon/carbon materials, dendrimers, and magnetic materials. These transport mechanisms allow the enclosing of various curing agents such as small molecule inhibitors, nucleic acids, peptides, and protein-based drugs, this increases the therapeutic index of the drug being used and its clinical significance (Rudzińska et al. 2021). Many nanocarriers have undergone broad research to find out the extent of their ability to combat drug resistance in cancer cures. These materials include dendrimers, liposomes, and polymeric NPs. The way these nanocarriers fight drug resistance is much more complex and comprises of:

1.6.4.1 Enhanced Permeability and Retention (EPR) Effect

The EPR effect allows nanocarriers to pile up in the tumor tissue. The EPR effect is possible due to low lymphatic drainage and broken vasculature of tumors. This enables us to transport drugs precisely to the tumor area. As a result, gain in drug absorption occurs in tumors which also interprets as a reduction in drug resistance.

1.6.4.2 Passive and Active Targeting

Passive Targeting

The EPR effect plays a significant part in the passive targeting mechanism of NPs, which uses the increased vascular permeability and enfeeble lymphatic drainage of cancer cells and allows NPs to target cancer cells without active resistance.

Active Targeting

Active targeting is obtained by the reaction between ligands and receptors. Nanocarriers can be engineered to recognize and bind to cancer cell's surface receptors that are overexpressed, such as folate receptors, transferrin receptors, lectin, and epidermal growth factor receptor (EGFR), which can enhance the selectivity and efficiency of drug delivery while minimizing toxicity to healthy cells (Yao et al. 2020).

1.6.4.3 Multidrug Resistance (MDR) Modulation

Nanocarriers have the potential to combat multidrug resistance (MDR) by hindering the activity of efflux pumps responsible for removing drugs from cancer cells, thereby diminishing drug resistance. This can be accomplished by integrating efflux pump inhibitors into the nanocarrier or by simultaneously administering the drug and inhibitor (Wong et al. 2015; Wu 2021).

1.6.4.4 Co-delivery of Multiple Drugs

One approach to overcoming drug resistance is through combination therapy, which involves the use of multiple drugs. Nanocarriers can facilitate this strategy by enabling the co-delivery of several drugs side by side, which can increase the effectiveness of treatment and reduce the risk of drug resistance development.

1.6.4.5 Controlled Drug Release

Nanocarriers can be engineered to achieve controlled drug release, which can decrease the probability of drug resistance development by sustaining therapeutic drug levels for a longer duration (Da Silva et al. 2017).

In conclusion, nanocarriers offer a range of mechanisms to address drug resistance in cancer therapy. They enhance drug accumulation in tumors, target exact cell surface receptors, obstruct efflux pumps, co-deliver various drugs, and control drug discharge. These characteristics make nanocarriers a promising approach to enhancing the effectiveness of cancer treatment and reducing drug resistance.

1.7 The Role of Nanoparticles in Immunotherapy

NPs possess distinguished physicochemical properties including size, shape, and charge, which can be customized to achieve various therapeutic goals such as cancer immunotherapy. For example, the size of NPs affects cellular uptake and endocytosis. As compared with the larger NPs (>100 nm), smaller ones (25–40 nm) have a greater potential for immune response activation, since smaller NPs are allowed to move to lymph nodes via dendritic cells, while the larger ones are retained at the target site (Hoshyar et al. 2016). Immunotherapy advancement has brought cancer treatment into a novel age. Nanoparticle properties are not limited to chemotherapy, but it has revealed extraordinary growth for immunotherapy applications. Activation of antitumor immune response gives rise to cancer immunotherapy. NPs that are linked to immunotherapy comprise artificial antigen-presenting cells (APC), nanovaccines, and immunosuppressed tumor microenvironment (TME) targeting. Likewise, NPs are usually altered with polyethylene glycol (PEG) for minimizing the interaction with the reticuloendothelial system (Zang et al. 2017). Additionally, the stimulation of cytotoxic T cell's antitumor properties is achieved by NPs which themselves act as adjuvants to endorse dendritic cell maturation and antigen-presenting cells (APC) presentation (Yang et al. 2018). The immune response over tumor cells is enhanced by dendrimers, polymeric micelles, gold NPs, and liposome NPs because they all have the ability of cytoplasmic delivery of tumor-associated antigens (TAAs) into dendric cells (DCs). Inorganic NPs like polymers such as acetylated dextran (AcDEX) and mesoporous silica have been known to act as an adjuvant in immunotherapy which results in immune response activation (Fontana et al. 2017a, b). Tumor-associated antigens (TAAs) and antigen-presenting cells (APC) are being delivered by nanovaccines such as dendritic cells. Nanovaccines and artificial APCs both function with MHC-antigen complexes but differ in their complexity, mechanisms of T cell activation, and primary applications. Nanovaccines and artificial APCs both utilize MHC antigen complexes but differ in their complexity, mechanisms of T cell activation, and primary applications. Nanovaccines focus on efficient antigen delivery, while artificial APCs offer more precise control over the immune response and are often used in specialized therapeutic settings. Targeting the myeloid-derived suppressor cells (MDSCs), regulatory T cells (Tregs), and tumor-associated macrophages (TAMs) that are significant cell types in the TME results in targeting of immunosuppressive TME (Shao et al. 2015). Although immunotherapy is efficient to treat different types of cancer, still there is a certain challenge in the delivery of immunotherapeutic agents which is expected to be resolved using NPs. A major goal of the utilization of NPs in cancer immunotherapy is to improve the therapeutic index by enhancing the delivery of immunotherapeutic agents directly to the site of interest only, enhancing accumulation and potency at a region of interest, while simultaneously minimizing the dose-dependent systemic toxicity (Patra et al. 2018). Despite the effectiveness of immunotherapy in treating various forms of cancer, there remains a particular obstacle in effectively delivering immunotherapeutic agents. This challenge is anticipated to be addressed through the utilization of NPs. A primary objective in employing NPs in cancer immunotherapy is to enhance the therapeutic index by facilitating the targeted delivery of immunotherapeutic agents exclusively to the desired site. This approach aims to increase accumulation and potency at the intended region while simultaneously reducing the potential for systemic toxicity associated with dose-dependent effects (Raza et al. 2023).

1.7.1 Nanoparticles as the Carrier of Immunotherapeutic Agents

NPs should be precisely designed to target regions of interest preferentially from the site of administration (common vaccine administration routes are mucosal or parenteral) to enhance the efficacy of immunotherapeutic agents. NPs targeting lymphoid tissues, where the majority of immune cells are concentrated, would enhance the efficacy of immunotherapeutic agents due to direct access to immune cells. To maximize the effectiveness of immunotherapeutic agents, it is crucial to intricately engineer NPs that exhibit a preference for targeting specific regions of interest. This is particularly important when considering the common routes of vaccine administration, such as mucosal or parenteral routes. By designing NPs to target lymphoid tissues, which house a significant concentration of immune cells, the efficacy of immunotherapeutic agents can be significantly enhanced. This direct access to immune cells provided by NPs would bolster the effectiveness of the immunotherapy (Ke et al. 2019). Depending on their physicochemical characteristics, including particle size, hydrophobicity, shape, and surface charge, NPs can directly drain to the nearest lymph node, or stay in the injection site and attract migratory DC or macrophages. Several kinds of literature show that NPs with particle sizes >100 nm tend to form depots and are taken up by APCs and then drained to lymph nodes. However, NPs with moderate particle sizes <100 nm drained to lymph nodes via lymphatics and retained relatively for a long time, while NPs with small particle sizes (<10 nm) drain to blood capillaries (Jia et al. 2018). NPs possess distinct physicochemical characteristics, such as particle size, hydrophobicity, shape, and surface charge, which determine their behavior in the body. Depending on these properties, NPs can follow different pathways. For instance, NPs with a size larger than 100 nm tend to accumulate and remain at the injection site, attracting migratory dendritic cells (DCs) or macrophages. These immune cells take up the NPs, leading to their subsequent drainage to the nearest lymph node. Conversely, NPs with a moderate particle size below 100 nm can directly access lymph nodes via lymphatic vessels, where they are retained for an extended period. Conversely, NPs with a small particle size below 10 nm are more likely to drain into the bloodstream through blood capillaries. These findings have been supported by several studies in the literature, highlighting the differential behaviors of NPs based on their size and emphasizing the importance of understanding these characteristics for effective NP-based drug delivery systems (Khan et al. 2019; Abbasi et al. 2023).

1.7.2 Antigens and Adjuvants Delivery to Antigen-Presenting Cells (APCs)

APCs are used as the link between innate and adaptive immune responses by interacting with T cells. APCs are primarily used to recognize and present tumorigenic antigens on their surface via MHC complexes to T cells to initiate an effective adaptive response (Gaudino and Kumar 2019). Antigen-presenting cells (APCs) play a crucial role in bridging innate and adaptive immune responses by engaging with T cells. Their primary function is to identify and display tumor-associated antigens on their surface using major histocompatibility complex (MHC) molecules, facilitating the activation of T cells and initiating a robust adaptive immune response. However, due to enzymes susceptible to antigens in the body, they are not easily transferred to APCs which decreases their immunogenicity. Hence by using NPs, it is possible to overcome this limitation. NPs can encapsulate and deliver cancer antigens to APCs without tumor antigen degradation by the intracellular enzyme. In addition, nano-formulated antigens are more efficiently taken up and processed by APCs than soluble vaccines to amplify T cell responses due to intradermally or subcutaneously injected NPs drain to lymph nodes, in which APCs are in closeness to T cells (Debele et al. 2020). Nevertheless, the immunogenicity of antigens can be compromised within the body due to the presence of enzymes that are capable of breaking them down, making it challenging to deliver intact antigens to antigen-presenting cells (APCs). This limitation can be overcome through the use of NPs. NPs can encapsulate cancer antigens and transport them to APCs, protecting them from degradation by intracellular enzymes. Furthermore, nano-formulated antigens exhibit enhanced uptake and processing by APCs compared to soluble vaccines. This leads to the amplification of T cell responses. When NPs are injected intradermally or subcutaneously, they can drain to the lymph nodes, where APCs reside near T cells, further enhancing the effectiveness of antigen presentation and immune activation (Kishimoto and Maldonado 2018).

1.7.3 Antigens and Adjuvants Delivery to Tumor Microenvironment (TME)

The TME is comprised of proliferating tumor cells, the tumor stroma, infiltrating inflammatory cells, apoptotic cancer cells, cancer-associated fibroblasts, myeloid-derived suppressor cells, tumor-associated macrophages, and a variety of associated tissue cells which are participating in the suppression of antitumor immunity. These immunosuppressive cells secrete numerous soluble mediators including transforming growth factor-beta (TGF-β), indoleamine 2,3-dioxygenase (IDO), arginase, prostaglandin E2, and nitric oxide synthase 2 (NOS2). By reducing the supply of indispensable amino acids (such as arginine (R) and tryptophan (W)), IDO and arginase directly suppress T cell proliferation and differentiation. The activity of arginase and IDO translates not only into amino acid deprivation but also into the production of metabolites (such as L-kynurenine and spermidine) capable of numerous physiologic effects (Debele et al. 2020). The tumor microenvironment (TME) consists of various components, including actively dividing tumor cells, the supportive tissue surrounding the tumor (tumor stroma), infiltrating inflammatory cells, dying cancer cells, cancer-associated fibroblasts, myeloid-derived suppressor cells, tumor-associated macrophages, and other related cells. All of these contribute to the suppression of the body's immune response against the tumor. These immunosuppressive cells release several soluble substances, including transforming growth factor-beta (TGF-β), indoleamine 2,3-dioxygenase (IDO), arginase, prostaglandin E2, and nitric oxide synthase 2 (NOS2). By limiting the availability of essential amino acids like arginine and tryptophan, IDO and arginase directly inhibit the growth and differentiation of T cells. The actions of arginase and IDO not only lead to amino acid deprivation but also result in the production of metabolites such as L-kynurenine and spermidine, which have various physiological effects. To summarize, NPs have the potential to boost the body's immune response against cancer by modulating the tumor microenvironment (TME). They can achieve this by either suppressing immune-suppressive factors or promoting immune activation. This ability to regulate the TME can complement the effects of existing immunotherapeutic treatments like immune checkpoint inhibitors (ICIs) and potentially work synergistically with them.

1.7.4 Nanoparticles as the Direct Immunomodulators

Compounds with immunomodulatory properties, including cytokines, monoclonal antibodies, and adjuvants, have been utilized to reshape the tumor microenvironment (TME) and initiate a response against cancer cells. However, the clinical application of these compounds faces certain limitations, such as concerns about their therapeutic effectiveness and the potential for undesirable side effects when administered systemically (Zhuang et al. 2019). The utilization of immunomodulatory NPs offers a promising approach to enhance therapeutic outcomes by promoting immune activation while minimizing the occurrence of off-target side effects. Consequently, there is a growing focus on research efforts aimed at comprehending the mechanisms underlying the interactions between NPs and the immune system. This understanding is crucial for determining the immunomodulatory potential of NPs, whether they exhibit immune-stimulating or immunosuppressive effects. The ability of NPs to modulate the immune response is influenced by various intrinsic factors, including surface chemistry, charge, size, and shape. Additionally, extrinsic factors such as the route of administration also play a significant role in the immunomodulatory function of NPs (Sushnitha et al. 2020). Among the different types of inorganic nanoparticles, such as gold nanoparticles (AuNPs), titanium nanoparticles (TiNPs), iron nanoparticles (FeNPs), zinc nanoparticles (ZnNPs), and silver nanoparticles (AgNPs), these particles are considered highly stable and hold significant promise for immune system modulation (Chugh et al. 2018).

1.7.5 Targeting Dendritic Cells

Dendritic cells (DCs) derived from bone marrow are recognized as highly potent antigen-presenting cells (APCs). A lipid-coated calcium phosphate nanoparticle was employed to deliver a specific tumor antigen, BRAFV600E peptide, to DCs via antigen presentation. The lipid bilayers of these NPs demonstrated efficient penetration of physical barriers, safeguarding the cargo drugs from degradation and aggregation. The findings revealed that the NPs substantially enhanced antigen-specific T cell responses and the production of interferon-gamma (IFN-γ). Notably, 20% of mice treated with these NPs achieved tumor-free survival. In another investigation, an innovative nano platform was developed by enveloping erythrocyte membranes with an antigenic peptide (hgp10025-33), a TLR4 agonist, and mannose. The use of erythrocyte membranes offered advantages such as easy isolation and biocompatibility. Interestingly, the erythrocyte membrane acted as an adjuvant and created a "depot effect" at the site of administration, prolonging the exposure time of antigens. By binding to DC receptors through mannose, this nanoparticle effectively activated DCs and significantly suppressed tumor growth. In addition to the incorporation of adjuvants into nanocomplexes, studies have also revealed that certain NPs possess inherent immunomodulatory properties (Yang et al. 2021).

1.8 Conclusion and Future Perspectives

NPs have shown great promise in cancer therapy, offering a wide range of benefits over traditional cancer therapies. The use of NPs in cancer therapy has significant potential to enhance the efficacy and reduce the toxicity of cancer therapies. NPs can be engineered to specifically target cancer cells, overcome drug resistance, and enhance the immune response against cancer. Continued research and development in this area have the potential to revolutionize cancer therapy and improve patient outcomes. The high-efficient, nontoxic, low-cost, and specific cancer therapy is a promising goal, which can be achieved through the development of nanotechnology. Nanometer-sized particles have novel optical, electronic, magnetic, or structural properties and are currently under intense development for applications in cancer, cardiovascular diseases, and degenerative neurological disorders. These integrated nanoparticle probes may allow simultaneous imaging and therapy of tumors and cardiovascular plaques in live animal models. Targeted nanoparticle drugs offer significant advantages in improving cancer therapeutic efficacy and simultaneously reducing drug toxicity. NPs have shown great promise in the field of cancer therapy, offering new and exciting possibilities for targeted drug delivery and improved treatment outcomes. From the extensive research conducted so far, it can be concluded that NPs hold significant potential for revolutionizing cancer treatment. One of the major advantages of NPs is their ability to specifically target cancer cells while sparing healthy cells, which reduces the side effects associated with traditional chemotherapy. They can be engineered to carry anticancer drugs, genes, or imaging agents, allowing for precise and controlled delivery to the tumor site. Additionally,

NPs can overcome biological barriers such as the blood–brain barrier, enabling the treatment of previously inaccessible tumors. NPs also offer the potential for combination therapies, where multiple therapeutic agents can be incorporated into a single nanoparticle system. This approach allows for synergistic effects and increased treatment efficacy. Moreover, NPs can be designed to respond to specific stimuli such as pH, temperature, or light, enabling triggered drug release at the tumor site, further enhancing therapeutic outcomes.

In the future, NPs in cancer therapy are expected to continue evolving and advancing. Researchers are exploring new materials, such as biodegradable polymers and inorganic NPs, to enhance drug delivery capabilities and reduce toxicity. Furthermore, efforts are being made to optimize the size, shape, and surface properties of NPs to improve their circulation time and targeting efficiency. The integration of nanotechnology with other emerging fields such as immunotherapy and precision medicine holds immense potential. NPs can be used to deliver immune-modulating agents, such as checkpoint inhibitors or vaccines, directly to the tumor microenvironment, enhancing the immune response against cancer cells. Additionally, personalized NPs could be developed to target specific mutations or biomarkers present in individual patients, enabling personalized treatment strategies.

Overall, the future of NPs in cancer therapy is bright. With ongoing advancements in nanotechnology, it is expected that NPs will play a pivotal role in improving treatment outcomes, reducing side effects, and ultimately contributing to the development of more effective and personalized cancer therapies. However, it is important to note that further research and rigorous clinical trials are necessary to ensure the safety, efficacy, and widespread implementation of nanoparticle-based therapies in clinical practice.

References

Abbasi R, Shineh G, Mobaraki M, Doughty S, Tayebi L (2023) Structural parameters of nanoparticles affecting their toxicity for biomedical applications: a review. J Nanopart Res 25(3):43. https://doi.org/10.1007/s11051-023-05690-w

Aghebati-Maleki A, Dolati S, Ahmadi M, Baghbanzhadeh A, Asadi M, Fotouhi A, Yousefi M, Aghebati-Maleki L (2020) Nanoparticles and cancer therapy: perspectives for application of nanoparticles in the treatment of cancers. J Cell Physiol 235(3):1962–1972. https://doi.org/10.1002/jcp.29126

Al-Zoubi MS, Al-Zoubi RM (2022) Nanomedicine tactics in cancer treatment: challenge and hope. Crit Rev Oncol 174(June):103677. https://doi.org/10.1016/j.critrevonc.2022.103677

Bae KH, Chung HJ, Park TG (2011) Nanomaterials for cancer therapy and imaging. Mol Cells 31(4):295–302. https://doi.org/10.1007/s10059-011-0051-5

Cao L, Zhu Y, Wang W, Wang G, Zhang S, Cheng H (2021) Emerging nano-based strategies against drug resistance in tumor chemotherapy. Front Bioeng Biotechnol 9:798882. https://doi.org/10.3389/fbioe.2021.798882

Cerna T, Stiborova M, Adam V, Kizek R, Eckschlager T (2016) Nanocarrier drugs in the treatment of brain tumors. Nanocarrier Drugs Treat Brain Tumors 2:407–416. https://doi.org/10.20517/2394-4722.2015.95

Cheng S, Li Z, Gao R, Xing B, Yunong Gao Y, Yang SQ et al (2021) A pan-cancer single-cell transcriptional atlas of tumor infiltrating myeloid cells. Cell 184(3):792–809. https://doi.org/10.1016/j.cell.2021.01.010

Chugh H, Sood D, Chandra I, Tomar V, Dhawan G, Chandra R (2018) Role of gold and silver nanoparticles in cancer nano-medicine. Artif Cells Nanomed Biotechnol 46(1):1210–1220. https://doi.org/10.1080/21691401.2018.1449118

Da Silva CG, Peters GJ, Ossendorp F, Cruz LJ (2017) The potential of multi-compound nanoparticles to bypass drug resistance in cancer. Cancer Chemother Pharmacol 80(5): 881–894. https://doi.org/10.1007/s00280-017-3427-1

Dadwal A, Baldi A, Narang RK (2018) Nanoparticles as carriers for drug delivery in cancer. Artif Cells Nanomed Biotechnol 46(2):295–305. https://doi.org/10.1080/21691401.2018.1457039

Debele TA, Yeh C-F, Wen-Pin S (2020) Cancer immunotherapy and application of nanoparticles in cancers immunotherapy as the delivery of immunotherapeutic agents and as the immunomodulators. Cancer 12(12):3773. https://doi.org/10.3390/cancers12123773

Dong P, Rakesh KP, Manukumar HM, Mohammed YHE, Karthik CS, Sumathi S, Mallu P, Qin H-L (2019) Innovative nano-carriers in anticancer drug delivery-a comprehensive review. Bioorg Chem 85(April):325–336. https://doi.org/10.1016/j.bioorg.2019.01.019

Ertas YN, Dorcheh KA, Akbari A, Jabbari E (2021) Nanoparticles for targeted drug delivery to cancer stem cells: a review of recent advances. Nanomaterials 11(7):1755. https://doi.org/10.3390/nano11071755

Fontana F, Liu D, Hirvonen J, Santos HA (2017a) Delivery of therapeutics with nanoparticles: what's new in cancer immunotherapy? WIREs Nanomed Nanobiotechnol 9(1):e1421. https://doi.org/10.1002/wnan.1421

Fontana F, Shahbazi M-A, Liu D, Zhang H, Mäkilä E, Salonen J, Hirvonen JT, Santos HA (2017b) Multistaged nanovaccines based on porous silicon@acetalated dextran@cancer cell membrane for cancer immunotherapy. Adv Mater 29(7):1603239. https://doi.org/10.1002/adma.201603239

Gaudino SJ, Kumar P (2019) Cross-talk between antigen presenting cells and T cells impacts intestinal homeostasis, bacterial infections, and tumorigenesis. Front Immunol 10:360. https://doi.org/10.3389/fimmu.2019.00360

Goodman, Louis S., Maxwell M. Wintrobe, William Dameshek, Morton J. Goodman, Alfred Gilman, And Margaret T. Mclennan. 1946. "Nitrogen mustard therapy: use of methyl-bis (beta-chloroethyl)amine hydrochloride and tris(beta-chloroethyl)amine hydrochloride for Hodgkin's disease, lymphosarcoma, leukemia and certain allied and miscellaneous disorders." J Am Med Assoc 132(3): 126–132. DOI: https://doi.org/10.1001/jama.1946.02870380008004.

Hoshyar N, Gray S, Han H, Bao G (2016) The effect of nanoparticle size on in vivo pharmacokinetics and cellular interaction. Nanomedicine 11(6):673–692. https://doi.org/10.2217/nnm.16.5

Hossen S, Khalid Hossain M, Basher MK, Mia MNH, Rahman MT, Jalal Uddin M (2019) Smart nanocarrier-based drug delivery systems for cancer therapy and toxicity studies: a review. J Adv Res 15(January):1–18. https://doi.org/10.1016/j.jare.2018.06.005

Hu N, Antoury L, Baran TM, Soumya Mitra C, Bennett F, Rigo F, Foster TH, Wheeler TM (2018) Non-invasive monitoring of alternative splicing outcomes to identify candidate therapies for myotonic dystrophy type 1. Nat Commun 9(1):5227. https://doi.org/10.1038/s41467-018-07517-y

Jain P, Kathuria H, Momin M (2021) Clinical therapies and nano drug delivery systems for urinary bladder cancer. Pharmacol Ther 226:107871. https://doi.org/10.1016/j.pharmthera.2021.107871

Jia J, Zhang Y, Xin Y, Jiang C, Yan B, Zhai S (2018) Interactions between nanoparticles and dendritic cells: from the perspective of cancer immunotherapy. Front Oncol 8:404. https://doi.org/10.3389/fonc.2018.00404

Joudeh N, Linke D (2022) Nanoparticle classification, physicochemical properties, characterization, and applications: a comprehensive review for biologists. J Nanobiotechnol 20(1):262. https://doi.org/10.1186/s12951-022-01477-8

Ke X, Howard GP, Tang H, Cheng B, Saung MT, Santos JL, Mao H-Q (2019) Physical and chemical profiles of nanoparticles for lymphatic targeting. Adv Drug Deliv Rev 151:72–93. https://doi.org/10.1016/j.addr.2019.09.005

Khan I, Saeed K, Khan I (2019) Nanoparticles: properties, applications and toxicities. Arab J Chem 12(7):908–931. https://doi.org/10.1016/j.arabjc.2017.05.011

Kishimoto TK, Maldonado RA (2018) Nanoparticles for the induction of antigen-specific immunological tolerance. Front Immunol 9:230. https://doi.org/10.3389/fimmu.2018.00230

Lu H, Guangzhao Yang F, Ran TG, Sun C, Zhao Q, Wang S (2020) Polymer-functionalized mesoporous carbon nanoparticles on overcoming multiple barriers and improving oral bioavailability of probucol. Carbohydr Polym 229:115508. https://doi.org/10.1016/j.carbpol.2019.115508

Mandriota G, Di Corato R, Benedetti M, De Castro F, Fanizzi FP, Rinaldi R (2019) Design and application of cisplatin-loaded magnetic nanoparticle clusters for smart chemotherapy. ACS Appl Mater Interfaces 11(2):1864–1875. https://doi.org/10.1021/acsami.8b18717

Marzi M, Osanloo M, Vakil MK, Mansoori Y, Ghasemian A, Dehghan A, Zarenezhad E (2022) Applications of metallic nanoparticles in the skin cancer treatment. Biomed Res Int 2022: 2346941. https://doi.org/10.1155/2022/2346941

Mu W, Chu Q, Liu Y, Zhang N (2020) A review on nano-based drug delivery system for cancer chemoimmunotherapy. Nano-Micro Lett 12(1):142. https://doi.org/10.1007/s40820-020-00482-6

Nasrollahzadeh M, Mohammad Sajadi S, Sajjadi M, Issaabadi Z (2019) Chapter 1 - an introduction to nanotechnology. In: Mahmoud Nasrollahzadeh S, Sajadi M, Sajjadi M, Issaabadi Z, Atarod M (eds) Interface science and technology, vol 28. Elsevier, Amsterdam, pp 1–27. https://doi.org/10.1016/B978-0-12-813586-0.00001-8

Nie S, Xing Y, Kim GJ, Simons JW (2007) Nanotechnology applications in cancer. Annu Rev Biomed Eng 9(1):257–288. https://doi.org/10.1146/annurev.bioeng.9.060906.152025

Nsairat H, Khater D, Sayed U, Odeh F, Al Bawab A, Alshaer W (2022) Liposomes: structure, composition, types, and clinical applications. Heliyon 8(5):e09394. https://doi.org/10.1016/j.heliyon.2022.e09394

Ozben T (2006) Mechanisms and strategies to overcome multiple drug resistance in cancer. FEBS Lett 580(12):2903–2909. https://doi.org/10.1016/j.febslet.2006.02.020

Patra JK, Das G, Fraceto LF, Campos EVR, del Pilar M, Rodriguez-Torres LS, Acosta-Torres LA, Diaz-Torres et al (2018) Nano based drug delivery systems: recent developments and future prospects. J Nanobiotechnol 16(1):71. https://doi.org/10.1186/s12951-018-0392-8

Pérez-Herrero E, Fernández-Medarde A (2015) Advanced targeted therapies in cancer: drug nanocarriers, the future of chemotherapy. Eur J Pharm Biopharm 93(June):52–79. https://doi.org/10.1016/j.ejpb.2015.03.018

Raj AT, Kheur S, Bhonde R, Mani VR, Baeshen HA, Patil S (2021) Assessing the effect of human dental pulp mesenchymal stem cell secretome on human oral, breast, and melanoma cancer cell lines. Saudi J Biol Sci 28(11):6556–6567. https://doi.org/10.1016/j.sjbs.2021.07.029

Raza F, Evans L, Motallebi M, Zafar H, Pereira-Silva M, Saleem K, Peixoto D et al (2023) Liposome-based diagnostic and therapeutic applications for pancreatic cancer. Acta Biomater 157(February):1–23. https://doi.org/10.1016/j.actbio.2022.12.013

Rodríguez F, Caruana P, De la Fuente N, Español P, Gámez M, Balart J, Llurba E et al (2022) Nano-based approved pharmaceuticals for cancer treatment: present and future challenges. Biomolecules 12(6):784. https://doi.org/10.3390/biom12060784

Rudzińska M, Daglioglu C, Savvateeva LV, Kaci FN, Antoine R, Zamyatnin AA (2021) Current status and perspectives of protease inhibitors and their combination with nanosized drug delivery systems for targeted cancer therapy. Drug Des Devel Ther 15:9–20. https://doi.org/10.2147/DDDT.S285852

Sahoo SK, Labhasetwar V (2003) Nanotech approaches to drug delivery and imaging. Drug Discov Today 8(24):1112–1120. https://doi.org/10.1016/S1359-6446(03)02903-9

Sajja KH, Michael East P, Mao H, Wang AY, Nie S, Yang L (2009) Development of multifunctional nanoparticles for targeted drug delivery and noninvasive imaging of therapeutic effect. Curr Drug Discov Technol 6(1):43–51. https://doi.org/10.2174/157016309787581066

Shao L, Gao Y, Yan F (2011) Semiconductor quantum dots for biomedical applications. Sensors 11(12):11736–11751. https://doi.org/10.3390/s111211736

Shao K, Singha S, Clemente-Casares X, Tsai S, Yang Y, Santamaria P (2015) Nanoparticle-based immunotherapy for cancer. ACS Nano 9(1):16–30. https://doi.org/10.1021/nn5062029

Shi J, Votruba AR, Farokhzad OC, Langer R (2010) Nanotechnology in drug delivery and tissue engineering: from discovery to applications. Nano Lett 10(9):3223–3230. https://doi.org/10.1021/nl102184c

Su S, Kang PM (2020) Recent advances in nanocarrier-assisted therapeutics delivery systems. Pharmaceutics 12(9):837. https://doi.org/10.3390/pharmaceutics12090837

Sung H, Ferlay J, Siegel RL, Laversanne M, Soerjomataram I, Jemal A, Bray F (2021) Global cancer statistics 2020: GLOBOCAN estimates of incidence and mortality worldwide for 36 cancers in 185 countries. CA Cancer J Clin 71(3):209–249. https://doi.org/10.3322/caac.21660

Sushnitha M, Evangelopoulos M, Tasciotti E, Taraballi F (2020) Cell membrane-based biomimetic nanoparticles and the immune system: immunomodulatory interactions to therapeutic applications. Front Bioeng Biotechnol 8:627. https://doi.org/10.3389/fbioe.2020.00627

Sutradhar KB, Amin ML (2014) Nanotechnology in cancer drug delivery and selective targeting. ISRN Nanotechnol 2014:939378. https://doi.org/10.1155/2014/939378

Wong AD, Ye M, Ulmschneider MB, Searson PC (2015) Quantitative analysis of the enhanced permeation and retention (EPR) effect. PLoS ONE 10(5):e0123461. https://doi.org/10.1371/journal.pone.0123461

Wu J (2021) The enhanced permeability and retention (EPR) effect: the significance of the concept and methods to enhance its application. J Personal Med 11(8):771. https://doi.org/10.3390/jpm11080771

Yadav N, Yadav RR, Dey KK (2022) Microwave assisted formation of trimetallic AuPtCu nanoparticles from bimetallic nano-islands: why it is a superior new age biocidal agent compared to monometallic & bimetallic nanoparticles. J Alloys Compd 896(March):163073. https://doi.org/10.1016/j.jallcom.2021.163073

Yang R, Jun X, Ligeng X, Sun X, Chen Q, Zhao Y, Peng R, Liu Z (2018) Cancer cell membrane-coated adjuvant nanoparticles with mannose modification for effective anticancer vaccination. ACS Nano 12(6):5121–5129. https://doi.org/10.1021/acsnano.7b09041

Yang M, Li J, Ping G, Fan X (2021) The application of nanoparticles in cancer immunotherapy: targeting tumor microenvironment. Bioactive Mater 6(7):1973–1987. https://doi.org/10.1016/j.bioactmat.2020.12.010

Yao Y, Zhou Y, Liu L, Yanyan X, Chen Q, Wang Y, Shijie W, Deng Y, Zhang J, Shao A (2020) Nanoparticle-based drug delivery in cancer therapy and its role in overcoming drug resistance. Front Mol Biosci 7:193. https://doi.org/10.3389/fmolb.2020.00193

Yuan F, Dellian M, Fukumura D, Leunig M, Berk DA, Torchilin VP, Jain RK (1995) Vascular permeability in a human tumor xenograft: molecular size dependence and cutoff size. Cancer Res 55(17):3752–3756

Zang X, Zhao X, Haiyang H, Qiao M, Deng Y, Chen D (2017) Nanoparticles for tumor immunotherapy. Eur J Pharm Biopharm 115(June):243–256. https://doi.org/10.1016/j.ejpb.2017.03.013

Zhang J, Zhang T, Gao J (2022) Biocompatible iron oxide nanoparticles for targeted cancer gene therapy: a review. Nanomaterials 12(19):3323. https://doi.org/10.3390/nano12193323

Zhuang J, Holay M, Park JH, Fang RH, Zhang J, Zhang L (2019) Nanoparticle delivery of immunostimulatory agents for cancer immunotherapy. Theranostics 9(25):7826–7848. https://doi.org/10.7150/thno.37216

Zitvogel L, Apetoh L, Ghiringhelli F, Kroemer G (2008) Immunological aspects of cancer chemotherapy. Nat Rev Immunol 8(1):59–73. https://doi.org/10.1038/nri2216

Synthesis of Different Types, Shapes, and Sizes of Nanocarriers Using Synthetic and Biological Approaches

2

Mohaddeseh Fatemi and Zohreh Bahrami

Abstract

The application of nanomaterials in different fields totally depends on their characteristics. Considering the impact of the synthesis method on the final product's characteristics such as size and morphology, selection of the proper method is critical. In general, the nanomaterial synthesis methods can be classified into two main categories: top-down and bottom-up. In the top-down methods, bulk materials are converted into nanomaterials using specific techniques such as mechanical milling. In the bottom-up methods, following the selection of a suitable precursor and creation of adequate conditions, atoms and molecules will lie next to each other and make nanomaterials consequently. Although most of the top-down synthesis methods are capable of mass production of nanomaterials, the diversity and ease of the bottom-up synthesis methods have made this category more interesting. Different types of nanomaterials can be used as carriers for cancer therapy, including zero-dimensional (0D), one-dimensional (1D), two-dimensional (2D), and three-dimensional (3D) nanomaterials. This chapter investigates the most important synthesis methods for the abovementioned nanomaterials and evaluates the corresponding effective parameters.

Keywords

Sol-gel · Co-precipitation · Polyol · Green synthesis · Sonochemistry · Hydrothermal · Template synthesis · Microemulsion

M. Fatemi · Z. Bahrami (✉)
Faculty of New Sciences and Technologies, Department of Nanotechnology, Semnan University, Semnan, Iran
e-mail: bahrami.zoh@semnan.ac.ir

F. A. Khan (ed.), *Nano Drug Delivery for Cancer Therapy*,
https://doi.org/10.1007/978-981-99-6940-1_2

2.1 Nanomaterials Synthesis Methods

2.1.1 0D Nanomaterials

Various categories of zero-dimensional nanomaterials serve as carriers in cancer therapy, encompassing metallic, metal oxide, polymer nanoparticles (NPs), quantum dots, and dendrimers. In the forthcoming sections, we discuss the important techniques employed for the synthesis of these nanocarriers.

2.1.1.1 Metal NPs

2.1.1.1.1 Co-precipitation Method

The fundamental concept behind this technique is that the final product is separated from the initial solution as a solid and insoluble phase. In this method, the synthesis takes place in both aqueous and non-aqueous media. First, the appropriate precursor is dissolved in the solvent, and then, with the help of a reducing agent, the metal ions in the solution are converted into the metal atoms. In this way, the initial nucleation happens and following the growth procedure, metal NPs will be produced. To convert metal ions into metal atoms, reducing agents such as H_2, hydrazine, sodium borohydride ($NaBH_4$), sodium citrate, electrochemical reduction, and/or high-energy beams are used. Gold nanoparticles (Au NPs) are among the most important types of metal NPs used as nanocarriers for anticancer drugs. They can be synthesized by co-precipitation method in both aqueous and non-aqueous media. Synthesis of Au NPs in aqueous medium begins with dissolving the $HAuCl_4$ precursor in water and continues using reducing agents like sodium citrate salt, which can also serve as a stabilizing agent. The presence of citrate anions around the Au NPs prevents them from agglomeration by electrostatic repulsion and thus their size can be controlled. This method, also known as the "citrate method" (see Fig. 2.1a) allows synthesis of Au NPs in the size range of 1–150 nm with different morphologies such as spherical, nanostar, nanocage, and nanoshell. The use of a strong reducing agent such as $NaBH_4$ in the nucleation process ensures that the resultant particles are small and monodisperse due to the simultaneous rapid nucleation of a large number of nanoparticles in the solution (Her et al. 2017; Li et al. 2021b; Personick and Mirkin 2013; Schletz et al. 2023; Stangherlin et al. 2020). In 2010, Ojea Jiménez et al. used heavy water instead of ordinary water to synthesize Au NPs in the presence of sodium citrate salt. Results showed that the nanoparticles synthesized in heavy water were smaller in size and had a narrower particle size distribution (Ojea-Jiménez et al. 2010). In order to synthesize Au NPs by co-precipitation method in a non-aqueous medium, we should be careful in choosing the precursor and reducing agent, because all the materials used must be soluble in a non-aqueous solvent such as tetrahydrofuran (THF). In this case, a thiolating agent is typically used as a stabilizer which can also control the growth rate of the initial nuclei and final NPs (see Fig. 2.1b). Electrochemical reduction of gold ions can occur on the surface of an electrode inside an electrochemical cell in the presence of an aqueous/organic solvent. In this technique, compounds such as

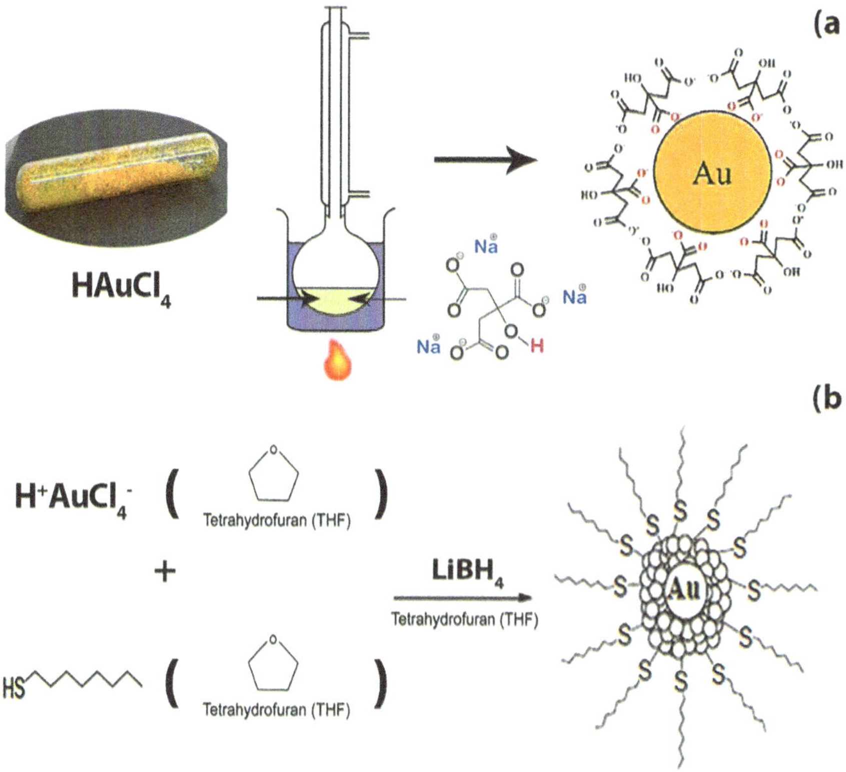

Fig. 2.1 Synthesis of gold nanoparticles by the co-precipitation method: (**a**) in aqueous medium using sodium citrate and (**b**) in non-aqueous THF medium in the presence of a thiolating agent

tetraalkylammonium salts, acetonitrile, cefazolin, polypyrrole, poly(N-vinylpyrrolidone), penicillin, and cationic surfactants are used as stabilizers (Haro-González et al. 2019). Synthesis of metal nanoparticles using high-energy beams is also possible. In this method, first an appropriate precursor is dissolved in water and then high-energy beams such as gamma, X-rays, and UV are irradiated to the solution. As a result of high-energy beams colliding the water molecules, free electrons are released for a short time in the reaction medium. Metal ions are reduced in the presence of free electrons and metal NPs are synthesized subsequently. This technique is only suitable for the synthesis of noble metal NPs. Use of stabilizing agents such as chitosan in the reaction medium allows control of particles size and morphology (Chen et al. 2007). In a study, Au NPs were synthesized using X-ray irradiation to an aqueous solution of two different gold precursors. First, complexes of gold and polymer compounds such as poly(1-vinyl-1,2,4-triazole) and poly (1-vinyl-1,2,4 triazole)-poly(acrylic acid) were prepared. After complete dissolution of precursors in water and before X-ray irradiation, a certain amount of ethanol was added to the solutions as a scavenger of hydroxyl radicals. Then, the solutions were irradiated by X-rays using a 5-BKhV-6 W tube with tungsten anode (applied voltage was 45 kVp with anode current of ca. 80 mA). Au NPs with very small size (1.5–3 nm), narrow particle size distribution, and spherical morphology were synthesized (Zharikov et al. 2022).

2.1.1.1.2 Polyol Method

In order to better control the size and morphology of metal nanoparticles, they can be synthesized in an alcoholic medium. Alcohol can be used as a solvent and a reducing agent, simultaneously. However, the main problem encountered when using alcohols is the final product agglomeration. To address this problem, polyols can be used. "Polyol method" is in fact a kind of co-precipitation in a non-aqueous medium. In this technique, an appropriate precursor along with additives like polyvinylpyrrolidone and sodium borohydride is dissolved in a polyol such as ethylene glycol. Then, the reaction is carried out under reflux in the temperature range of 80–200 °C. Finally, the product is separated as a solid phase (Ammar and Fiévet 2020). Copper nanoparticles (Cu NPs) can be used as a drug delivery system, the same way as gold, silver, and palladium NPs. It has been shown that selection of the initial precursor, polyol used, additives and synthesis temperature can affect the size of Cu NPs. Using various precursors like anhydrous copper sulfate, pentahydrate copper sulfate, copper hydroxide, copper acetate, and copper chloride in solvents such as ethylene glycol and diethylene glycol along with temperature control in the range of 80–200 °C, 2–220 nm Cu NPs are synthesized (Chandrakala et al. 2022; Gande and Pushpavanam 2021). In the synthesis method used by Wolf et al. in 2022, silver (Ag) ions were reduced in hot ethylene glycol in the presence of polyacrylic acid (PAA) and Ag NPs were synthesized. In this study, 3 nm spherical NPs with a narrow particle size distribution were obtained. Due to the presence of a PAA layer around the Ag NPs, they were completely stable in an aqueous media at pH 10. In this method, ethylene glycol was used both as a solvent and a reducing agent (Wolf et al. 2022).

2.1.1.1.3 Green Synthesis

Although lots of physical and chemical routes have been approved for the synthesis of metal NPs, such methods are not considered as environmentally friendly which has limited their applications in food and medical fields. The main idea behind the green synthesis method is to use biological systems such as microbes, fungi, and plant extracts for NPs synthesis. Extracts are advantageous in that they can act as mild, renewable, and nontoxic reducing and stabilizing agents, eliminating the need for chemical reducing agents such as sodium borohydride and expensive polymer stabilizers and capping agents. Generally, the three important requirements for the synthesis of NPs are selections of: (1) a good solvent, (2) a suitable reducing agent, and (3) a proper stabilizer. Frequently, the chemical methods used are too costly and incorporate the use of hazardous and toxic chemicals causing environmental risks, but the green synthesis technique is considered as a safe and eco-friendly approach to synthesize NPs for biomedical applications. In the biomedical field, the used NPs should have very little or zero toxicity and high biocompatibility and

Table 2.1 Green synthesis of metal NPs using plant extracts

Type of NPs	Plant	Morphology	Particle size (nm)	References
Ag	*Flaxseed*	Spherical	46.98 ± 12.45	Alzubaidi et al. (2023)
Ag	*Lawsonia inermis*	Spherical	28–60	Said et al. (2023)
Ag	*Carrot*	Spherical	17–30	Fareed et al. (2023)
Au	*Capsicum annum* fruit	Spherical	20–30	Patil et al. (2023)
Au	*Mentha longifolia*	Spherical	36.4	Li et al. (2021a)
Au	*Spinacia oleracea L*	Spherical	16.7	Zhu et al. (2022)
Au	*Tecoma capensis*	Spherical	10–35	Hosny et al. (2022)
Au	*Jatropha integerrima Jacq*	Spherical	38.8	Suriyakala et al. (2022)
Au	*Solidago canadensis L*	Spherical, triangular, hexagonal	8–200	Mariychuk et al. (2020)
Cu	*Jatropha curcas*	Spherical	12 ± 1	Ghosh et al. (2020)
Cu	*Cotton*	Spherical	13–35	Pérez-Alvarez et al. (2021)
Cu	*Hagenia abyssinica (Brace)*	Spherical, hexagonal, triangular	34.76	Murthy et al. (2020)

biodegradability. Use of plants for the synthesis of metallic nanoparticles has attracted more attention in recent years due to its rapid, environmentally friendly, pathogenic and economical protocol, providing a single-step mechanism for the biological process (Chandrakala et al. 2022). Different parts of a plant can be used such as roots, stems, leaves, flowers, fruits, and seeds. Plants are rich in metabolites like alkaloids, phenoloids, and phenolic acids. These compounds can act as reducing agents and stabilizers and prevent particles from agglomeration while controlling their size. Table 2.1 is a summary of the latest works done in the field of metal NPs synthesis from plants. Biological synthesis of NPs by microorganisms is also feasible. Using their enzymes, bacteria are capable of converting metal ions into metal atoms. This method is promising for synthesizing NPs because it requires low energy and is capable of controlling the synthesis process. For example, Ag NPs were synthesized from the source of *Streptacidiphilus durhamensis HGG16n* with a size range of 8–48 nm. Similarly, *Bacillus endophyticus* and *Deinococcus radiodurans* were capable of producing Ag NPs with different morphologies and sizes (Chandrakala et al. 2022). In different studies, *Psychrophilic bacteria*, *Phaeocystis antarctica*, *Pseudomonas meridiana*, *Arthrobacter kerguelensis*, *Arthrobacter gangotriensis*, *Bacillus indicus*, *Bacillus subtilis*, *Klebsiella pneumonia,* and *Bacillus cecembensis* were employed to biosynthesize Ag NPs with sizes ranging from 6 to 13 nm. It seemed that the generation and stability of Ag NPs depended on temperature, pH, or kind of bacteria from which the supernatant was obtained. Several shapes and sizes of Ag NPs were found in the final product. These included hexagonal, spherical, triangular, circular, and cuboidal which depended on the culture medium utilized for the growth of bacteria (Alsamhary 2020; Chopra et al. 2022; Ibrahim et al. 2019; Ibrahim et al. 2021; Saleh and Alwan 2020). The same way as Ag NPs, synthesis of Au NPs from bacteria is also possible. Various types of bacteria were used for this purpose including *Bacillus subtilis 168*, *Micrococcus yunnanensis strain J2*, *Shewanella algae*, *Rhodopseudomonas capsulate*, *Plectonema*, *Rhodobacter*, *Delftia* sp, and *Marinobacter algicola*. By controlling the synthesis parameters such as temperature, pH, and concentration, NPs with different sizes (100–5 nm) and various morphologies (spherical, octahedral, cubic, periplasmic) were obtained (El-Bendary et al. 2022; Feng et al. 2008; He et al. 2007; Jafari et al. 2018; Sayadi et al. 2021; Wu and Ng 2017). Kumar et al. prepared Au NPs using culture supernatant of *Delftia* sp. *strain KCM-006* and used them as a carrier for the anticancer drug Resveratrol. Spherical shaped Au NPs with an approximate size of 11.3 nm and a zeta potential of -25 mV showed good stability. The designed release system was pH-sensitive and showed the highest release rate (95%) in the acidic medium (pH = 5.2) (Kumar et al. 2014).

2.1.1.2 Metal Oxide NPs

2.1.1.2.1 Sol-Gel Method

Sol-gel is one of the common and well-known bottom-up methods in the liquid media which is used to produce various types of nanomaterials. Fundamentals of sol-gel method are: production of a sol and its conversion into a gel. Sol which has been derived from the word "Solution" is a colloidal mixture, a blend of solid particles less than 100 nm in a liquid phase. Due to the Brownian motion of nanoparticles, they do not precipitate and would be dispersed in the liquid phase. Generally, two types of precursors are used in the sol-gel process: metal salts and metal alkoxides. Partial or complete hydrolysis also occurs in this technique. For example, in the synthesis of silica (SiO_2) NPs, tetraethyl orthosilicate (TEOS) is used as a precursor. This compound has 4 Si-OR groups. If all four OR groups convert into OH groups, the hydrolysis will be complete, but if the number of conversions is less than 4, the hydrolysis will be partial (see Fig. 2.2).

At this stage, the sol is formed and then it must be converted into a gel based on condensation reactions. These reactions are reversal of the hydrolysis in such a way that in condensation, two small molecules bond to each other in order to make a larger molecule, while in hydrolysis, the molecules are decomposed. During the condensation process, temperature conditions and pH changes are the effective parameters. There are two types of condensation reactions: alcoholic and aqueous. If the water is removed in the condensation reaction, it is called aqueous, and if the alcohol is removed, it is called alcoholic. The result of both is the same, formation of a bond between two silicon groups through an oxygen bridge. Such a structure is called a gel. The initial gel is very voluminous. It fills the entire volume of the reaction chamber and traps the solvent inside it. For this, the initial gel is called a wet gel. As long as the solvent is inside the gel, it will not have a strong structure, so it must be dried. Before the drying stage, aging is a must. At this step, the gel has a

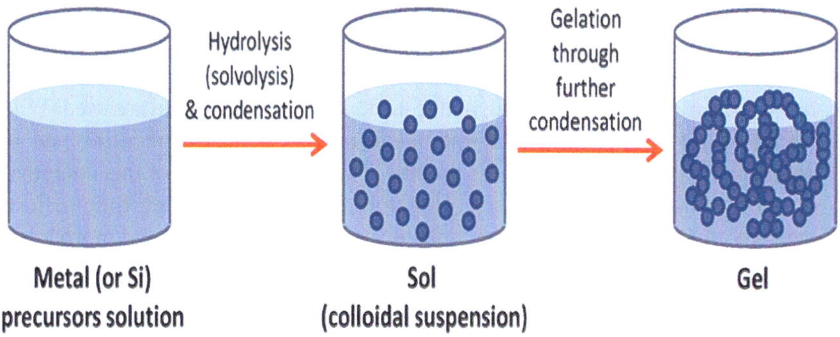

Fig. 2.2 Schematics of sol-gel synthesis method

chance to form new bonds and thus becomes stronger. This process may last a few hours to several days (Navas et al. 2021). The last step in the sol-gel method is gel drying. This process can occur in three ways: evaporation, supercritical drying, and freeze drying. Depending on the drying procedure used, the final product will be different. During the evaporation, the gel is dried at room temperature, and solvents such as water and ethanol are easily evaporated and removed. In this condition, the capillary pressure within the gel will increase and cause the network to shrink. As a result, the gel will collapse and its volume will reduce 5–10 times the wet gel. The final product obtained this way is called xerogel, which has an integrated structure without any porosity. In supercritical drying, an autoclave reactor is used. The critical temperature is determined according to the phase diagram of the used solvent. First of all, the gel is put inside the autoclave chamber and then the autoclave is placed inside the oven. When the solvent's temperature reaches the critical temperature, the pressure inside the chamber increases gradually. After the supercritical conditions is reached, the solvent within the gel turns into a supercritical fluid. By making the temperature fixed and gradual reduction of the pressure, the supercritical fluid turns into gas and leaves the system. In this case, the porous structure of the wet gel is preserved and this causes the formation of an aerogel. Aerogel is a very lightweight porous structure which is considered as a category of 3D nanomaterials. Use of aerogel as a drug delivery system is one of the significant applications of this nanomaterial. Due to the high critical temperature of water (374 °C) and ethanol (240 °C), other solvents with lower critical temperatures are also used in the supercritical drying. Carbon dioxide is one of the potential solvents for replacing water or ethanol, because it has a critical temperature close to the ambient temperature (~31 °C). In this case, the gel's drying method is called cold supercritical drying. Freeze drying is the counterpoint of supercritical drying. In this technique, first the solvent within the wet gel freezes and then is sublimed away slowly at low temperatures. The solvent removal does not cause network shrinkage and thus preserves the structural porosity. The final product of this technique is called cryogel, which is similar to the aerogel regarding the structural properties (Bokov et al. 2021). The important influential parameters in the sol-gel method are: type of the precursor, hydrolysis rate, pH, and calcination temperature. If the precursor's hydrolysis rate is very high, precipitation will occur before the sol is formed. In order to control the hydrolysis rate and thus the sol formation, chelating agents such as acetic acid, maleic acid, citric acid, tartaric acid, oxalic acid, and polyacrylic acid are used. When chelating agents are present, a compound forms between the precursor and the chelating agent before the hydrolysis process. The hydrolysis rate of this compound is slower than that of the precursor. Therefore, the sol has enough time to form. Conversely, the chelating agent may influence controlling the size and morphology of the final product. The effect of the chelating agent's concentration change can be studied using a parameter called complex coefficient (X). This parameter is obtained by dividing the chelating agent's concentration by the concentration of metal ions. In the synthesis of a certain type of NPs, various chelating agents will work differently. This fact was investigated in the synthesis of zirconia (ZrO_2) nanoparticles (Kazemi et al. 2020). In this study, citric acid, sucrose, and fructose were used as the chelating agents and zirconium hydroxide was the precursor. Compared with the other two chelating agents, when the fructose was used, ZrO_2 nanoparticles had the lowest size (39 nm), the highest surface area (27 $m^2\ g^{-1}$), and a tetragonal crystal phase. But, in the presence of citric acid and sucrose, a mixture of monoclinic and tetragonal crystalline phases was obtained. In order to remove the chelating agent, rinsing or calcination can be used. Calcination temperature affects the particle size and crystallinity. In a study in 2021, TiO_2 nanoparticles were synthesized using titanium tetraisopropoxide, isopropanol, and acetic acid as precursor, solvent, and chelating agent, respectively. The resulting sol turned into a gel after 12 h. Following the drying at 110 °C, the obtained gel was calcined at 400 °C, 600 °C, and 800 °C for 5 h, in order to investigate the effect of calcination temperature. XRD results showed that as the temperature increases, the thermodynamically more stable crystalline phase of rutile is the main phase in the synthesized samples. In addition, with the temperature rise, more growth is possible due to the increase in the amount of energy available for the initial nuclei, and the particles size also increases as a result (Saikumari et al. 2021). The effect of pH on size, morphology, and crystallinity of zinc oxide (ZnO) nanoparticles was investigated by Alias et al. (2010). In order to synthesize ZnO nanoparticles, a certain amount of zinc acetate was dissolved in methanol. After a clear solution was obtained, the pH was adjusted in the range of 6–11 using NaOH. In this pH range, particles with a size range of 50–36 nm were synthesized. Increasing the pH from 6 to 11, increased the crystallinity while decreased the size and agglomeration of the particles.

2.1.1.2.2 Solvothermal/Hydrothermal Method

Solvothermal method is among the most important liquid phase techniques used for the synthesis of nanomaterials. In general, high temperature synthesis can reduce the particles size. However, most of safe, cost-effective, and readily available solvents such as water have a relatively low boiling point and cannot be used in high temperatures under normal conditions. In order to address this limitation, an autoclave reactor is used. This way, the solvents can be used at temperatures higher than their boiling point without their volume change. Sealing the reaction chamber's cap prevents the solvent's vapor from releasing and at the same time increases the vapor pressure within the chamber. Therefore, the synthesis is possible at temperatures higher than the boiling point of the solvent and under pressures higher than the ambient pressure. Simplicity is another advantage of this method. Depending on the final product, an appropriate precursor is selected and dissolved in a proper solvent. Then the resulting solution is transferred to the reaction's chamber and placed in an oven with predetermined temperature for a specific time period. After this period and when the autoclave is cool, the reaction chamber's contents are removed and centrifuged. The final product will be in a solid phase and will not need any other purification. However, sometimes calcination is done in order to achieve a certain crystal phase. If water is used as the solvent, the technique will be called hydrothermal, while in the solvothermal technique non-aqueous solvents such as ethanol, methanol, ethylene glycol, 1,4-butanediol, and toluene are used (Choudhury et al. 2022). In this technique, factors such as type of the precursor, reaction's time and temperature, and additives can affect the final product characteristics, namely size, morphology, crystallinity, and crystal phase. Lots of metal oxide nanoparticles have been synthesized with this method, including Al_2O_3, CuO, Fe_2O_3, NiO, ZrO_2, TiO_2, $BaTiO_3$, $SrTiO_3$, and Fe_3O_4. These nanoparticles can be synthesized by both hydrothermal and solvothermal methods (Li et al. 2016). Following is a summary of related studies. Ahmadi et al. (2012) synthesized superparamagnetic Fe_3O_4 NPs using $FeCl_2$, $FeCl_3$, and NaOH with a molar ratio of 1:2:8 under hydrothermal conditions at temperatures 100 °C, 150 °C, and 200 °C for 30 min. Results showed that increasing the synthesis temperature caused particles to grow. In addition, saturation magnetization and coercivity of the obtained NPs also increased. Temperature rise did not have any effects on the NPs morphology and all the synthesized samples were spherical in shape. In 2018, Kim et al. (2018) synthesized superparamagnetic Fe_3O_4 NPs with solvothermal method using different concentrations of iron (III) chloride, trisodium citrate dihydrate, and sodium acetate in ethylene glycol as the solvent at 200 °C and for 10 h. Resulted NPs were spherical in shape and the particles size increased to ~300 nm with increasing the iron salt concentration. The synthesized NPs showed superparamagnetic properties, high saturation magnetization, high stability, and hydrophilicity. Temperature and time are the most significant parameters in the solvothermal/hydrothermal method. In 2020, Mohan et al. (2020) investigated the effect of these two parameters on the synthesis of ZnO NPs. They used zinc nitrate as the precursor. In order to investigate the effect of temperature, time was fixed and the synthesis was carried out at temperatures 100 °C, 125 °C, and 150 °C for 2 h. To study the effect of time, the reaction temperature was fixed at 120 °C and the synthesis was performed for 1, 3, and 5 h, respectively. XRD results showed that increasing the temperature decreased the crystallite sizes while increasing the time caused the crystallites to grow. All of the synthesized samples had the same wurtzite crystal phase. SEM micrographs showed that the particles size and morphology were strongly influenced by the synthesis temperature and time. At the temperatures 100 °C and 125 °C, ZnO flower-like nanoparticles were synthesized. By increasing the temperature to 150 °C, the morphology changed and spherical nanoparticles were obtained. ZnO nanorods were synthesized at a fixed temperature (120 °C) and different times. Time increase caused the initial nuclei to grow and the rods to get longer. In recent years, use of supercritical water for synthesis of metal oxide NPs by hydrothermal method has attracted many attention. In the critical point of water (374 °C and 22.1 MPa) its characteristics change, and this allows the synthesis of nanomaterials. Batchwise and continuous methods are the two potential techniques for the synthesis of NPs using supercritical water. The batchwise method is very similar to the conventional hydrothermal method. The only difference is that the temperatures needed for the synthesis are higher than the critical temperature of water. The autoclave used must be equipped with a pressure gauge. In order to avoid corrosion of the inner surface of the autoclave, which is in contact with supercritical water, it can be coated with a thin layer of nickel or platinum (Yoko et al. 2018). In continuous method, first the precursor is dissolved in water and heated a bit. In a separate step, water is converted into the supercritical water stream and then it is mixed with the precursor solution. The resulting mixture is transferred to the autoclave chamber and synthesis is performed under supercritical conditions. It has been proved that mixing the supercritical water with a lower temperature solution influences the final product properties strongly. The mixing rate must be higher than the rate of the chemical reaction that leads to the final product. Following the synthesis process termination, the contents within the reaction chamber are quickly cooled down and after that the precipitate is separated. Compared with the batchwise method, the NPs obtained from the continuous method are smaller and have a narrower particle size distribution (Yoko et al. 2020). Using the supercritical water, synthesis of a wide range of metal oxides including AlOOH, α-Fe_2O_3, Fe_3O_4, Co_3O_4, NiO, ZrO_2, and TiO_2 with a size range of 20–200 nm is achievable (Yoko et al. 2018).

2.1.1.2.3 Co-precipitation Method

The same way as metal NPs, metal oxide NPs can also be synthesized using co-precipitation method in aqueous and non-aqueous media. Fe_3O_4 NPs are among the most significant metal oxide NPs synthesized with this method in an aqueous medium. There are many reports regarding their synthesis (Ba-Abbad et al. 2022; Dheyab et al. 2020; LaGrow et al. 2019; Panariello et al. 2022; Slimani et al. 2021; Wen and Jin-Ming 2014). Typically, iron (II) and iron (III) salts with a molar ratio of 1:2 in the presence of an alkalizing agent such as NaOH or ammonia are used. The reaction medium's pH is usually adjusted in the range of 9–14. In order to avoid the oxidation of iron ions and synthesis of Fe_2O_3 instead of Fe_3O_4, the reaction is carried out under an inert atmosphere such as N_2 gas. The final product is a dark brown or black precipitate. Although the co-precipitation method is a simple and relatively fast technique for synthesis of Fe_3O_4 nanoparticles, reaching suitable conditions for mass production of NPs with controlled size and morphology is still a challenge and needs more detailed studies. Therefore, in recent years, there have been many studies related to the improvement of the co-precipitation method effective parameters. In 2022, the effect of additives on Fe_3O_4 particles morphology and size was investigated. In this study, ammonium hydroxide and sodium hydroxide were used as additives. The results showed that in the presence of ammonium hydroxide, the synthesized NPs were the same in size with a spherical shape, while when using sodium hydroxide, the morphology changed from spherical to cubic. In addition, compared with the ammonium hydroxide, sodium hydroxide caused the growth of initial nuclei and formation larger NPs. Size of the synthesized NPs in the presence of NH_4OH and NaOH was 10 ± 2 nm and 32 ± 2 nm, respectively. Although both samples showed ferromagnetic properties, NPs with cubic morphology had a higher saturation magnetization. The reason is the effect of structural characteristics on magnetic properties of NPs (Ba-Abbad et al. 2022). In another study conducted by Dheyab et al. in 2020, the effect of citric acid as a surface modification agent on size, morphology, magnetic properties, stability, and hydrophilicity of Fe_3O_4 NPs was investigated. Surface modification was implemented in a single step simultaneous with NPs synthesis. The operating temperature was 65 °C with a relatively short time interval (30 min) compared with the conventional co-precipitation method. In this study, surface modified Fe_3O_4 NPs ~19 nm in size, with spherical morphology and a narrow particle size distribution were synthesized. Zeta potential increase from −31 mV in pristine NPs to −45 mV in the surface modified NPs indicating the stability was increased. The latter had a relatively good saturation magnetization and due to the reduced agglomeration and increased stability, was suggested for biomedical applications (Dheyab et al. 2020). With the aim of achieving a scalable method, Gavriilidis et al. used microwaves with co-precipitation method for the synthesis of Fe_3O_4 NPs. For this, a flow reactor was used and the effect of residence time and microwave power was investigated. An increase in both parameters led to NPs size increase. By controlling the synthesis conditions, NPs with a size range of 35–70 nm were synthesized. The presence of stabilizing agents such as citric acid and dextran reduced the synthesis time. Capacity of mass production in this method was evaluated by increasing the dimensions of the reactor used. The best production rate obtained was 3.3 g/h (Panariello et al. 2022).

2.1.1.2.4 Combustion Method

Combustion synthesis (CS) is a technique for achieving a rapid and energy-efficient synthesis. A typical CS procedure utilizes a self-sustained exothermic reaction among well-mixed reactants to achieve the rapid and economical synthesis of particulate products. CS is classified into self-propagating high temperature synthesis (SHS) and solution combustion synthesis (SCS). In conventional SHS, the reactants are all in a solid state, still suffering from an inhomogeneous mixing. In order to reduce the particles size, additional treatments such as mechanical activation, intensive milling, or chemical dispersion are required after SHS, which increase the preparation cost and steps. On the contrary, SCS begins with an aqueous solution, which guarantees a molecular-level mixing of reactants. As a result, SCS serves as a promising route for rapid and direct synthesis of nanomaterials, especially for oxides, due to its simplicity, energy- and time-effectiveness, and low cost. Controllability over the final product phases and morphologies is the general difficulty encountered in SCS, which is the result of the rapid and uncontrollable combustion procedure (Wen and Jin-Ming 2014). This synthesis method starts with dehydration and thermal decomposition of the homogeneous solution and involves several thermally coupled exothermic reactions, which result in the formation of at least one solid product and a large amount of gas. It is worth nothing that this process allows for the synthesis of a variety of nanomaterials, including oxides, metals, alloys, and sulfides, which can be used in many important applications (Varma et al. 2016). SCS systems are classified based on the chemical composition of fuel, oxidizer, and solvent. Various types of organic fuels or their mixtures are typically dissolved in a solvent with metal nitrate hydrates as oxidizer. The prevalence of this class of oxidizer is probably explained by their good solubility in water and relatively low decomposition temperature that leads to the formation of active oxygen. Organic fuels such as urea, glycine, sucrose, and citric acid are the source of carbon and hydrogen and, in many cases, form complexes with metal ions, thus facilitating homogeneous mixing of cations in solutions or gels. The ideal fuel should have high solubility in the solvent and low decomposition temperature (below 400 °C), yield no other residual mass, be compatible with metal nitrates (reaction should not lead to explosion), and be readily available or easy to prepare. In most works related to SCS, water is used as a solvent. However, occasionally kerosene, alcohols, or formaldehyde are also utilized (Varma et al. 2016). Masoudpanah et al. investigated the role of fuel on morphologies of ZnO via solution combustion route. Poly(vinyl-pyrrolidone) (PVP) was used as fuel at different ratios to oxidant. The increase in fuel led to the transformation of giant pyramid-like particles to hexagonal nano-disk ones. N_2 adsorption–desorption isotherms showed that the specific surface area increased from 23 to 41 m^2/g at the higher fuel contents. By considering the decrease in particle size, the band gap energy value was turned from 3.02 to 3.13 eV (Vasei et al. 2020). Nanocrystalline MgO particles were prepared through combustion method using magnesium nitrate as oxidizer and hexamine as a fuel. The materials obtained by combustion method were subsequently annealed at 800 °C for 3 h to improve the crystallinity and phase purity. The cubic crystal structure with average crystalline size of 22 nm was obtained for the MgO NPs. The SEM micrographs showed that the MgO nanomaterials are seemingly porous and highly agglomerated with fine particles (Balamurugan et al. 2014). Copper oxide NPs were produced using the solution combustion technique with ascorbic acid as the fuel. The synthesized NPs were characterized by XRD, SEM, FT-IR, and UV–Vis spectrophotometer. The XRD results confirmed the monoclinic crystal phase. FT-IR absorption exhibited the existence of metal oxygen stretching bond at 422 cm^{-1}. SEM results showed that the particles were highly porous in nature with irregular shape. The band gap was found to be 1.65 eV from UV–Vis studies (Harini et al. 2022). A facile method for synthesis of ZnO NPs by rapidly heating a paste of zinc nitrate and sucrose on the hot plate at 500 °C was reported by Islam et al. (2019). In detail, the desired amounts of sucrose and $Zn(NO_3)_2{\cdot}6H_2O$ salt were dissolved in distilled water. The beaker was put on the preheated hot plate. Within 1–2 min, the mixture turned into a viscous solution and immediately after that the viscous solution started to decompose and generated brownish yellow gases. After 3–4 min, a blackish brown foam was obtained from the decomposition of zinc nitrate salt. Afterward, the blackish brown hot foam/powder was transferred from the beaker to the same bare hot plate. The blackish powder turned into yellowish-white powder of ZnO NPs. The transmission electron microscopy images revealed the spherical shape of the NPs with an average size of 35 nm. The band gap and the specific surface area of the ZnO were measured to be ~3.32 eV and 80.11 m^2/g, respectively (Islam et al. 2019).

2.1.1.2.5 Sonochemical Synthesis

High intensity ultrasound can be used for the production of novel materials. Sonochemistry originates from the extreme transient conditions induced by ultrasound, which produces unique hot spots that can achieve temperatures above 5000 K, pressures exceeding 1000 atmospheres, and heating and cooling rates in excess of 10^{10} K s^{-1}. Several phenomena are responsible for sonochemistry and in specific the production or modification of nanomaterials during ultrasonic irradiation. The most notable effects are consequences of acoustic cavitation (the formation, growth, and implosive collapse of bubbles) and can be categorized as primary sonochemistry (gas-phase chemistry occurring inside collapsing bubbles), secondary sonochemistry (solution-phase chemistry occurring outside the bubbles), and physical modifications (caused by high-speed jets or shock waves derived from bubble collapse) (see Fig. 2.3) (Xu et al. 2013).

Fe_3O_4 NPs can easily be synthesized from an inorganic iron precursor through decomposing iron salts using a sonochemical method via very high pressures and temperatures generated by ultrasonic irradiation. Due to the strong influence of NPs size on magnetic and structural properties of Fe_3O_4 NPs, the size selection proportional to the desired magnetization is very significant (Ali Dheyab et al. 2021). Synthesis of Fe_3O_4 NPs with ethylenediamine as a novel precipitating agent by ultrasonic treatment (40 kHz, 150 W) was reported by Boustani et al. (2020). Different parameters such as the effect of reaction temperature, reaction time, and precipitating agent were investigated. It was demonstrated that ethylenediamine

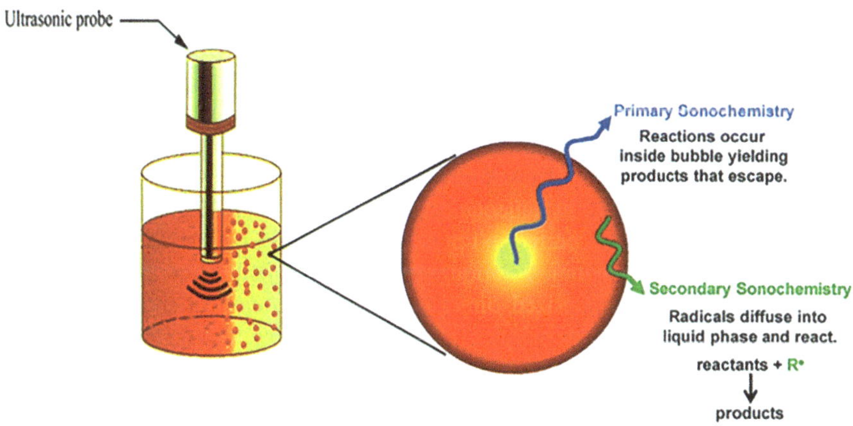

Fig. 2.3 Primary and secondary sonochemistry for preparation of nanomaterials

leads to the synthesis of larger magnetite NPs with the desired saturation magnetization (Ms). In another study, monodisperse magnetite nanocubes with a uniform particle size of about 80 nm were synthesized in an aqueous medium by single reaction sonochemical method using inexpensive and nontoxic metal salt ($FeSO_4$ $7H_2O$) as a reactant. The crystallinity of the magnetite nanocube enhanced by annealing treatment up to 600 °C with uniform shape and size distribution. The magnetic characterization of the NPs revealed saturation magnetization of 91 emu/g at 5 K for the as-synthesized sample and 94.8 emu/g for the sample annealed at the temperature of 600 °C in a vacuum chamber (Abbas et al. 2013). Due to the high magnetization, these NPs can be used for various applications, such as MRI and drug delivery (Ali Dheyab et al. 2021). In another work a pH-triggered release system based on core@shell mesoporous magnetic nanoparticles (MNP@$mSiO_2$) obtained using a simple and rapid ultrasound-assisted method was reported. A solution containing $MnSO_4$ and $FeSO_4$ in deionized water was prepared. The prepared solution was sonicated (130 W, 20 kHz). After 10 min of continuous irradiation, 10 mL of NaOH 2N solution was poured into the reactor, and the appearance of a dark brown color indicated the formation of magnetic nanoparticles. The coating of silica was performed based on the hydrolysis and condensation of TEOS. Performed characterizations revealed magnetic cores of $Fe_{2.9}Mn_{0.1}O_4$ (38 ± 6 nm) with specific loss power values adequate for hyperthermia (463 W/g) were surrounded by a hydroxyl group (–OH) functionalized mesoporous silica shell (10 ± 2 nm) with a large surface area (269 m^2 g^{-1}). MNP@$mSiO_2$ were loaded with DOX and aminosilane groups, providing pH-triggered DOX release at acidic media, driven by dipolar intermolecular interactions (Fuentes-García et al. 2021).

2.1.1.2.6 Green Synthesis

Another approach for producing metal oxide NPs is the green synthesis method. Numerous studies have addressed this approach, and some of the most recent findings are summarized in Table 2.2.

2.1.1.3 Polymeric NPs

Polymeric NPs are solid colloidal particles that span sizes from 10 to 1000 nm (1 μm). They are widely utilized due to their exceptional characteristics and high surface areas. These nanoparticles competently take drugs, DNA, and proteins to their target cells or organs. They are generally synthesized from preformed polymers or by direct polymerization of monomers. Regardless of the synthesis method, the final products are obtained as aqueous colloidal suspensions. The methods used to synthesize these NPs from preformed polymers are emulsification-solvent evaporation, salting-out, dialysis, ionic gelation, and solvent emulsification-diffusion. Conversely, they can be synthesized directly by polymerization of monomers using various polymerization techniques such as microemulsion, nanoemulsion, surfactant-free emulsion, and interfacial polymerization (Castro et al. 2022; Khan et al. 2023). The most important technique for synthesis of polymeric NPs is the microemulsion method, which will be discussed in the following.

Table 2.2 Green synthesis of metal oxide NPs using plant extracts

Type of NPs	Plant	Morphology	Particle size (nm)	References
ZnO	*Capsicum annum* fruit	Spherical	20–30	Patil et al. (2023)
ZnO	*Mangifera indica* seed-wrapped	Spherical, cylindrical, rectangle, and triangle shape clustered	40–70	Rajeshkumar et al. (2023)
ZnO	*Ocimum tenuiflorum*	Spherical	19	Elsamra et al. (2023)
ZnO	Aqueous leaf	Spherical	100	Dangana et al. (2023)
CuO	*Portulaca oleracea*	Spherical	5–30	Eid et al. (2023)
CuO	*Tanacetum parthenium*	Spherical	28	Ranjbar et al. (2023)
TiO_2	*Manihot esculenta*	Spherical	15–28	Bopape et al. (2023)
TiO_2	*Commelina benghalensis*	Spherical	~200	Al Masoudi et al. (2023)
TiO_2	*Juniperus phoenicea*	Spherical	10–30	Rathi and Jeice (2023)

2.1.1.3.1 Microemulsion Method

The term microemulsion applies to a mixture with at least three components: an oily phase, an aqueous phase, and a surface-active species, which is called surfactant. Sometimes the fourth component, i.e., co-surfactant can/must be present. Depending on the ratios between the components, in the two extremes, the microstructure of the microemulsions varies from very tiny water droplets dispersed in the oil phase (w/o microemulsion) to oil droplets dispersed in the water phase (o/w microemulsion). The microstructure of the mixture changes continuously from one to another extreme, namely from a spherical to cylindrical, tubular, and interconnected continuous oil and water phases separated with a very thin layer of surfactant molecules, in the middle, which is defined as a continuous microemulsion. The microemulsions of each kind are thermodynamically stable and transparent solutions. There are main differences between emulsions and microemulsions in terms of structure and stability (Asgari et al. 2019; Najjar 2012).

Surfactants have a critical role in microemulsion formation. They are molecules with at least two parts: one part soluble in polar solvents (hydrophilic) and the other part insoluble in the polar solvent (hydrophobic). Because of this double character, the term amphiphile is also used for surfactants. The polar part of the surfactant molecule is referred to as the head, and the non-polar part of the molecule as the tail. Having these two parts with opposing solubilization abilities gives the surfactant molecules unique capabilities, such as the tendency to adsorb at the surfaces and interfaces, which results in the decrease in the surface tension, and also formation of the aggregates inside the solutions, resulting in the formation of the microemulsions.

This double character of the surfactant molecule enables it to orient in the desired way while in contact with the two phases with different hydro/lipophilic properties, or to make aggregates inside of the solution with hydro- or lipophilic parts directed toward the media. Such aggregates can solubilize an oil in the aqueous phase (micelles) or water in the oily phase (reversed micelles). Usually, many surfactants can be used to form microemulsion, including cationic surfactants such as cetyltrimethylammonium bromide (CTAB), anionic surfactants such as bis (2-ethylhexyl) sulfosuccinate (AOT), sodium dodecyl benzene sulfonate (SDBS) and lauryl sodium sulfate (SDS), and nonionic surfactants such as Triton X-100 and sorbitan monooleate Span 80, nonylphenyl ether (NP-5) or polyoxyethylene (9) nonylphenyl ether (NP-9) (Asgari et al. 2019; Najjar 2012). With this introduction in mind, subsequent section presents some examples of the latest works done on synthesis of polymeric NPs by microemulsion method. Using reverse microemulsion method and zoledronic acid (Zol) and calcium as model bisphosphonate and metal molecules, respectively, Cui et al. prepared DOPA (an anionic lipid)-coated Zol-Ca nanocomposites (Zol-Ca@DOPA). Results showed that these composites are spherical, around 15 nm in size, monodispersed, and uniform. Based on such Zol-Ca@DOPA nanocomposites, developing a novel nanoparticle formulation for targeting delivery of Zol to subcutaneous tumors in order to explore the potential of using the Zol to directly inhibit tumor cell growth in bone diseases is possible (Li et al. 2017). In another study, Luo et al. prepared 1-hexadecyl-3-methylimidazolium bromide C16mimBr/butan-1 ol/cyclohexane/water ionic liquid microemulsion. They studied the effects of n-alkyl alcohols, alkanes, water content and temperature on the properties of microemulsion by dilution experiment. The microregion of microemulsion was identified by pseudo-ternary phase diagram and conductivity measurement. Then starch NPs were synthesized by water in oil (W/O) microemulsion-cross-linking methods with C16mimBr as surfactant. Synthesis of starch NPs with a mean diameter of 94.3 nm and narrow particle size distribution (SD = 3.3) was confirmed by dynamic light scattering (DLS). Scanning electron microscope (SEM) micrographs revealed that starch NPs were spherical granules with a size of ~60 nm. In addition, the results of Fourier transform infrared spectroscopy (FT-IR) and X-ray diffraction (XRD) verified the formation of cross-linking bonds in starch molecules. Finally, the drug loading and releasing properties of starch NPs were investigated with methylene blue (MB) as a drug model (Wang et al. 2016). In the work done by Zhao et al. bisphosphonate functionalized coordination polymer nanoparticles (CPNs) were prepared by W/O microemulsion for the treatment of bone metastatic breast cancer. CPNs are a class of burgeoning materials which are self-assembled from metal ions and organic bridging ligands. They have shown a great potential in drug delivery systems, because of their tunable compositions, sizes, shapes, biological compatibility, stability in physiological environments, and intrinsic biodegradability. In this study, the particle size was regulated by adjusting the volume ratio of water phase to oil phase in microemulsion. Herein, the ultra-small CPNs, ~55 nm were used for enhancing the targeting efficiency of nanoparticles by regulating its sizes (He et al. 2017). Luo et al. used a [C_3OHmim]Ac/TX-100 + 1-butanol/cyclohexane ionic liquid microemulsion cross-linking technique which allowed the formation of spherical octenyl succinic anhydride (OSA) starch-based nanoparticles with concentrated size distribution and nanosize diameters. This was done by tailor-made polar phase [C3OHmim]Ac with high dissolving capacity that was conducive to the uniform distribution of starch molecules, and composite surfactant TX-100/1-butanol that lowered the interfacial tension much more than the traditional surfactants. Regarding the drug loading properties of starch nanoparticles, it was found that the drug loading and encapsulation efficiency were influenced by loading time, loading temperature, and particle size as well as drug concentration. OSA-modified starch nanoparticles as drug carriers showed an increased initial release rate and total release amount in the release experiment of indomethacin (Qi et al. 2017).

2.1.2 1D Nanomaterials

1D nanomaterials possess unique shape-dependent properties and can be classified, according to their shapes, to nanorods, nanotubes, nanowires, self-assembled nanochains, etc. The biological characteristics of these nanomaterials, including high drug loading efficiency, prolonged blood circulation, the ability to capture cancer cells, unique cellular uptake mechanisms, efficient photothermal conversion, and material tunability have aided in extending their potential to biomedical applications, particularly in cancer therapy and diagnosis (Shin et al. 2023). Considerable research has been conducted on new routes to controllable synthesis of 1D nanomaterials. Template synthesis is the most considerable method for this goal which we will discuss it in the following sections.

2.1.2.1 Template Synthesis Method

Template synthesis is a very effective method for synthesis of nanomaterials that has been widely used in recent years. Synthesis using a pre-existing nanostructured template is one of the most effective strategies for achieving high degrees of synthetic control. Typically, nanomaterials synthesized by templating strategies hold a well-defined size, shape, and configuration, which usually benefits from the directing effect of the templates. The general route for the template synthesis of nanostructured materials includes the following steps: (1) template preparation, (2) directed synthesis of target materials using the template, and (3) template removal (if necessary). In nanoscale synthesis, theoretically, a template can be any substance with nanostructured features. Thus, there are abundant types of materials that can be taken into consideration. Some are naturally occurring (such as nanostructured minerals, biological molecular assemblies, and organs), while others are sourced from synthetic nanostructures which usually benefit from pre-existing synthesis methods (both direct synthesis and template synthesis), including top-down and bottom-up approaches. Their physical properties can vary greatly, from soft materials such as molecular assemblies to hard materials such as colloidal particles, porous membranes, and patterned solid surfaces. With regard to their chemical properties, they can either behave as a pure physical directing agent, which we refer to as a physical template, or as a physical director which is also involved in a chemical transformation during the synthesis, which we refer to as a chemical template. For the synthesis of target materials, synthetic strategies based on physical templates include a variety of methods such as surface coating and channel replication. In cases of chemical template synthesis, chemical transformations during the synthesis may include isomerization, addition, substitution, and elimination of chemical species. As a result, nearly all types of materials can be utilized to form 1D nanomaterials. In order to remove the template after synthesis, as long as the physical and chemical properties of the resultant materials can be preserved, both physical methods such as dissolution, and chemical methods including calcination and etching can be applied according to the nature of the templates (Liu et al. 2013b; Xie et al. 2016). The hard template is a rigid material, whose stable structure directly determines the size and morphology of the sample particle. A wide choice of hard templates is available, such as polymer microspheres, porous membranes, plastic foam, ion exchange resin, carbon fiber, and porous anodic aluminum oxide (AAO). Because of their special structures and the effect on particle size restriction, they play an important role in many fields. The soft template does not have a fixed rigid structure. Common soft templates are surfactants, polymers, and biopolymers. The soft template has broad prospects for development in the synthesis of 1D nanomaterials because of advantages such as good repeatability and simplicity of the process (Li et al. 2022; Xie et al. 2016). As an example, Zhao et al. clarified the template-induced mechanism for hydroxyapatite (HAp) nanomaterials synthesized using a three-phase reaction system. HAP nanocrystals were initially grown via an emulsion-hydrothermal synergistic strategy in an aliphatic-ethanol (EtOH)-water three-phase mixture system. Four kinds of eighteen-carbon (C18) aliphatic amphiphiles were used including stearic acid, oleic acid, stearyl alcohol, and oleyl alcohol. The two fatty acids both formed entangled long wormlike micelle structures, and the fatty alcohols formed globular or ellipsoidal micelles in the mixture solution. Ultra-long HAp nanowires were successfully fabricated using fatty acids and nanorods were obtained using stearyl alcohol (Zhao et al. 2022). In another work, monodisperse SiO_2 nanorods with hollow structure were prepared using FeOOH as a hard template. With a $FeCl_3$ concentration of 0.1M, the average length and width of the prepared hollow SiO_2 nanorods were 397 nm and 116 nm, respectively, and the average length-diameter ratio was equal to 3.4. The average size of the nanopores of the amorphous SiO_2 shell was measured as about 2.5 nm. Increasing the concentration of Cl ions improved length-diameter ratios of the FeOOH nanorod template and hollow SiO_2 nanorods (Yang et al. 2021). Pure silica nanotubes with controlled diameter and length were also synthesized using a novel and modified template-sol-gel method. The consistency and order of silica nanotubes mainly depended on the AAO template that was used during the preparation process. The AAO membrane was chemical etched using different concentrations of phosphoric acid. The obtained silica nanotubes had diameters in the range of 220–280 nm and thicknesses about 70–90 nm. Scanning transmission electron microscopy (STEM) images showed the arrangement of the hollow cylindrical silica nanotubes (Ananth et al. 2016). Hu et al. successfully synthesized, for the first time, mesoporous TiO_2 nanotubes (m-TiO2-NTs) through a dual-template method. This method used multi-walled carbon nanotube (CNT) as a hard template to control the morphology and cetyltrimethylammonium bromide (CTAB) as a soft template to form the mesoporous structure. The m-TiO_2-NTs had a multiple channel tubular structure and higher specific surface area than TiO_2 nanoparticles (TiO_2-NPs) (He et al. 2019). In a study by Zeng et al., silver nanowires (Ag NWs) were successfully synthesized using a polyvinylpyrrolidone (PVP)-free hydrothermal method with an Alpinia zerumbet leaf chunk as a reducing agent and template. The Ag NWs had a diameter of ~77 nm and a length of ~10 μm. During the hydrothermal process, the biomass initially served as a reducing agent in order to reduce silver ions. As the reaction proceeded, the biomass formed a pipe-shaped soft template by hydrothermal carbonization. Silver ions were adsorbed and reduced along the pipe-shaped soft templates to form silver nanorods, and adjacent nanorods were merged to Ag NWs. Thus, Ag NWs were grown along the pipeline soft template based on the oriented attachment mechanism (Li et al. 2023). Apart from the template synthesis method, there are also other routes for synthesis of 1D nanomaterials that are summarized in Table 2.3.

2.1.3 2D Nanomaterials

The biomedical applications of 2D nanomaterials such as graphene-based materials, including drug delivery, have grown rapidly in the past few years. Graphene and graphene oxide (GO) have been extensively explored as some of the most promising materials for biomedical applications due to their unique properties like two-dimensional planar structure, large surface area, chemical and mechanical stability, superb conductivity, and good biocompatibility. These properties result in potential applications for the design of advanced drug delivery systems and delivery of a broad range of therapeutics. Despite these, graphene is challenging to employ in biomedical applications due to its high hydrophobic nature; therefore, its derivatives, like graphene oxide are frequently utilized as preferable materials. GO possesses many oxygen-based moieties compared to graphene, namely epoxy and hydroxyl groups on the sheet's basal plane and carbonyl, carboxyl, hydroxyl, lactone, and phenol structures toward the sheet edge (Liu et al. 2013a). The synthesis of GO can essentially be divided into two main categories: "bottom-up" methods where simple carbon molecules are used to construct pristine graphene, and "top-down" methods where layers of graphene derivatives are extracted from a carbon source, typically graphite. It has been shown that bottom-up synthesis techniques (such as chemical vapor deposition, epitaxial growth on silicon carbide wafers, etc.) is time-consuming and faces challenges to scalability (Smith et al. 2019).

The most common GO synthesis method is the "Hummer method." In this method oxidation of graphite using the concentrated sulfuric acid with potassium permanganate and sodium nitrate as catalysts leads to the synthesis of graphene oxide. It was demonstrated that the oxidation using sodium nitrate would emit toxic gasses such as NO_2 and N_2O_4. Therefore, the replacing use of $NaNO_3$ with H_2SO_4,

Table 2.3 Different routes for synthesis of nanomaterials

Type of 1D nanomaterial	Synthesis method	Morphology	Size	References
Au	Microwave–polyol method	Nanorod	Diameter: 20–100 nm Length: 0.6–5 μm	Tsuji et al. (2004)
Au	Green synthesis	Nanorod	Diameter: 21–66 nm Length: 8–33 nm	Lage et al. (2021)
CuO	Solution combustion	Nanorod	Diameter: 20–25 nm Length: 100–250 nm	Christy et al. (2013)
ZnO	Hydrothermal	Nanotube	Diameter: 17 nm Length: 2 μm	Aal et al. (2015)
ZnO	Sol-gel/hydrothermal	Nanorod	Diameter: 22–30 nm Length: 200–500 nm	Torres et al. (2022)
Hydroxyapatite	Sol-gel	Nanotube	Diameter: 140–350 nm Length: 1 μm	Chandanshive et al. (2013)
Fe/Fe_3O_4	Electrochemical deposition	Nanowire	Diameter: 30–40 nm Length: 6.4 ± 1.3 μm	Martínez-Banderas et al. (2019)

H_3PO_4, and double the amount of $KMnO_4$ was proposed in the improved Hummer's method. In addition, the obtained GO has more hydrophilicity and conductivity (Zaaba et al. 2017). Gu et al. reported a new graphene-based co-delivery nanosystem of a cell-membrane-targeted anticancer protein and a chemotherapeutic agent for combination cancer treatment. This designed nanocarrier was composed of graphene oxide (GO), polyethylene glycol (PEG) linker, and a furin cleavable peptide, which was achieved by step-by-step site-specific conjugations. For carboxylation of GO, sodium hydroxide, sodium chloroacetate ($ClCH_2COONa$), and HCl were used step by step under ultrasonication in an ice-water bath. A nanosized GO sheet was obtained after ultrasonication and purification, which had an average size of 152 nm. Carboxylated GO functionalized by PEG and peptide and used as a carrier for DOX delivery (Jiang et al. 2015). In another work, Cao et al. reported the synthesis of near-infrared (NIR)-responsive PEGylated graphene oxide incorporated with core-shell gold nanorods/mesoporous silica nanoparticles. The modified Hummer method was used for the fabrication of graphene oxide. The core-shell nanoparticles were obtained in two steps: seed-mediated method and hydrolysis of TEOS via sol-gel. The nanocarriers were fabricated by adding core-shell nanoparticles into PEGylated graphene oxide suspension in a dropwise manner after stirring for 24 h at room temperature (Qi et al. 2019). Other 2D nanomaterials that can be used for drug delivery are nanosheets. Synthesis of folic acid-functionalized polyethylene glycol coated ZnO nanosheet was reported by Kannan et al. for the first time. The PEG-ZnO nanosheet was fabricated by Borassus flabellifer fruit extract via green synthesis at 160 °C for 6–7 h. The surface functionalization of the ZnO nanosheet was carried out using folic acid, 1-ethyl-3-(3-(dimethylamino) propyl)-carbodiimide, and N-hydroxysuccinimide in DMSO as a solvent. The average size of the functionalized nanosheet was about 100 nm (Vimala et al. 2017). Molybdenum disulfide (MoS_2) has been extensively used for biomedical applications due to its excellent photothermal conversion ability. In the work done by Williams et al., they reported a nanoplatform based on folic acid (FA) targeted dual-stimuli responsive MoS_2 nanosheets and explored this for the treatment of FA-receptor positive human breast cancer. MoS_2 was exfoliated from MoS_2 flakes in hexane under a N_2 atmosphere and ultrasonication. The nanosheets had a uniform diameter (133 nm) and could be loaded with the anticancer drug doxorubicin (DOX) to a high capacity (151.4 mg/g). The release of DOX was greatly accelerated at pH 5.0 as compared to pH 7.4 (Zhang et al. 2019).

2.1.4 3D Nanomaterials

Different types of 3D nanostructures such as aerogels, mesoporous silica materials, metal-organic frameworks, flower-like structures, and nanocomposites can be used as drug release systems. Generally, by changing the effective parameters in common synthetic methods such as sol-gel, polyol, hydrothermal, etc. such structures can be produced. Sometimes, a combination of different synthetic methods is used. Table 2.4 represents a summary of some widely used routes for the synthesis of 3D nanomaterials. As an example, Moghaddas et al. reported the synthesis of carboxylic acid-functionalized silica aerogel nanostructure as a carrier for celecoxib delivery. The simple ambient pressure drying method was used for preparing hydrophobic silica aerogel. The silica sol was obtained using sodium silicate solution and tartaric acid as the starting materials at pH 4–5. The gelation of the sol was carried out in the sealed reactor at 55 °C. Water was exchanged with 2-propanol and n-hexane in the final gel. The silica aerogel was prepared after calcination at 500 °C for 3 h. The carboxylic acid functionalization was done by N-(2-aminoethyl)-3 aminopropyltrimethoxysilane and succinic anhydride at 60 °C for 24 h. Silica aerogel possessed uniform nanoparticles in a porous nanostructure with an average size of 24 ± 4 nm and 37 ± 5 nm before and after functionalization, respectively (Jabbari-Gargari et al. 2021).

Table 2.4 Some of the widely used routes for synthesis of 3D nanomaterials

Type of 3D nanomaterial	Method	Morphology	Size	References
Chitosan	Supercritical drying	Aerogel	13–59 μm	Menshutina et al. (2022)
Hydroxyapatite	*Template synthesis*	Flower-like	4 μm	Liu et al. (2020)
ZnO	Hydrothermal	Flower-like	1.4 ± 0.6 μm	Paino et al. (2016)
Fe_2O_3	Co-precipitation	Flower-like	23.4 nm	Tabassum et al. (2023)
Mesoporous silica (SBA-15)	Sol-gel Hydrothermal/ template	Nanorod	Diameter: 100 nm and length: 1 μm	Bahrami et al. (2014, 2015a, b)
Mesoporous silica (MCM-41)	Sol-gel hydrothermal/ template	Spherical	300 nm	Popova et al. (2020)
Metal-organic framework (ZIF-8)	Wet chemical synthesis	Hexagonal nanocrystals	80–100 nm	Kaur et al. (2017)
Magnetite/silica nanocomposites	Co-precipitation/ sol-gel	Spherical	8/2–13/2 nm	Taufiq et al. (2020)

References

Aal NA, Al-Hazmi F, Al-Ghamdi AA, Al-Ghamdi AA, El-Tantawy F, Yakuphanoglu F (2015) Novel rapid synthesis of zinc oxide nanotubes via hydrothermal technique and antibacterial properties. Spectrochim Acta A Mol Biomol Spectrosc 135:871–877

Abbas M, Takahashi M, Kim CG (2013) Facile sonochemical synthesis of high-moment magnetite (Fe3O4) nanocube. J Nanopart Res 15:1–12

Ahmadi S, Chia C-H, Zakaria S, Saeedfar K, Asim N (2012) Synthesis of Fe3O4 nanocrystals using hydrothermal approach. J Magn Magn Mater 324(24):4147–4150

Ali Dheyab M, Aziz AA, Jameel MS (2021) Recent advances in inorganic nanomaterials synthesis using sonochemistry: a comprehensive review on iron oxide, gold and iron oxide coated gold nanoparticles. Molecules 26(9):2453

Alias SS, Ismail AB, Mohamad AA (2010) Effect of pH on ZnO nanoparticle properties synthesized by sol–gel centrifugation. J Alloys Compd 499:231–237

Alsamhary KI (2020) Eco-friendly synthesis of silver nanoparticles by Bacillus subtilis and their antibacterial activity. Saudi J Biol Sci 27(8):2185–2191

Alzubaidi AK, Al-Kaabi WJ, Al Ali A, Albukhaty S, Al-Karagoly H, Sulaiman GM, Asiri M, Khane Y (2023) Green synthesis and characterization of silver nanoparticles using flaxseed extract and evaluation of their antibacterial and antioxidant activities. Appl Sci 13(4):2182

Ammar S, Fiévet F (2020) Polyol synthesis: a versatile wet-chemistry route for the design and production of functional inorganic nanoparticles. Nanomaterials 10(6):1217

Ananth KP, Jose SP, Joseph Nathanael A, Tae Hwan O, Mangalaraj D, Ballamurugan AM (2016) A novel modified sol-gel template synthesis of high aspect ratio silica nanotubes in the presence of phosphoric acid. J Nano Res 35:27–38

Asgari S, Saberi AH, McClements DJ, Lin M (2019) Microemulsions as nanoreactors for synthesis of biopolymer nanoparticles. Trends Food Sci Technol 86:118–130

Ba-Abbad MM, Benamour A, Ewis D, Mohammad AW, Mahmoudi E (2022) Synthesis of Fe3O4 nanoparticles with different shapes through a co-precipitation method and their application. JOM 74(9):3531–3539

Bahrami Z, Badiei A, Atyabi F (2014) Surface functionalization of SBA-15 nanorods for anticancer drug delivery. Chem Eng Res Des 92(7):1296–1303

Bahrami Z, Badiei A, Atyabi F, Darabi HR, Mehravi B (2015a) Piperazine and its carboxylic acid derivatives-functionalized mesoporous silica as nanocarriers for gemcitabine: adsorption and release study. Mater Sci Eng C 49:66–74

Bahrami Z, Badiei A, Ziarani GM (2015b) Carboxylic acid-functionalized SBA-15 nanorods for gemcitabine delivery. J Nanopart Res 17:1–12

Balamurugan S, Ashna L, Parthiban P (2014) Synthesis of nanocrystalline MgO particles by combustion followed by annealing method using hexamine as a fuel. J Nanotechnol 2014: 841803

Bokov D, Jalil AT, Chupradit S, Suksatan W, Ansari MJ, Shewael IH, Valiev GH, Kianfar E (2021) Nanomaterial by sol-gel method: synthesis and application. Adv Mater Sci Eng 2021:1–21

Bopape DA, Tetana ZN, Mabuba N, Motaung DE, Hintsho-Mbita NC (2023) Biosynthesis of TiO2 nanoparticles using Commelina benghalensis for the photodegradation of methylene blue dye and antibiotics: effect of plant concentration. Results Chem 5:100825

Boustani K, Shokri A, Shayesteh SF, Jafari A (2020) Ultrasound-assisted synthesis and tuning the magnetic and structural features of superparamagnetic Fe3O4 nanoparticles by using ethylenediamine as a precipitating agent. J Supercond Nov Magn 33:1879–1887

Castro KC, Costa JM, Campos MGN (2022) Drug-loaded polymeric nanoparticles: a review. Int J Polym Mater Polym Biomater 71(1):1–13

Chandanshive BB, Rai P, Rossi AL, Ersen O, Khushalani D (2013) Synthesis of hydroxyapatite nanotubes for biomedical applications. Mater Sci Eng C 33(5):2981–2986

Chandrakala V, Aruna V, Angajala G (2022) Review on metal nanoparticles as nanocarriers: current challenges and perspectives in drug delivery systems. Emerg Mater 5(6):1593–1615

Chen P, Song L, Liu Y, Fang Y-e (2007) Synthesis of silver nanoparticles by γ-ray irradiation in acetic water solution containing chitosan. Radiat Phys Chem 76(7):1165–1168

Chopra H, Bibi S, Singh I, Hasan MM, Khan MS, Yousafi Q, Baig AA, Rahman MM, Islam F, Emran TB (2022) Green metallic nanoparticles: biosynthesis to applications. Front Bioeng Biotechnol 10:548

Choudhury S, Paul S, Goswami S, Deb K (2022) Methods for nanoparticle synthesis and drug delivery. In: Advances in nanotechnology-based drug delivery systems. Elsevier, Amsterdam, pp 21–44

Christy AJ, Nehru LC, Umadevi M (2013) A novel combustion method to prepare CuO nanorods and its antimicrobial and photocatalytic activities. Powder Technol 235:783–786

Dangana RS, George RC, Agboola FK (2023) The biosynthesis of zinc oxide nanoparticles using aqueous leaf extracts of Cnidoscolus aconitifolius and their biological activities. Green Chem Lett Rev 16(1):2169591

Dheyab MA, Aziz AA, Jameel MS, Noqta OA, Khaniabadi PM, Mehrdel B (2020) Simple rapid stabilization method through citric acid modification for magnetite nanoparticles. Sci Rep 10(1): 10793

Eid AM, Fouda A, Hassan SE-D, Hamza MF, Alharbi NK, Elkelish A, Alharthi A, Salem WM (2023) Plant-based copper oxide nanoparticles; biosynthesis, characterization, antibacterial activity, tanning wastewater treatment, and heavy metals sorption. Catalysts 13(2):348

El-Bendary MA, Afifi SS, Moharam ME, Abo Elsoud MM, Gawdat NA (2022) Optimization of Bacillus subtilis NRC1 growth conditions using response surface methodology for sustainable biosynthesis of gold nanoparticles. Sci Rep 12(1):20882

Elsamra RMI, Masoud MS, Zidan AA, El Zokm GM, Okbah MA (2023) Green synthesis of nanostructured zinc oxide by Ocimum tenuiflorum extract: characterization, adsorption modeling, cytotoxic screening, and metal ions adsorption applications. Biomass Convers Biorefinery. https://doi.org/10.1007/s13399-022-03709-1

Fareed N, Nisa S, Bibi Y, Fareed A, Ahmed W, Sabir M, Alam S, Sajjad A, Kumar S, Hussain M (2023) Green synthesized silver nanoparticles using carrot extract exhibited strong antibacterial activity against multidrug resistant bacteria. J King Saud Univ Sci 35(2):102477

Feng Y, Lin X, Wang Y, Wang Y, Hua J (2008) Diversity of Aurum bioreduction by Rhodobacter capsulatus. Mater Lett 62(27):4299–4302

Fuentes-García JA, Alavarse AC, Eduardo C, de Castro F, Giacomelli C, Ibarra MR, Bonvent J-J, Goya GF (2021) Sonochemical route for mesoporous silica-coated magnetic nanoparticles towards pH-triggered drug delivery system. J Mater Res Technol 15:52–67

Gande VV, Pushpavanam S (2021) Continuous synthesis of copper nanoparticles using a polyol process in a milli-channel reactor. J Flow Chem 11(3):661–674

Ghosh MK, Sahu S, Gupta I, Ghorai TK (2020) Green synthesis of copper nanoparticles from an extract of Jatropha curcas leaves: characterization, optical properties, CT-DNA binding and photocatalytic activity. RSC Adv 10(37):22027–22035

Harini HV, Nagaraju G, Nirmala B (2022) Combustion synthesis CuO nanoparticles: application to photocatalytic activity. Mater Today Proc 49:860–864

Haro-González PG, Ramírez-Rico DS, Larios-Durán ER (2019) Synthesis of gold nanoparticles in aqueous solutions by electrochemical reduction using poly (ethylene glycol) as stabilizer. Int J Electrochem Sci 14(10):9704–9710

He S, Zhirui Guo Y, Zhang S, Wang J, Ning G (2007) Biosynthesis of gold nanoparticles using the bacteria Rhodopseudomonas capsulata. Mater Lett 61(18):3984–3987

He Y, Huang Y, Huang Z, Jiang Y, Sun X, Shen Y, Chu W, Zhao C (2017) Bisphosphonate-functionalized coordination polymer nanoparticles for the treatment of bone metastatic breast cancer. J Control Release 264:76–88

He G, Zhang J, Yun H, Bai Z, Wei C (2019) Dual-template synthesis of mesoporous TiO2 nanotubes with structure-enhanced functional photocatalytic performance. Appl Catal B Environ 250:301–312

Her S, Jaffray DA, Allen C (2017) Gold nanoparticles for applications in cancer radiotherapy: mechanisms and recent advancements. Adv Drug Deliv Rev 109:84–101. https://doi.org/10.1016/j.addr.2015.12.012

Hosny M, Fawzy M, El-Badry YA, Hussein EE, Eltaweil AS (2022) Plant-assisted synthesis of gold nanoparticles for photocatalytic, anticancer, and antioxidant applications. J Saudi Chem Soc 26(2):101419

Ibrahim E, Fouad H, Zhang M, Zhang Y, Qiu W, Yan C, Li B, Mo J, Chen J (2019) Biosynthesis of silver nanoparticles using endophytic bacteria and their role in inhibition of rice pathogenic bacteria and plant growth promotion. RSC Adv 9(50):29293–29299

Ibrahim S, Ahmad Z, Manzoor MZ, Mujahid M, Faheem Z, Adnan A (2021) Optimization for biogenic microbial synthesis of silver nanoparticles through response surface methodology, characterization, their antimicrobial, antioxidant, and catalytic potential. Sci Rep 11(1):770

Islam MT, Dominguez A, Alvarado-Tenorio B, Bernal RA, Montes MO, Noveron JC (2019) Sucrose-mediated fast synthesis of zinc oxide nanoparticles for the photocatalytic degradation of organic pollutants in water. ACS Omega 4(4):6560–6572

Jabbari-Gargari A, Moghaddas J, Hamishehkar H, Jafarizadeh-Malmiri H (2021) Carboxylic acid decorated silica aerogel nanostructure as drug delivery carrier. Microporous Mesoporous Mater 323:111220

Jafari M, Rokhbakhsh-Zamin F, Shakibaie M, Moshafi MH, Ameri A, Rahimi HR, Forootanfar H (2018) Cytotoxic and antibacterial activities of biologically synthesized gold nanoparticles assisted by Micrococcus yunnanensis strain J2. Biocatal Agric Biotechnol 15:245–253

Jiang T, Sun W, Zhu Q, Burns NA, Khan SA, Mo R, Zhen G (2015) Furin-mediated sequential delivery of anticancer cytokine and small-molecule drug shuttled by graphene. Adv Mater 27(6):1021–1028

Kaur H, Mohanta GC, Gupta V, Kukkar D, Tyagi S (2017) Synthesis and characterization of ZIF-8 nanoparticles for controlled release of 6-mercaptopurine drug. J Drug Deliv Sci Technol 41: 106–112

Kazemi F, Arianpour F, Taheri M, Saberi A, Rezaie HR (2020) Effects of chelating agents on the sol-gel synthesis of nano-zirconia: comparison of the Pechini and sugar-based methods. Int J Miner Metall Mater 27:693–702

Khan S, Wong A, Zeb S, Mortari B, Villa JEL, Sotomayor MDPT (2023) Characterization of polymeric nanoparticles. In: Smart polymer nanocomposites. Elsevier, Amsterdam, pp 141–163

Kim J, Tran VT, Sangjin O, Kim C-S, Hong JC, Kim SI, Joo Y-S, Mun S, Kim M-H, Jung J-W (2018) Scalable solvothermal synthesis of superparamagnetic Fe3O4 nanoclusters for bioseparation and theragnostic probes. ACS Appl Mater Interfaces 10(49):41935–41946

Kumar CG, Poornachandra Y, Mamidyala SK (2014) Green synthesis of bacterial gold nanoparticles conjugated to resveratrol as delivery vehicles. Colloids Surf B: Biointerfaces 123:311–317

Lage AC, Pinheiro LO, Ladeira LM, Paniago RM, Castilho RO, Amorim JM, Pessoa ES, Nuncira J, Faraco AAG (2021) Synthesis and characterization of gold nanorods using the natural products resveratrol, gallic acid, and a purified fraction of Stryphnodendron obovatum by seedless method. Environ Nanotechnol Monit Manage 16:100473

LaGrow AP, Besenhard MO, Hodzic A, Sergides A, Bogart LK, Gavriilidis A, Thanh NTK (2019) Unravelling the growth mechanism of the co-precipitation of iron oxide nanoparticles with the aid of synchrotron X-ray diffraction in solution. Nanoscale 11(14):6620–6628

Li J, Wu Q, Ji W (2016) Synthesis of nanoparticles via solvothermal and hydrothermal methods. In: Handbook of nanoparticles. Springer, Cham

Li X, Naguib YW, Cui Z (2017) In vivo distribution of zoledronic acid in a bisphosphonate-metal complex-based nanoparticle formulation synthesized by a reverse microemulsion method. Int J Pharm 526(1-2):69–76

Li S, Al-Misned FA, El-Serehy HA, Yang L (2021a) Green synthesis of gold nanoparticles using aqueous extract of Mentha Longifolia leaf and investigation of its anti-human breast carcinoma properties in the in vitro condition. Arab J Chem 14(2):102931

Li W, Tong W, Yadav A, Bladt E, Bals S, Funston AM, Etheridge J (2021b) Shape control beyond the seeds in gold nanoparticles. Chem Mater 33(23):9152–9164

Li H, Wang L, Wei Y, Yan W, Feng J (2022) Preparation of templated materials and their application to typical pollutants in wastewater: a review. Front Chem 10:882876

Li Y, Wang Y, Junqing W, Pan Y, Ye H, Zeng X (2023) Synthesis of silver nanowires using a polyvinylpyrrolidone-free method with an Alpinia zerumbet leaf based on the oriented attachment mechanism. ACS Omega 8(2):2237–2242

Liu J, Cui L, Losic D (2013a) Graphene and graphene oxide as new nanocarriers for drug delivery applications. Acta Biomater 9(12):9243–9257

Liu Y, Goebl J, Yin Y (2013b) Templated synthesis of nanostructured materials. Chem Soc Rev 42(7):2610–2653

Liu Y, Tang Y, Jiang W, Sun J, Liao X, Teng Z, Guangming L (2020) Facile synthesis of biodegradable flower-like hydroxyapatite for drug and gene delivery. J Colloid Interface Sci 570:402–410

Mariychuk R, Grulova D, Grishchenko LM, Linnik RP, Lisnyak VV (2020) Green synthesis of non-spherical gold nanoparticles using Solidago canadensis L. extract. Appl Nanosci 10(12): 4817–4826

Martínez-Banderas AI, Aires A, Quintanilla M, Holguín-Lerma JA, Lozano-Pedraza C, Teran FJ, Moreno JA, Perez JE, Ooi BS, Ravasi T (2019) Iron-based core–shell nanowires for combinatorial drug delivery and photothermal and magnetic therapy. ACS Appl Mater Interfaces 11(47): 43976–43988

Masoudi A, Luluah M, Alqurashi AS, Zaid AA, Hamdi H (2023) Characterization and biological studies of synthesized titanium dioxide nanoparticles from leaf extract of Juniperus phoenicea (L.) growing in Taif Region, Saudi Arabia. Processes 11(1):272

Menshutina N, Majouga A, Uvarova A, Lovskaya D, Tsygankov P, Mochalova M, Abramova O, Ushakova V, Morozova A, Silantyev A (2022) Chitosan aerogel particles as nasal drug delivery systems. Gels 8(12):796

Mohan S, Vellakkat M, Aravind A, Reka U (2020) Hydrothermal synthesis and characterization of zinc oxide nanoparticles of various shapes under different reaction conditions. Nano Exp 1(3): 030028

Murthy HC, Desalegn T, Kassa M, Abebe B, Assefa T (2020) Synthesis of green copper nanoparticles using medicinal plant Hagenia abyssinica (Brace) JF. Gmel. leaf extract: antimicrobial properties. J Nanomater 2020:3924081

Najjar R (2012) Microemulsions: an introduction to properties and applications: BoD–books on demand. IntechOpen, London

Navas D, Fuentes S, Castro-Alvarez A, Chavez-Angel E (2021) Review on sol-gel synthesis of perovskite and oxide nanomaterials. Gels 7(4):275

Ojea-Jiménez I, Romero FM, Bastús NG, Puntes V (2010) Small gold nanoparticles synthesized with sodium citrate and heavy water: insights into the reaction mechanism. J Phys Chem C 114(4):1800–1804

Paino IMM, Gonçalves FJ, Souza FL, Zucolotto V (2016) Zinc oxide flower-like nanostructures that exhibit enhanced toxicology effects in cancer cells. ACS Appl Mater Interfaces 8(48): 32699–32705

Panariello L, Besenhard MO, Damilos S, Sergides A, Sebastian V, Irusta S, Tang J, Thanh NTK, Gavriilidis A (2022) Microwave-assisted flow synthesis of multicore iron oxide nanoparticles. Chem Eng Process Process Intensif 182:109198

Patil TP, Vibhute AA, Patil SL, Dongale TD, Tiwari AP (2023) Green synthesis of gold nanoparticles via Capsicum annum fruit extract: characterization, antiangiogenic, antioxidant and anti-inflammatory activities. Appl Surf Sci Adv 13:100372

Pérez-Alvarez M, Cadenas-Pliego G, Pérez-Camacho O, Comparán-Padilla VE, Cabello-Alvarado CJ, Saucedo-Salazar E (2021) Green synthesis of copper nanoparticles using cotton. Polymers 13(12):1906

Personick ML, Mirkin CA (2013) Making sense of the mayhem behind shape control in the synthesis of gold nanoparticles. J Am Chem Soc 135(49):18238–18247. https://doi.org/10.1021/ja408645b

Popova M, Koseva N, Trendafilova I, Lazarova H, Mitova V, Mihály J, Momekova D, Momekov G, Koleva IZ, Aleksandrov HA (2020) Tamoxifen delivery system based on PEGylated magnetic MCM-41 silica. Molecules 25(21):5129

Qi L, Ji G, Luo Z, Xiao Z, Yang Q (2017) Characterization and drug delivery properties of OSA starch-based nanoparticles prepared in [C3OHmim] Ac-in-oil microemulsions system. ACS Sustain Chem Eng 5(10):9517–9526

Qi Z, Shi J, Zhang Z, Cao Y, Li J, Cao S (2019) PEGylated graphene oxide-capped gold nanorods/silica nanoparticles as multifunctional drug delivery platform with enhanced near-infrared responsiveness. Mater Sci Eng C 104:109889

Rajeshkumar S, Parameswari RP, Sandhiya D, Al-Ghanim KA, Nicoletti M, Govindarajan M (2023) Green synthesis, characterization and bioactivity of mangifera indica seed-wrapped zinc oxide nanoparticles. Molecules 28(6):2818

Ranjbar M, Khakdan F, Mukherjee A (2023) In vitro analysis of green synthesized CuO nanoparticles using Tanacetum parthenium extract for multifunctional applications. Environ Sci Pollut Res 30(21):60180–60195

Rathi VH, Jeice AR (2023) Green fabrication of titanium dioxide nanoparticles and their applications in photocatalytic dye degradation and microbial activities. Chem Phys Impact 6: 100197

Said A, Abu-Elghait M, Atta HM, Salem SS (2023) Antibacterial activity of green synthesized silver nanoparticles using Lawsonia inermis against common pathogens from urinary tract infection. Appl Biochem Biotechnol. https://doi.org/10.1007/s12010-023-04482-1

Saikumari N, Monish Dev S, Avinaash S, Dev. (2021) Effect of calcination temperature on the properties and applications of bio extract mediated titania nano particles. Sci Rep 11(1):1734
Saleh MN, Alwan SK (2020) Bio-synthesis of silver nanoparticles from bacteria Klebsiella pneumonia: their characterization and antibacterial studies. J Phys Conf Ser 1664:012115
Sayadi K, Akbarzadeh F, Pourmardan V, Saravani-Aval M, Sayadi J, Chauhan NPS, Sargazi G (2021) Methods of green synthesis of Au NCs with emphasis on their morphology: a mini-review. Heliyon 7(6):e07250
Schletz D, Breidung M, Fery A (2023) Validating and utilizing machine learning methods to investigate the impacts of synthesis parameters in gold nanoparticle synthesis. J Phys Chem C 127(2):1117–1125
Shin J, Kang N, Kim B, Hong H, Le Y, Kim J, Kang H, Kim JS (2023) One-dimensional nanomaterials for cancer therapy and diagnosis. Chem Soc Rev 52:4488–4514
Slimani S, Meneghini C, Abdolrahimi M, Talone A, Murillo JPM, Barucca G, Yaacoub N, Imperatori P, Illés E, Smari M (2021) Spinel iron oxide by the co-precipitation method: effect of the reaction atmosphere. Appl Sci 11(12):5433
Smith AT, LaChance AM, Zeng S, Liu B, Sun L (2019) Synthesis, properties, and applications of graphene oxide/reduced graphene oxide and their nanocomposites. Nano Mater Sci 1(1):31–47
Stangherlin S, Cathcart N, Sato F, Kitaev V (2020) Gold nanoprisms: synthetic approaches for mastering plasmonic properties and implications for biomedical applications. ACS Appl Nano Mater 3(8):8304–8318. https://doi.org/10.1021/acsanm.0c01741
Suriyakala G, Sathiyaraj S, Babujanarthanam R, Alarjani KM, Hussein DS, Rasheed RA, Kanimozhi K (2022) Green synthesis of gold nanoparticles using Jatropha integerrima Jacq. flower extract and their antibacterial activity. J King Saud Univ Sci 34(3):101830
Tabassum N, Singh V, Chaturvedi VK, Vamanu E, Singh MP (2023) A facile synthesis of flower-like iron oxide nanoparticles and its efficacy measurements for antibacterial, cytotoxicity and antioxidant activity. Pharmaceutics 15(6):1726
Taufiq A, Nikmah A, Hidayat A, Sunaryono S, Mufti N, Hidayat N, Susanto H (2020) Synthesis of magnetite/silica nanocomposites from natural sand to create a drug delivery vehicle. Heliyon 6(4):e03784
Torres FCG, López JLC, Rodríguez ASL, Gallardo PS, Morales ER, Hernández GP, Guillen JCD, Flores LLD (2022) Sol–gel/hydrothermal synthesis of well-aligned ZnO nanorods. Bol Soc Esp Cerám Vidrio 62:348–356
Tsuji M, Hashimoto M, Nishizawa Y, Tsuji T (2004) Synthesis of gold nanorods and nanowires by a microwave–polyol method. Mater Lett 58(17-18):2326–2330
Varma A, Mukasyan AS, Rogachev AS, Manukyan KV (2016) Solution combustion synthesis of nanoscale materials. Chem Rev 116(23):14493–14586
Vasei HV, Masoudpanah SM, Habibollahzadeh M (2020) Different morphologies of ZnO via solution combustion synthesis: the role of fuel. Mater Res Bull 125:110784
Vimala K, Shanthi K, Sundarraj S, Kannan S (2017) Synergistic effect of chemo-photothermal for breast cancer therapy using folic acid (FA) modified zinc oxide nanosheet. J Colloid Interface Sci 488:92–108
Wang X, Chen H, Luo Z, Xiong F (2016) Preparation of starch nanoparticles in water in oil microemulsion system and their drug delivery properties. Carbohydr Polym 138:192–200
Wen W, Jin-Ming W (2014) Nanomaterials via solution combustion synthesis: a step nearer to controllability. RSC Adv 4(101):58090–58100
Wolf JB, Stawski TM, Smales GJ, Thünemann AF, Emmerling F (2022) Towards automation of the polyol process for the synthesis of silver nanoparticles. Sci Rep 12(1):5769
Wu J-W, Ng I-S (2017) Biofabrication of gold nanoparticles by Shewanella species. Bioresour Bioprocess 4(1):1–9
Xie Y, Kocaefe D, Chen C, Kocaefe Y (2016) Review of research on template methods in preparation of nanomaterials. J Nanomater 2016:2302595
Xu H, Zeiger BW, Suslick KS (2013) Sonochemical synthesis of nanomaterials. Chem Soc Rev 42(7):2555–2567

Yang QQ, Chen XL, He Y, Lai JQ, Xiong Y, Shen Q, Zhou XC, Shao B, Ma YL, Guo DL (2021) Preparation of monodisperse SiO2 nanorods with hollow structure and parameters affecting the length-diameter ratio. J Dispers Sci Technol 42(8):1146–1151

Yoko A, Aida T, Aoki N, Hojo D, Koshimizu M, Ohara S, Seong G, Takami S, Togashi T, Tomai T (2018) Supercritical hydrothermal synthesis of nanoparticles. In: Nanoparticle technology handbook. Elsevier, Amsterdam, pp 683–689

Yoko A, Seong G, Tomai T, Adschiri T (2020) Continuous flow synthesis of nanoparticles using supercritical water: process design, surface control, and nanohybrid materials. KONA Powder Particle J 37:28–41

Zaaba NI, Foo KL, Hashim U, Tan SJ, Liu W-W, Voon CH (2017) Synthesis of graphene oxide using modified hummers method: solvent influence. Proc Eng 184:469–477

Zhang X, Jianrong W, Williams GR, Niu S, Qian Q, Zhu L-M (2019) Functionalized MoS2-nanosheets for targeted drug delivery and chemo-photothermal therapy. Colloids Surf B: Biointerfaces 173:101–108

Zhao J, Chu T, Qin H, Lei Y, Liu L, Zhang G, Gao C, Zhang T, Song W (2022) The preparation of hydroxyapatite nanowires and nanorods via aliphatic micelles as soft templates. Cryst Eng Commun 24(25):4593–4598

Zharikov AA, Vinogradov RA, Zezina EA, Pozdnyakov AS, Feldman VI, Vasiliev AL, Zezin AA (2022) The radiation-induced preparation of ultrasmall gold nanoparticles in Au (III) complexes with units of poly (1-vinyl-1,2,4-triazole) and poly (1-vinyl-1,2,4-triazole)–poly (acrylic acid). Colloid Interf Sci Commun 47:100602

Zhu B, Xie N, Yue L, Wang K, Bani-Fwaz MZ, Osman H-EH, El-kott AF, Bai X (2022) Formulation and characterization of a novel anti-human endometrial cancer supplement by gold nanoparticles green-synthesized using Spinacia oleracea L. Leaf aqueous extract. Arab J Chem 15(3):103576

3 Synthesis of Nanocarriers Loaded with Anti-Cancer Drugs by Using Functionalization or Conjugations, Encapsulation Methods

Vijaya Ravinayagam and B. Rabindran Jermy

Abstract

Recently, inorganic porous nanocomposites are gaining importance in biomedical applications. This is due to several advantages such as high surface area, designable pore structures, and robust mechanical/chemical stability with adjustable optical and electrical properties. Inorganic nanoparticles are interesting as nanocarriers for drug delivery due to the chemical, mechanical strength, microbial resistance, and biocompatibility. The presence of large surface area, structured pores of tunable sizes, ability to functionalize with different chain length of chemical moieties has been an attractive feature of nanoparticles in nanotherapeutic area of research. Drug loading on porous structured silica requires ambient condition that minimize the drug disintegration. An effective nanocarrier needs to exhibit both spatial placement (target specificity) and temporal delivery (sustainability and controllability). In this chapter we have discussed several nanocarriers based on structured silica, metal oxides, carbon, liposomes, nucleic acids, and polymeric species based therapeutic tools has been reported for treating chronic diseases such as cancer, cardiovascular, and diabetics. In addition, hexagonal shaped mesoporous silica (SiM41 and SBA-15), cubic shaped mesosilica (SBA-16 and MCM-48) with high textural properties are reported for multifunctional drug delivery applications are also discussed.

V. Ravinayagam
Deanship of Scientific Research & Department of Nanomedicine Research, Institute for Research and Medical Consultations, Imam Abdulrahman Bin Faisal University, Dammam, Saudi Arabia

B. R. Jermy (✉)
Department of Nanomedicine Research, Institute for Research and Medical Consultations, Imam Abdulrahman Bin Faisal University, Dammam, Saudi Arabia
e-mail: rJermy@iau.edu.sa

F. A. Khan (ed.), *Nano Drug Delivery for Cancer Therapy*,
https://doi.org/10.1007/978-981-99-6940-1_3

Keywords

Nanocarriers loaded with anti-cancer drugs · Functionalization · Conjugations · Encapsulation methods

3.1 Introduction

Engineering a stimuli responsive drug delivery system offers a controlled drug release, improves the therapeutic efficiency, and reduces the drug dose. Three class of materials explored are classified into organic (Dendrimer, micelles, polymers, and liposomes), inorganic (mesoporous silica, metal oxides, graphene oxide, carbon nanotubes), and biological systems (Viosomes, extracellular vesicles) (Tian et al. 2022). Recently, inorganic porous nanocomposites are gaining importance in biomedical applications. This is due to several advantages such as high surface area, designable pore structures, and robust mechanical/chemical stability with adjustable optical and electrical properties (Khakpour et al. 2023). Inorganic nanoparticles are interesting as nanocarriers for drug delivery due to the chemical, mechanical strength, microbial resistance, and biocompatibility. The presence of large surface area, structured pores of tuneable sizes, ability to functionalize with different chain length of chemical moieties has been an attractive feature of nanoparticles in nanotherapeutic area of research. Drug loading on porous structured silica requires ambient condition that minimize the drug disintegration. An effective nanocarrier needs to exhibit both spatial placement (target specificity) and temporal delivery (sustainability and controllability) (Di Stefano 2023). Several nanocarriers based on structured silica, metal oxides, carbon (Debnath and Srivastava 2021), liposomes, nucleic acids (Guo et al. 2022), and polymeric species based therapeutic tools has been reported for treating chronic diseases such as cancer, cardiovascular, and diabetics (Li et al. 2019). Hexagonal shaped mesoporous silica (SiM41 and SBA-15), cubic shaped mesosilica (SBA-16 and MCM-48) with high textural properties (surface area ranging between 500 and 1100 m^2/g) are reported for multifunctional drug delivery applications. The nanotherapeutics transformed from single drug delivery system to multifunctional system (Fig. 3.1).

Amidoxime ligand modified mesoporous SBA-15 as nanocarrier platform has been reported for doxorubicin drug loading and release (Fig. 3.2). The grafting of amidoxime on the surface of SBA-15 had little impact on the morphology of nanocarrier, while increases the binding functional moieties such as amine and hydroxyl groups to facilitate effective interaction with DOX. The presence of such reactive functional moieties enables a pH responsive drug delivery and effective against MCF-7 cells for potential application in cancer therapy (Ulagesan et al. 2022). Atiyah et al. (2022) reported a comparative study of *curcumin loading on SBA-15 and amine (aminopropyltrimethoxysilane) functionalized SBA-15*. Several parameters such as nanocarrier concentrations, curcumin concentration, contact time, and pH on curcumin loadings were studied. The curcumin loading efficiency almost doubled from 45% to 89.7% with amine functionalization on SBA-15.

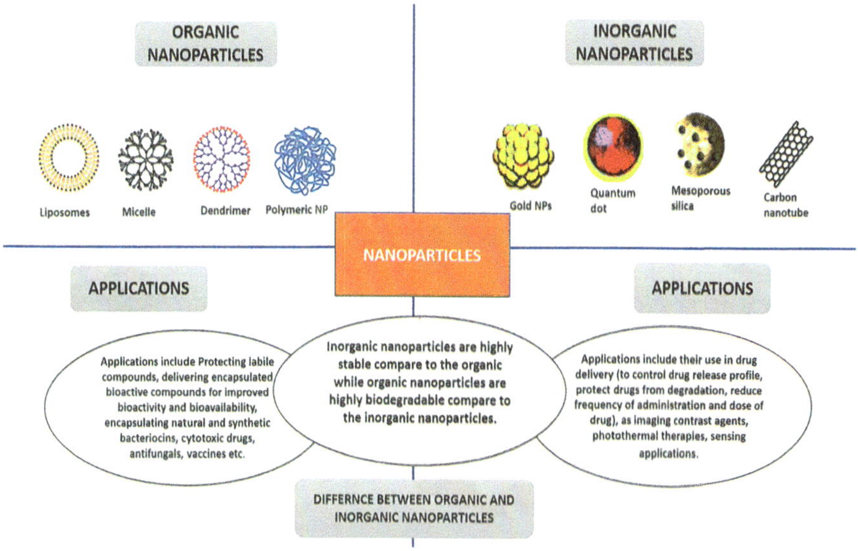

Fig. 3.1 Nanoparticles and their classifications (Seidu et al. 2022)

Kinetics study revealed a higher correlation coefficient using Langmuir isotherm. Drug delivery study showed that SBA-15 exhibits high curcumin release of 82.6%, while amine functionalization regulates the curcumin in a sustained manner to lower release percentage of 41.2%. Three-dimensional porous *SBA-16 silica* loaded with *biocompatible ZnO* was reported to be an effective carrier for chemo drug temozolomide. Characterization of nanoformulation using various techniques shows the deposition of nanosized ZnO alongside retainment of mesoporous characteristics with high percentage of temozolomide loading (Fekri et al. 2022). *SBA-16* functionalized with silane coupling agent *3-aminopropyltriethoxysilane* was found to be effective for drug delivery of nonsteroidal anti-inflammatory drug indomethacin ($C_{19}H_{16}ClNO_4$). Importantly, the wrapping of polymer *gelatin and glutaraldehyde* through condensation process exhibited a pH release behavior corresponding to the pH environment of colon cancer (Yu et al. 2023).

Three-dimensional Ia3d cubic structured silica designated as Korea Advanced Institute of Science and Technology-6 material (KIT-6) containing guanidine and curcumin has been reported for potential breast cancer treatment. In the first step, KIT-6 was hydrothermally synthesized using Pluronic P123 template. After calcination, the sample was pegylated (20%) and treated with aminopropyltriethyoxysilane 1-H-pyrazole-1-carboxamidine hydrochloride and diisopropyl ethylamine to form 25% guanidine (~2.0 mmol/g). Then acetone containing curcumin was added under inert atmosphere, rinsed with excess curcumin and dried. The final loading of curcumin was found to be 1.67 mmol/g. The KIT-6 based formulation exhibited high loading capacity (50%), efficiency (100%) and colloidal stability with zeta potential value 24 ± 5 mV. Textural analysis using nitrogen adsorption technique

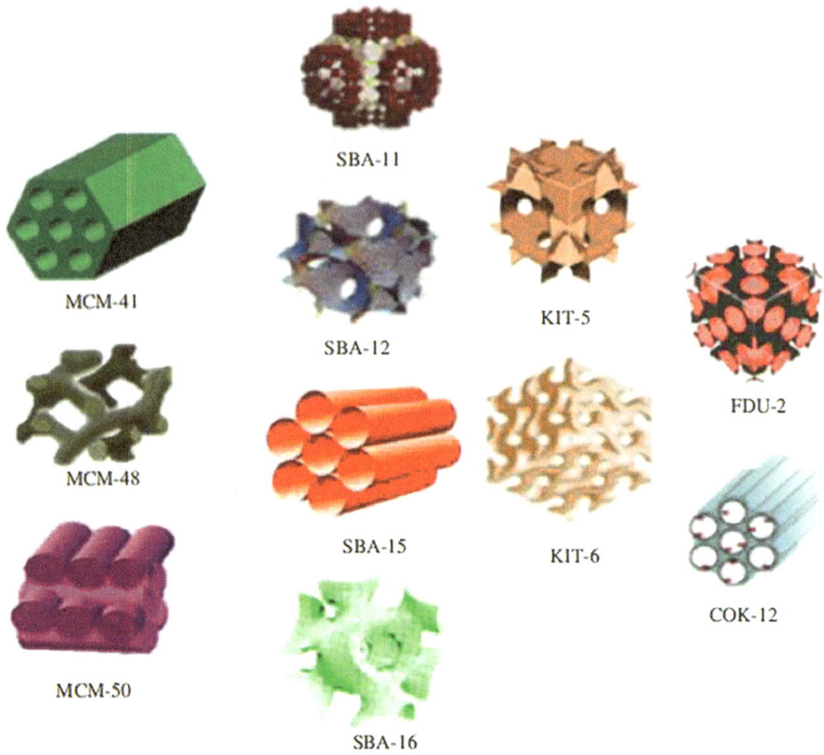

Fig. 3.2 Different types of structured mesoporous silica with different shaped pore architectures (Tella et al. 2022)

revealed a major change in the surface area, pore volume and pore size of KIT-6 to final formulation. For instance, the surface area of KIT-6 decreases from 834 m^2/g to 234 m^2/g, while pore volume decreases from 1.81 cm^3/g to 0.219 cm^3/g, respectively. The nano formulation exhibited a high pH sensitive curcumin release of 82.26% at pH 5.4 than 41.19% at pH 7.4 at 96 h in comparison to raw curcumin. The nanoformulation showed a strong cytotoxicity against MCF-7 cells (live cells of only 7.62% after 48 h) than normal cells (Ma'mani et al. 2014).

Spherical mesoporous silica (~150 nm) functionalized with pH sensitive polypeptide *Poly-L-Histidine*/tamoxifen has been reported for effective delivery of doxorubicin (DOX). The study relates the effect of DOX release through pre-surface and post-surface grafting technique. In the pre-surface grafting, the DOX was diffused into the pores of structured silica and then grafted with polypeptide and tamoxifen. In post-surface grafting, the silica was functionalized with the polypeptide and tamoxifen in the first step and followed by loaded with DOX. The polypeptide with pK_a value of 6.5 has been reported to act as an effective gate keeper due to solubility transition (Najafi Rad and Farzad 2022). At the normal pH 7.4, the peptide condenses, block the silica pore, and safeguard the DOX. However, at the tumor acidic pH condition of 5.0, the polypeptide gets protonated and due to repulsive forces, an expansion occurs leading to the release of DOX. In addition, the solubility of DOX improves due to acidic environment and targeted release improves the therapeutic efficiency. The drug release study of DOX loaded silica without modification exhibited high release pattern (33.6% at pH 7.4) and (42.3% at pH 5.0). In case of functionalized mesoporous silica, a less DOX release (18% at pH 7.4) and (23% at pH 5.0) was observed (Table 3.1; Fig. 3.3). This confirms the effective gate keeper role of grafted Poly-L-Histidine (Day et al. 2021).

AlMofty et al. (2023) has reported the targeted therapeutics based on cerium impregnated spherical silica and halloysite nanoformulations. Cerium nanoparticles are interesting, rare metals that show antioxidant and anti-inflammatory properties. The presence of variable oxidation states such as Ce^{3+} and Ce^{4+} helps to reactive with reactive oxygen species. Characterization tools such as diffuse reflectance spectroscopy and transmission electron microscope revealed the presence of Ce^{3+} and Ce^{4+} species (particle size ~5–10 nm) at 250 nm and 300 nm, respectively. The dual drugs (cisplatin and curcumin) functionalization was established using Fourier transform infrared (FTIR) spectroscopy. Drug release study was performed using dialysis membrane for 72 h. The result showed halloysite based formulation exhibits a high release of cisplatin of about 26% than spherical silica (21.4%).

Tumor specific MSN has been developed through conjugation with siRNA multidrug resistant protein 1 (MRP1) and myricetin with loading efficiency of 93.3% and 36.7%, respectively. The presence of folic acid receptor facilitates endocytosis via folate receptor. The nano formulation with hydrodynamic diameter of 109.8 nm shown to improve the therapeutic efficiency in treating non-small cell lung cancer. The presence of folic acid and MRP-1 facilitated apoptosis and upregulate Caspase-3 and PARP. A reduction in the colony formation and cell death was observed lung cancer cells (A549 and NCI-H1299). In vivo study has indicated a tumor shrinkage from 1800 mm^3 to about 450 mm^3 (Yinxue et al. 2020).

Biocompatible mesoporous silica with controlled particle size using tri-ethanolamine has been reported to deliver DOX to targeted therapeutics of breast cancer. For high drug loading, the surface of silica was functionalized with silane chemistry (APTES, EDC/NHS), further modified with fluorescence Rhodamine B isothiocyanate and cancer drug Trastuzumab. The surface functionalization led to high loading of DOX (57.4%) and exhibited therapeutic effect on breast cancer cells HER2 overexpressed SKBR3 (Mozafarinia et al. 2021). Glutathione responsive MSN linked with disulfide bonds was shown to load DOX effectively along with photosensitizer chlorine e6. The wrapping with biocompatible carboxymethyl chitosan on the nano formulation has shown to exhibit particle monodispersion, robust dimensional stability, and anti-cancer efficacy against MCF-7 cells. The photodynamic therapy of nano formulation well exhibited with laser irradiation ($<$ 660 nm). An effective antitumor effect was observed both in in vitro and in vivo (Zhu et al. 2022). Glycosylated-MSN with prodrug capecitabine capable of delivering 5-fluorouracil has been reported with high textural characteristics (particle size of 245.24 nm, pore diameter of 8.12 nm and pore volume of 0.73 cm^3/g). Surface modification with capping (chitosan-glucuronic acid) and receptor (lectin)

Table 3.1 Nanocarriers with different types of functionalization/conjugation for drug loadings

Sl. No	Nanocarrier	Material Source	Functionalization/Conjugation	Drug	In vitro/in vivo	Reference
1	SBA-15	Pluronic (P123), tetraethylorthosilicate (TEOS), triethylamine	Silane (3-chloropropyltriethoxysilane), Amidoxime	Doxorubicin hydrochloride (DOX)	Human breast adenocarcinoma cells (MCF-7)	Ulagesan et al. (2022)
2	SBA-15	P123, TEOS	(3-aminopropyl) trimethoxysilane (APTMS)	Curcumin	–	Atiyah et al. (2022)
3	SBA-16	P123, Pluronic F127, TEOS, glutaraldehyde	3-aminopropyltriethoxy silane, gelatin	Indomethacin	–	Yu et al. (2023)
4	KIT-6	P123, TEOS	Guanidine, polyethylene glycol	Curcumin	MCF-7, mouse breast cancer cells (4 T1), human mammary epithelial cells (MCF10A)	Ma'mani et al. (2014)
5	Graphene nanosheet	–	Poly (L-histidine)	Gemcitabine	–	Najafi Rad and Farzad (2022)
6	MSN	Cetylpyridinium bromide (CTAB), TEOS	(3-mercaptopropyl) trimethoxysilane, Tetraethylene glycol	DOX	–	Day et al. (2021)
13	Spherical silica, Halloysite nanotube	Cerium nitrate, wet impregnation technique	Polyethylene glycol	Cisplatin, curcumin	MCF-7, HFF-1	Almofty et al. (2023)
7	MSN	CTAB, NaOH, TEOS	siRNA, folic acid	Myricetin	Lung cancer cell lines A594 and NCI-H1299	Yinxue et al. (2020)
8	MSN	TEOS, Triethylamine, rhodamine B isothiocyanate, CTAB	(3- Aminopropyl) tri-ethoxysilane (APTES),	Trastuzumab and	Breast cancer cells	

			1-Ethyl-3-(3-dimethylaminopropyl)-carbodiimide (EDC)/N-hydroxysuccinimide (NHS)	doxorubicin hydrochloride (DOX)		Mohammad Mozafarinia et al. (2021)
9	MSN	CTAB, TEOS, APTES, EDC/NHS, amylin hydrochloride	Glutathione, chlorin e6, disulfide bonds, carboxymethyl chitosan	DOX	Breast cancer cells (MCF-7)	Zhu et al. (2022)
10	MSN-Capecitabine	CTAB, TEOS, glucuronic acid, 1,2-dimethylhydrazine dihydrochloride, chitosan,	Chitosan-glucuronic acid, 1-Ethyl-3-(3-dimethylaminopropyl)-carbodiimide/N-hydroxysuccinimide	5-fluorouracil	Colorectal cancer HCT 116 cell lines	Narayan et al. (2021)
11	MSN	CTAB, dimethylformamide, triethylamine, ethylenediamine, glutaric dialdehyde, 5,5′-dithiobis-(2-nitrobenzoic acid), glutathione	Pluronic F127, PEO_{100}-PPO_{65}-PEO_{100}, ε-caprolactone, disulfide bond	DOX	–	Long et al. (2022)
12	MOS	CTAB, TEOS, triethanolamine, bis [3 (triethoxysilyl)propyl] disulfide, 1′-Dioctadecyl-3,3,3′,3′-tetramethylindocarbocyanine perchlorate	Glutathione-S-transferase (GST)-fused cell-specific affibody, 3-(trimethoxysilyl) propyl acrylate, human epidermal growth factor 2 (HER2), epidermal growth factor receptor	DOX	–	Yang et al. (2023)
	Graphene oxide					
13	GO	Graphite exfoliation process	Folic acid, polyethylene glycol, EDC/NHS	Camptothecin	J774, HepG2	de Sousa et al. (2018)
14	$MnFe_2O_4$/GO $MnFe_2O_4$/silica	Dry impregnation technique	Polyethylene glycol	Cisplatin	MCF-7, human foreskin fibroblast (HFF-1)	Almohazey et al. (2023)

(continued)

Table 3.1 (continued)

Sl. No	Nanocarrier	Material Source	Functionalization/Conjugation	Drug	In vitro/in vivo	Reference
15	GO	Layer by layer self-assembly technique	Chitosan, dextran	DOX	MCF-7	Meng Xie et al. (2016)
16	GN/FA-ZnO	Precipitation technique	Folic acid	Ehrlich ascites carcinoma/ bioimaging	Swiss albino mice	Wanas et al. (2023)
17	Metal oxide/ GO	ZnO, TiO_2, GO	Hydrophilic polymer polyethylenimine	Curcumin, quercetin	Sulforhodamine B assay	Monika Matiyani et al. (2023)
18	PEI-CTC	GO, topotecan, distilled water, sonication	Polyethylenimine	Cisplatin, Topotecan	HeLa cervical cells	Mallick et al. (2018)
19	GO	Copper (II) sulfate pentahydrate, ascorbic acid	EDC/NHS, folic acid, poly (ethylene glycol) bis(amine)	DOX	MCF-7, human fibroblasts (HDFa)	Mauro et al. (2020)

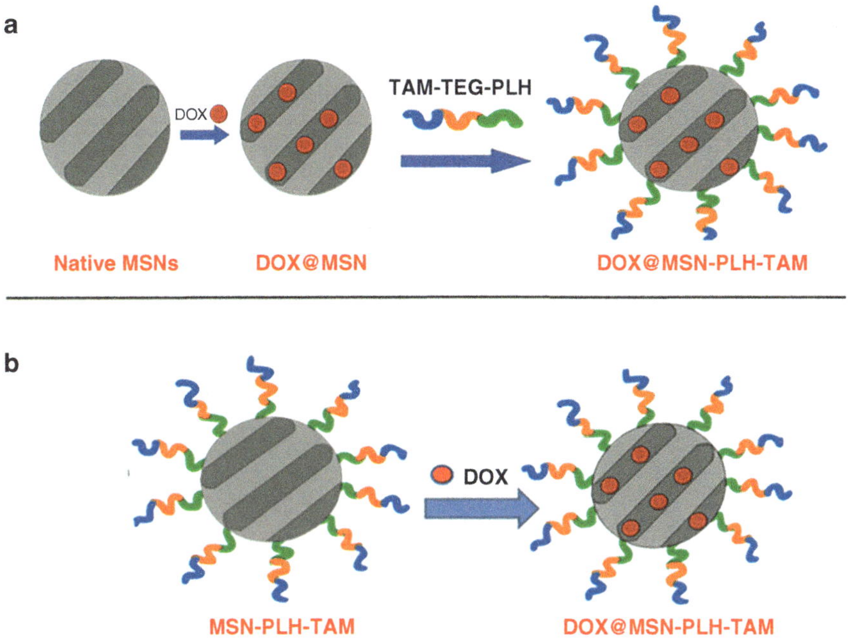

Fig. 3.3 Effect of pre- and post-grafting loading of mesoporous based nanocarrier for DOX release (Day et al. 2021)

demonstrated high pH stimuli prodrug release and uptake in HCT 116 cells. In vivo study revealed a tumor shrinkage and reduction of clusters of abnormal tube-like glands (aberrant crypt foci), abnormality in cells and inflammation (Narayan et al. 2021). Triple responsive MSN to delivery DOX was reported using disulfide bond interaction with polycaprolactone modified non-ionic copolymer Pluronic F127 micelle PEO_{100}-PPO_{65}-PEO_{100} (thermos sensitive gatekeeper). Glutathione stimuli assists the uncapping of micelle from MSN, while the tumor acidic pH combined with hyperthermia (40 °C) facilitated the quick release of drug (DOX) for anti-cancer activity (Long et al. 2022). In a similar stance, the presence of disulfide bond incorporated into organosilica bound mesoporous nanoparticles exhibited a highly targeted therapeutic efficacy of DOX. Glutathione-S-transferase fused cell-specific affibody at the external surface and presence of DOX at the internal pores facilitated prompt disintegration and release of DOX (Yang et al. 2023) (Fig. 3.4).

3.1.1 Functionalized Graphene Oxide

3.1.1.1 Conjugation and Functionalization of Graphene Oxide Based Nanocarriers

Graphene oxide (GO), a two-dimensional nanomaterial in the oxygenated form has attracted huge interest in the field of biomedical applications (Fig. 3.5). The presence

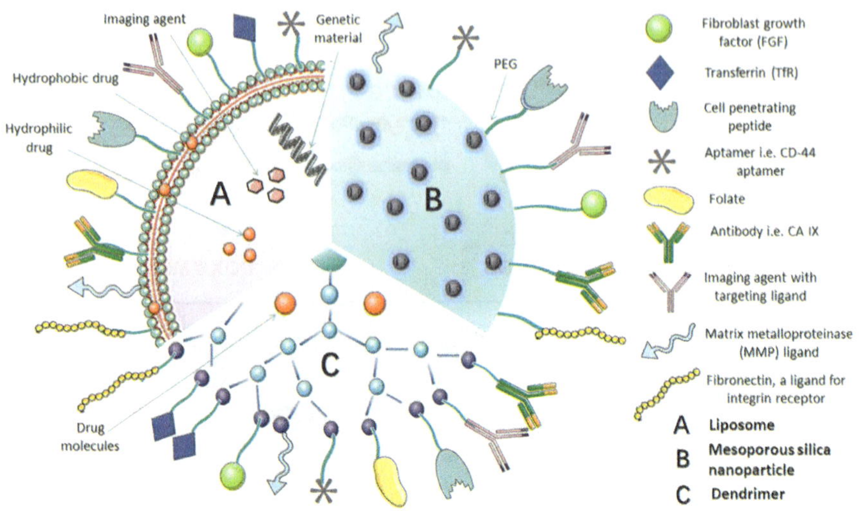

Fig. 3.4 Types of functionalization available for liposome, mesoporous silica, and dendrimer using different chemical moieties (Seidu et al. 2022)

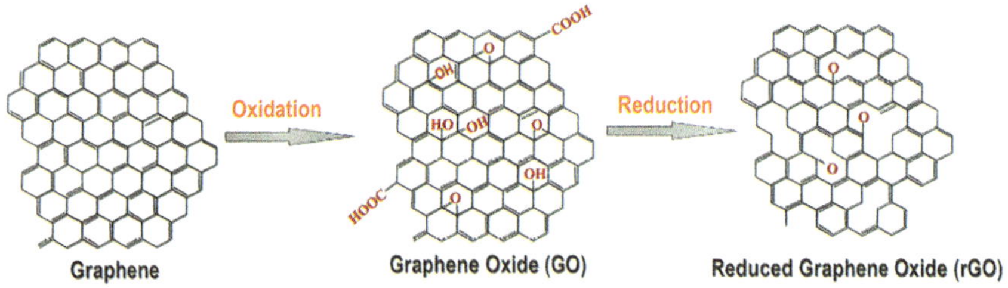

Figure 2 Structure of graphene and its oxidized derivatives.
Notes: Reproduced with permission from Priyadarsini S, Mohanty S, Mukherjee S, Basu S, Mishra M. Graphene and graphene oxide as nanomaterials for medicine and biology application. *Journal of Nanostructure in Chemistry*. 2018;8:123–137.[41] Copyright © 2018, The Authors. Creative Commons CC BY (https://creativecommons.org/licenses/by/4.0/legalcode).

Fig. 3.5 Graphene oxide for targeted anti-cancer applications

of hydrophobic large surface area and noncovalent interactions (π-π stacking) can be exploited for hydrophobic drug encapsulation, loading, and controlled drug delivery (Naline et al. 2022).

Generation of oxygenation through exfoliation and oxidation process offers exceptional scope for chemical functionalization and polymer wrapping. Furthermore, unlike liposomes and lipid nanoparticles, coupling drugs with GO was effective due to the presence of carboxylic functional moieties at the surface of GO. n-(3-(dimethylamino)-propyl)-n-ethylcarbodiimide (EDC), n-hydroxysuccinimide (NHS), and thionyl chloride ($SOCl_2$) can be used as coupling agents and improve the functionalization properties. Interestingly, a certain cytotoxicity of GO was reported to reduce significantly by the polymer (polyethylene glycol) wrapping around GO (Fig. 3.6).

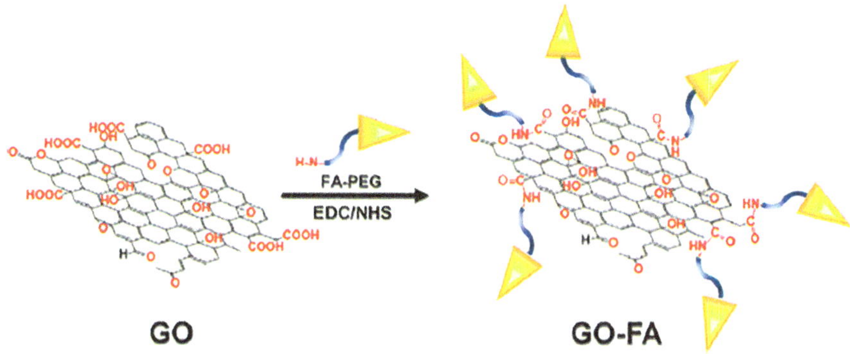

Fig. 3.6 GO functionalization using folic acid and EDC/NHS agents (de Sousa et al. 2018)

In the first step, conjugation of folic acid with polyethylene glycol (FA-PEG) was carried out. For functionalization, GO was crosslinked with EDC/NHS, then treated with FA-PEG. Then the hydrophobic drug camptothecin was loaded on the nanocarrier using DMSO solvent, followed by washing with methanol/deionized water mixture (de Sousa et al. 2018). Aromatic rings of GO tend to favor the adsorption of camptothecin via hydrophobic adsorption process and stacking through π–π interactions between aromatic rings. The presence of folic acid reduces the loading capacity of camptothecin from 52.3% to 37.8%, while loading efficiency decreases from 53.2% to 32.6%, respectively. This shows the functionalization with folic acid reduces the interaction of camptothecin. In the drug release study, the nanoformulation demonstrated a pH sensitive drug release. For instance, the lactone ring of camptothecin hydrolyze readily in normal physiological condition to carboxylate form and exhibit higher release than at pH 5.0. Nevertheless, the carboxylate form was inactive and non-toxic to tumors. In the tumor acidic pH, the carboxylate transforms into lactone form and becomes active against cancer cells. In case of folic acid functionalization, high camptothecin release of about 83% was observed, while it reaches plateau after 100 h. In case of tumor pH of 5.0, a slower and sustained release of about 48% was observed after 48 h. The release reached to about 60% even after 100 h. Functionalization of camptothecin to GO without folic acid strengthen the drug interaction, reducing the release to 56% in 48 h at pH 7.4. At pH 5.0, the formulation further reduced the drug release to about 29% and 41% after 48 h and 200 h, respectively. The study indicates that GO exert a strong interaction with drug, while folic acid functionalization triggers high drug release and promote cell death by apoptosis (de Sousa et al. 2018).

GO has some of the limitations including aggregation with physiological sources (salt and proteins) and non-specific protein binding in physiological conditions. Wrapping with hydrophilic polymer around GO have been reported to overcome such limitation partly and also improve the drug solubility. Some of nanocarriers reported includes pegylated $MnFe_2O_4$/GO (Almohazey et al. 2023), 3-aminopropyltriethoxysilane, folic acid, caffeic acid and chitosan wrapped GO

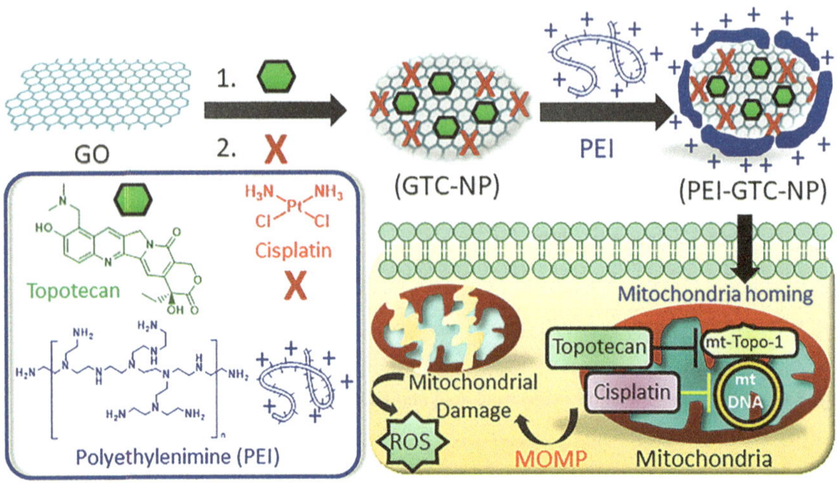

Figure 4 GO with PEI coverage coloaded with cisplatin and topotecan for mitochondria targeting of cancer cells.
Notes: Reproduced with permission from Mallick A, Nandi A, Basu S. Polyethylenimine coated graphene oxide nanoparticles for targeting mitochondria in cancer cells. *ACS Applied Bio Mater.* 2018;2:14–19.[116] Copyright 2019, American Chemical Society.

Fig. 3.7 GO/PEI/cisplatin/topotecan for mitochondria targeting cancer cells (Mallick et al. 2018)

(Tousian et al. 2022), GO/chitosan/dextran/DOX (Xie et al. 2016), hyperbranched polyglycerol functionalized GO via glycidol polymerization for curcumin release (Paknia et al. 2022), etc. Recently, biocompatible metal oxides (Fe_3O_4, ZnO, TiO_2, CuO) based nanocarriers have received attention for nanotherapeutic applications. Such nanoformulations shows biodegradable character and exhibit high colloidal stability. Wanas et al. (2023) have shown that folic acid functionalized metal oxide such as ZnO has been effectively used in vivo tumor imaging. Functionalization of metal oxides (ZnO or TiO_2)/GO nanocomposite with cationic hydrophilic *branched polyethylenimine* followed by loading of quercetin (49.39%) and curcumin (48.89%) exhibited a pH sensitive site-specific drug release and anti-cancer activity against breast cancer cell line (Matiyani et al. 2023). Combination of GO with cationic hydrophilic polymer such as polyethylenimine and chemodrugs such as cisplatin and topotecan enabled a mitochondria target specific drug delivery system. The presence of polymer induces the positive charge and able to reach into mitochondria at 6 h duration and generate reactive oxygen species to impart cancer cells death (Mallick et al. 2018) (Fig. 3.7).

GO based nano formulation conjugated with mitochondria targeting ligands and external coverage of polyglycerol for delivery DOX. The formulation exhibited a synergistic hyperthermia and chemotherapy effect. The presence of positive charge prompted the cellular internalization, while near-infrared irradiation combined with the cellular level release of DOX into inhibited both biochemical energy molecule "adenosine triphosphate" process and cell processes governed by mitochondria (Tu et al. 2018, 2020). Photothermal application of pegylated GO has been reported with physiological fluid stability and biocompatibility. Folic acid/pegylation/GO

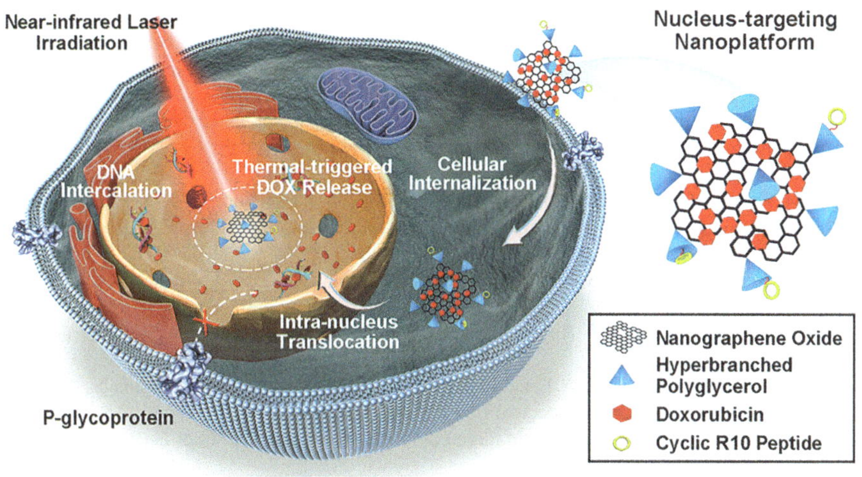

Fig. 3.8 Multifunctional drug delivery system GO/Polyglycerol/DOX for multiple drug resistance cancer treatment (Tu et al. 2020)

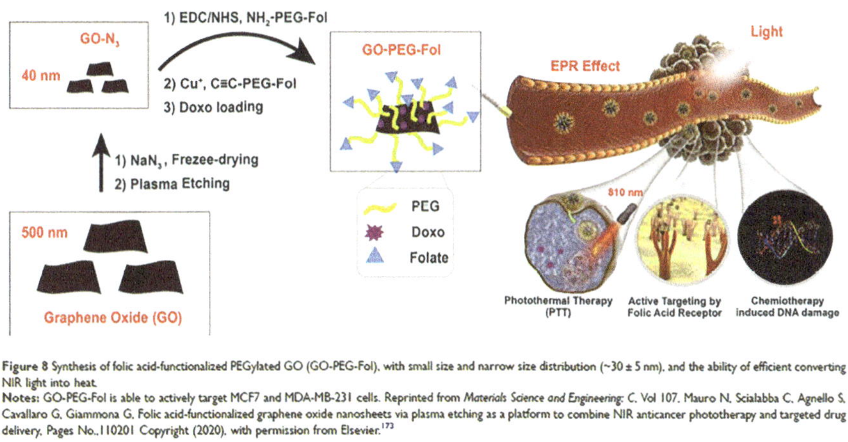

Fig. 3.9 Folic acid/pegylation/GO formulation for phototherapy

formulation with particle size ranging at about 30 nm ± 5 nm shown to convert the NIR light in to heat suitable for breast cancer therapy (Figs. 3.8 and 3.9).

References

AlMofty S, Ravinayagam V, AlGhamdi N, AlGhamdi W, AlBazroun Z, AlMulla L, Akhtar S et al (2023) Effect of CeO2/spherical silica and halloysite nanotubes engineered for targeted drug delivery system to treat breast cancer cells. OpenNano 13:100169

Almohazey D, Ravinayagam V, Alamoudi W, Sultan Akhtar H, Dafalla HN, AlSuwaidan S, Almutairi T, Alghamdi HS, Aldamen SA, Almessiere MA et al (2023) Insights of platinum drug interaction with spinel magnetic nanocomposites for targeted anti-cancer effect. Cancers 15(3):695. https://doi.org/10.3390/cancers15030695

Atiyah NA, Albayati TM, Atiya MA (2022) Interaction behavior of curcumin encapsulated onto functionalized SBA-15 as an efficient carrier and release in drug delivery. J Mol Struct 1260: 132879. https://doi.org/10.1016/j.molstruc.2022.132879

Day CM, Sweetman MJ, Song Y, Plush SE, Garg S (2021) Functionalized mesoporous silica nanoparticles as delivery systems for doxorubicin: drug loading and release. Appl Sci 11:6121. https://doi.org/10.3390/app11136121

de Sousa M, Visani de Luna LA, Fonseca LA, Giorgio S, Luiz AO (2018) Folic-acid-functionalized graphene oxide nanocarrier: synthetic approaches, characterization, drug delivery study, and antitumor screening. ACS Appl Nano Mater 1:922–932. https://doi.org/10.1021/acsanm.7b00324

Debnath S, Srivastava R (2021) Drug delivery with carbon-based nanomaterials as versatile nanocarriers: progress and prospects. Front Nanotechnol 3:1

Di Stefano A (2023) Nanotechnology in targeted drug delivery. Int J Mol Sci 24(9):8194. https://doi.org/10.3390/ijms24098194

Fekri MH, Soleymani S, Mehr MR, Akbari-adergani B (2022) Synthesis and characterization of mesoporous ZnO/SBA-16 nanocomposite: its efficiency as drug delivery system. J Non-Cryst Solids 591:121512. https://doi.org/10.1016/j.jnoncrysol.2022.121512

Guo Y, Cao X, Zheng X, Abbas SKJ, Li J, Tan W (2022) Construction of nanocarriers based on nucleic acids and their applications in nanobiology delivery systems. Natl Sci Rev 9(5): nwac006

Khakpour E, Salehi S, Naghib SM, Ghorbanzadeh S, Zhang W (2023) Graphene-based nanomaterials for stimuli-sensitive controlled delivery of therapeutic molecules. Front Bioeng Biotechnol 11:1129768

Li Y, Zhang T, Liu Q, He J (2019) PEG-derivatized dual-functional nanomicelles for improved cancer therapy. Front Pharmacol 10:808

Long H, Tian W, Jiang S, Zhao J, Zhou J, He Q, Tang Z, Shen W, Wang J (2022) A dual drug delivery platform based on thermo-responsive polymeric micelle capped mesoporous silica nanoparticles for cancer therapy. Microporous Mesoporous Mater 338:111943

Mallick A, Nandi A, Basu S (2018) Polyethylenimine coated graphene oxide nanoparticles for targeting mitochondria in cancer cells. ACS Appl Biomater 2(1):14–19

Ma'mani L, Nikzad S, Kheiri-Manjili H, Al-Musawi S, Saeedi M, Askarlou S, Foroumadi A, Shafiee A (2014) Curcumin-loaded guanidine functionalized PEGylated I3ad mesoporous silica nanoparticles KIT-6: practical strategy for the breast cancer therapy. Eur J Med Chem 83:646–654

Matiyani M, Rana A, Karki N, Garwal K, Pal M, Sahoo NG (2023) Development of multi-functionalized graphene oxide based nanocarrier for the delivery of poorly water soluble anticancer drugs. J Drug Deliv Sci Technol 83:104412

Mauro N, Scialabba C, Agnello S, Cavallaro G, Giammona G (2020) Folic acid-functionalized graphene oxide nanosheets via plasma etching as a platform to combine NIR anticancer phototherapy and targeted drug delivery. Mater Sci Eng C 107:110201

Mozafarinia M, Karimi S, Farrokhnia M, Esfandiari J (2021) In vitro breast cancer targeting using trastuzumab-conjugated mesoporous silica nanoparticles: towards the new strategy for decreasing size and high drug loading capacity for drug delivery purposes in MSN synthesis. Microporous Mesoporous Mater 316:110950

Najafi Rad Z, Farzad F, Razavi L (2022) Surface functionalization of graphene nanosheet with poly (l-histidine) and its application in drug delivery: covalent vs non-covalent approaches. Sci Rep 12(1):19046

Naline B, Baipaywad P, Ryu N, Lee JY, Park H (2022) Recent biomedical advancements in graphene oxide-and reduced graphene oxide-based nanocomposite nanocarriers. Biomater Res 26(1):65

Narayan R, Gadag S, Cheruku SP, Raichur AM, Day CM, Garg S, Manandhar S et al (2021) Chitosan-glucuronic acid conjugate coated mesoporous silica nanoparticles: a smart pH-responsive and receptor-targeted system for colorectal cancer therapy. Carbohydr Polym 261:117893

Paknia F, Mohabatkar H, Ahmadi-Zeidabadi M, Zarrabi A (2022) The convergence of in silico approach and nanomedicine for efficient cancer treatment; in vitro investigations on curcumin loaded multifunctional graphene oxide nanocomposite structure. J Drug Deliv Sci Technol 71: 103302

Seidu TA, Kutoka PT, Asante DO, Farooq MA, Alolga RN, Bo W (2022) Functionalization of nanoparticulate drug delivery systems and its influence in cancer therapy. Pharmaceutics 14: 1113. https://doi.org/10.3390/pharmaceutics14051113

Tella JO, Adekoya JA, Ajanaku KO (2022) Mesoporous silica nanocarriers as drug delivery systems for anti-tubercular agents: a review. R Soc Open Sci 9(6):220013

Tian H, Zhang T, Qin S et al (2022) Enhancing the therapeutic efficacy of nanoparticles for cancer treatment using versatile targeted strategies. J Hematol Oncol 15:132. https://doi.org/10.1186/s13045-022-01320-5

Tousian B, Ghasemi MH, Khosravi AR (2022) Targeted chitosan nanoparticles embedded into graphene oxide functionalized with caffeic acid as a potential drug delivery system: new insight into cancer therapy. Int J Biol Macromol 222:295–304

Tu Z, Qiao H, Yan Y, Guday G, Chen W, Adeli M, Haag R (2018) Directed graphene-based nanoplatforms for hyperthermia: overcoming multiple drug resistance. Angew Chem 130(35): 11368–11372

Tu Z, Donskyi IS, Qiao H, Zhu Z, Unger WES, Hackenberger CPR, Chen W, Adeli M, Haag R (2020) Graphene oxide-cyclic R10 peptide nuclear translocation nanoplatforms for the surmounting of multiple-drug resistance. Adv Funct Mater 30(35):2000933

Ulagesan S, Santhamoorthy M, Phan TTV, Alagumalai K, Thirupathi K, Kim SC, Nam TJ, Choi YH (2022) Mesoporous silica (SBA-15) with enriched amidoxime functionalities for pH-controlled anticancer drug delivery. Inorg Chem Commun 146:110132. https://doi.org/10.1016/j.inoche.2022.110132

Wanas W, Abd El-Kaream SA, Ebrahim S et al (2023) Cancer bioimaging using dual mode luminescence of graphene/FA-ZnO nanocomposite based on novel green technique. Sci Rep 13:27. https://doi.org/10.1038/s41598-022-27111-z

Xie M, Lei H, Zhang Y, Yuanguo X, Shen S, Ge Y, Li H, Xie J (2016) Non-covalent modification of graphene oxide nanocomposites with chitosan/dextran and its application in drug delivery. RSC Adv 6(11):9328–9337

Yang G, Kim S, Jun Yong O, Kim D, Jin S, Choi E, Ryu J-H (2023) Surface protein-retractive and redox-degradable mesoporous organosilica nanoparticles for enhanced cancer therapy. J Colloid Interface Sci 649:1014

Yinxue S, Zhou B, Xiangyang D, Wang Y, Zhang J, Ai Y, Xia Z, Zhao G (2020) Folic acid (FA)-conjugated mesoporous silica nanoparticles combined with MRP-1 siRNA improves the suppressive effects of myricetin on non-small cell lung cancer (NSCLC). Biomed Pharmacother 125:109561

Yu B, Shi R, Chen X, Zhang Y, Hu J, Khan S (2023) Gelatin-coated indomethacin drug-loaded SBA-16 silica-based composites: pH-responsive slow-release performance. Inorg Chem Commun 150:110469. https://doi.org/10.1016/j.inoche.2023.110469

Zhu L, Zhang W, Song P, Li W, Chen X, Ge F, Gui L, Yang K, Tao Y, Guocheng D (2022) Redox-responsive mesoporous silica nanoparticles for chemo-photodynamic combination cancer therapy. Mater Res Express 9(4):045401

Nano-Drug Carriers for Chemotherapeutic Agents Delivery in Cancer Disease Treatment

4

Priyanca Ahlawat, Rohit Kumar, Akhilesh Kumar, and Piyush Kumar Gupta

Abstract

Cancer is a leading cause of mortality in the non-communicable disease category. Researchers are on a constant quest to identify novel drug targets as well as therapeutic molecules for its treatment. However, innumerable chemotherapeutics that are experimentally evaluated are unable to reach bench to bedside due to many drawbacks. These include poor bioavailability, lower bioretention, and non-specific effects along with systemic toxicity. Additionally, these caveats lead to a long and intense treatment regimen that leads to poor patient compliance. Active efforts are underway to identify bioactives that do not fall short on these points. However, catering to such limitations, the introduction

Priyanca Ahlawat and Rohit Kumar are contributed equally as first authors.

P. Ahlawat
Translational Health Science and Technology Institute, Faridabad, Haryana, India

R. Kumar
Department of Life Sciences, Sharda School of Basic Sciences and Research, Sharda University, Greater Noida, India

Centre for Development of Biomaterials, Sharda University, Greater Noida, India

A. Kumar
Division of Medicine, ICAR Indian Veterinary Research Institute, Izatnagar, Bareilly, Uttar Pradesh, India

P. K. Gupta (✉)
Department of Life Sciences, Sharda School of Basic Sciences and Research, Sharda University, Greater Noida, India

Department of Biotechnology, Graphic Era Deemed to be University, Dehradun, Uttarakhand, India

Centre for Development of Biomaterials, Sharda University, Greater Noida, India
e-mail: piyush.kumar1@sharda.ac.in

F. A. Khan (ed.), *Nano Drug Delivery for Cancer Therapy*,
https://doi.org/10.1007/978-981-99-6940-1_4

of nanomedicine in the cancer chemotherapeutic regimen has opened many avenues and has helped in accelerating the cancer treatment. Nanomedicine has encompassed all the aforementioned lacunae and turned them into opportunities for the development of novel and versatile nanoformulations. In this chapter, we extensively discuss the recent advancements in the development of many varieties of nanosystems that have been evaluated for their anti-cancerous efficacy and propose bright future prospects for such nanoplatforms.

Keywords

Nano-carriers · Chemotherapeutic agents · Drug Delivery · Cancer treatment

4.1 Introduction

Cancer is a continued world health problem that has lasted centuries. With advancements in science and technology, the biology of cancer is being studied in depth. The scientific community has reached a consensus that this disorder has showcased manipulative behavior over the years, and hence its treatment has been explored from a versatile approach. Many compounds that target all the vital cell organelles and cell viability and proliferation mechanisms have been identified or synthesized. Although the overall survival of cancer patients has increased, most remain precarious and difficult to treat. Oncologists have struggled to identify the neoplasm and its metastasis to prescribe the best possible chemotherapeutic combinations for tumor regression. The situation remains grim in the twenty-first century, and newer and finer interventions are being sought.

Cancer researchers have been on an unending quest to identify the ultimate molecule or a compound that holds therapeutic efficacy to destroy tumor progression completely. Although innumerable novel drug candidates have been developed, their certain biological shortcomings, such as bioretention, elimination, or metabolism, discourage their progression through clinical trials.

That is why so many experimentally proven compounds are sieved to reach US FDA approval. Even the compounds that have been relatively fortunate to reach the bench to bedside have been highly patient and non-compliant. This is because chemotherapeutics tend to unleash systemic organ toxicities ranging from mild manifestations like skin rashes, alopecia, and fatigue that affect the emotional and social quotient of the patient to severe and fatal effects such as hematopoietic suppression, thrombocytopenia, neurological issues, immunological and renal suppression, among others. These major side effects restrict the continuation of the treatment regimen and burn a hole in the pockets of cancer patients to an unknown extent during the therapy.

These lacunae in conventional therapy have been a cue to the emergence of nanotechnology in cancer management and to their rescue comes the nanoformulations that eliminate such limitations and broaden chemotherapeutics' prospects. Elucidation of tumor biology fundamentals immensely supported this field's advancement and led to the development of a variety of nanoplatforms.

At present, cancer therapy lacks specificity in its ability to selectively target cellular proteins and other molecules due to the variability of these molecular fingerprints concerning the mutational state of the tumor. Nanomedicine has taken care of the disadvantages mentioned above of chemotherapeutics in terms of the fact that it has been developed to improve bioactive/drugs bioavailability in the systemic circulation, especially for the hydrophobic compounds, which is provided by the sustained and controlled release profile of the encapsulated drug. Secondly, it has selectivity over the free drugs and hence can be used as per the tumor histopathology type. Thirdly, nanoformulations can render longer chemotherapeutic effects with minimum side effects due to their higher mean retention time and negligible drug leaching. Finally, they can enhance patient compliance and lower the economic burden.

Therefore, this chapter deals with the advancements in the field of cancer-specific nanomedicines. The discussion elaborates on the rationale of the nanosystem development, the selection of experimental assays for their anticancer capacity evaluation, and the extent to which combinatorial formulations synergistically enhance the antiproliferative effects of the chemotherapeutics.

4.2 Lipid-Based Nanotherapy

Many pharmaceutically relevant compounds are generally insoluble or partially water soluble. Not only that, they are physiologically and chemically unstable or render severe systemic toxicities. Among other therapeutic bioactive compound carriers are lipid-based nanoparticles (LBNPs). Their utilization in chemotherapeutic cancer regimens as a drug carrier has enhanced the antitumor efficacy of many chemotherapeutics. LBNPs can be developed using naturally available lipids with great biological stability, etc. LBNPs were widely studied against a variety of cancer, both in vitro and in vivo, with encouraging outcomes in certain clinical trials. The different types used are enlisted below (Fig. 4.1). Lipids play a crucial role in the composition of cell membranes in the human body. Lipids encompass several organic compounds, namely fatty acids, waxes, neutral fats, and steroids, including cortisone. Compound lipids include several types of molecules, including glycolipids, lipoproteins, and phospholipids, which are intricately associated with other chemical entities. Lipids can be characterized as hydrophobic or amphipathic compounds and their amphiphilic nature allows for the construction of vesicles, massive uniflagellar liposomes, or a membrane. A biological lipid generally comprises two types of building subunits: ketoacyl and isoprene groups.

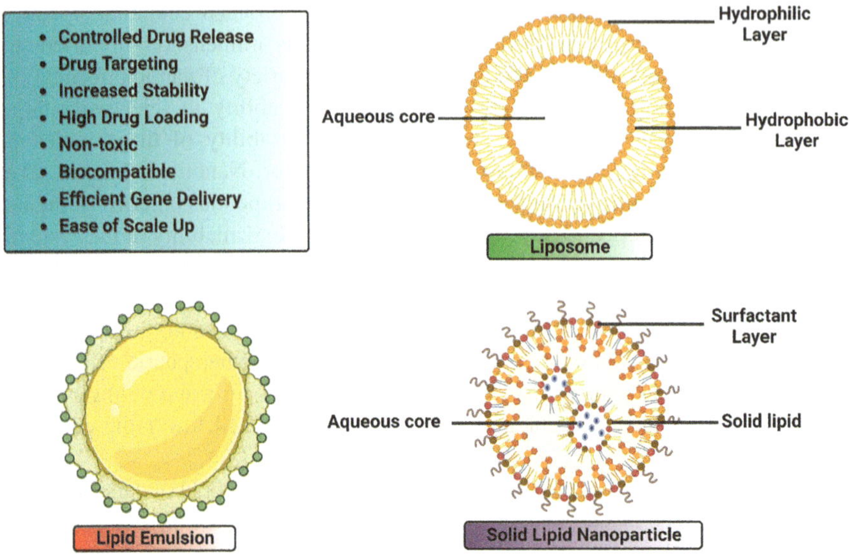

Fig. 4.1 The overall arrangement of solid lipid NPs, which have benefits over liposomes and lipid emulsions, is represented schematically (adopted from Tekade et al. 2017)

4.2.1 Solid Lipid Nanoparticles

Innumerable studies have indicated the beneficial use of solid lipid NPs (SLNs) as drug delivery vehicles for chemotherapeutic anticancer compounds. The drug carriers under consideration are colloidal NPs, with sizes ranging from 50 to 1000 nm. SLNs consist of a solid matrix core made up of biodegradable lipids, which are enveloped by a hydrophilic surfactant. They have demonstrated desirable biocompatibility and low systemic toxicity. SLNs also show sustained drug release by lipid matrix erosion, degradation, or diffusion. Thus, they have improved the limits of conventional anticancer compounds' therapeutic efficacy and bioavailability.

For example, a recent study conducted by Alajami et al. (2022) showcased the advantages of utilizing a colon delivery system consisting of SLNs loaded with celecoxib (CXB) to further enhanced its anticancer properties. The ultrasonic melt-emulsification technique was utilized in the fabrication of SLNs. The cytotoxic effects were assessed by employing an MTT test in three cancer cell lines: HT29 (colon cancer), Daoy (medulloblastoma), and HepG2 (hepatocellular carcinoma) cells. The utilization of SLNs resulted in a significant enhancement of CXB activity against a colon cancer cell line that exhibits resistance to COX-2, as well as two additional cancer cell lines that also display resistance. This development facilitated the administration of increased dosages of CXB without the occurrence of undesirable side effects associated with the COX-2 inhibitor. Therefore, it is apparent that this particular system possesses the capacity to be transformed into a viable therapeutic alternative for both preventing and treating colon cancer. The anticancer effect of andrographolide (ADG) against many cancer cell lines is extensively acknowledged. However, the compound's elevated hydrophobicity, coupled with its limited bioavailability, imposes constraints on its potential therapeutic use. Li et al. (2022) demonstrated that ADG-SLN has enhanced inhibitory efficacy against head and neck cancer and precancerous cells in comparison to free ADG. The observed result can be related to the increased cellular uptake and intracellular absorption of ADG-SLN. Moreover, ADG-SLN promoted cell cycle arrest and apoptosis in HIOEC, Leuk1, HN6, and HN30 cells.

The administration of cytotoxic drugs for brain cancer treatment is associated with an inadvertent impact on healthy cells, owing to the nonselective nature of the therapeutic strategy. Hence, a research investigation was undertaken to formulate tween 80-coated SLNs containing a folic acid-doxorubicin (FAD) mixture. The purpose of this formulation was to enable the targeted delivery of drugs to brain cancer cells. The U87 MG (Uppsala 87 Malignant Glioma) cell line exhibited both cytotoxic effects and internalization of SLN-C in U87 MG brain cancer cells. The findings reaffirmed that SLNs coated with tween 80 demonstrate promise in selectively delivering doxorubicin to brain cancer cells (Jain et al. 2022). Ahmed et al. (2020) employed a solvent emulsification process to fabricate SLNs loaded with brigatinib (BG). Subsequently, their team conducted an MTT experiment utilizing the optimized SLNs denoted as BS5. The study revealed that the BG-loaded SLNs containing BS5 exhibited enhanced cytotoxic effects on the A549 lung cancer cell line in comparison to both the BG solution and the blank SLN. Therefore, it was determined that the modified BS5 formulation exhibited a comparatively higher level of cytotoxicity and was able to achieve its effectiveness with a 74.91% lower dosage in comparison to the drug suspension (BG) used alone. The results provide evidence in favor of the proposition that drug delivery through sentinel lymph nodes (SLN) might potentially augment cytotoxic effects. Another group successfully formulated a dry powder inhaler dose form that incorporates gefitinib-loaded SLNs (Gef-G-SLNs) targeted to lung cancers using glucosamine. They exhibited the enhanced internalization of G-SLNs in A549 cells in comparison to non-targeted SLNs using flow cytometry and fluorescence microscopy techniques (Satari et al. 2020).

4.2.2 Liposomes

Liposomes are nano-sized synthetic vesicles with a spherical shape that can be developed using cholesterol and non-toxic phospholipids derived from natural sources. Liposomes, due to their distinct size, hydrophobic and hydrophilic properties, as well as their biocompatibility, have emerged as one of the potential drug delivery mechanisms.

Xu et al. (2022) conducted a study whereby they fabricate a codelivery system with extended circulation. This system used a combination of doxorubicin (DOX) and schizandrin A (SchA) to target MCF-7/ADR cells. The DOX-SchA long-circulating liposome (DOX-SchA-Lip) was synthesized via the ammonium sulfate gradient technique. Both drugs were simultaneously encapsulated into the distearoyl phosphatidylethanolamine-polyethylene glycol (DSPE-mPEG2000) liposome. The liposome showed efficacy in inducing early apoptosis in HepG2/ADR cells, and the administration of DOX-Sc resulted in a blockage of the cell cycle in the S-phase. The efficacy of DOX-SchA- DOX-SchA-Lip in inducing tumor cell death and overcoming multidrug resistance (MDR) was demonstrated in a way that was dependent on the dosage administered. Another research group explored the possibilities of the liposomal delivery system against resistant cancer cells. In this study, a liposome system was developed to simultaneously encapsulate a chemotherapeutic drug, namely doxorubicin hydrochloride (DOX), and a commonly used autophagy inhibitor, namely chloroquine phosphate (CQ), at a weight ratio of 1:2. The objective of this study was to examine the mechanism by which this liposome system may reverse drug resistance. The results of the MTT assay revealed that liposomes containing both DOX and CQ, when administered to DOX-resistant human breast cancer cells (MCF7/ADR), exhibited a half-maximal inhibitory concentration (IC50) of 4.7 ± 0.2 μM. This value was much lower, about 5.7 times than the IC50 observed for free DOX (26.9 ± 1.9 μM). The liposomal co-encapsulation of DOX/CQ exhibited enhanced antitumor effectiveness in both the MCF-7/ADR multicellular tumor spheroid model and transgenic zebrafish models as compared to the administration of liposomal DOX or DOX alone. The results of this study indicate that the co-encapsulation of chloroquine (CQ) with doxorubicin (DOX) into liposomes resulted in a notable enhancement of DOX sensitivity in cancer cells that had developed resistance to DOX (Gao et al. 2017).

In a study conducted by Zhu et al. in 2023, it was shown that Ginsenosides Rg3-based liposomes containing Paclitaxel (referred to as Rg3-PTX-LPs) could selectively target MCF7/T cancer cells and the tumor microenvironment (TME) concurrently. This targeted delivery mechanism was principally facilitated by the recognition of GLUT-1. The drug resistance reversal capacity and in vivo antitumor efficacy of Rg3-PTX-LPs exhibited notable enhancements in comparison to standard cholesterol liposomes. The remodeling processes of Rg3-PTX-LPs inside the TME encompass the suppression of IL-6/STAT3/p-STAT3 pathway activation, resulting in the conversion of protumor M2 macrophages into an anticancer M1 phenotype. Furthermore, these processes encompass the inhibition of myeloid-derived suppressor cells (MDSCs), the decrease in tumor-associated fibroblasts (TAFs) and collagen fibers inside the TME, and the facilitation of apoptosis in neoplastic cells. Hence, by the concurrent targeting of tumor cells and modulation of the TME, Rg3-PTX-LPs exhibited a noteworthy tumor inhibition rate of 90.3%.

In another study conducted by George et al. (2022b), a polyethylene glycol-modified nanoscale liposomal formulation (referred to as LipoRV) of a novel anthraquinone derivative was synthesized. The study demonstrated the therapeutic efficacy of LipoRV on several triple-negative breast cancer (TNBC) cell lines. The use of LipoRV has shown a high level of effectiveness in suppressing the progression of the cell cycle, triggering programmed cell death, and diminishing the extended capacity for proliferation in triple-negative breast cancer (TNBC) cells.

4.2.3 Lipid Emulsions

Lipid emulsions are generally used to enhance the therapeutic window of hydrophobic drugs. However, their lower retention and bioavailability restrict their wide applications. Therefore, scientists have further explored them to form nanoemulsions to support the purpose of cancer treatment.

In a study conducted in 2022, Bo et al. synthesized lipid nanoemulsions (LNs) that could simultaneously transport paclitaxel (PTX) and docosahexaenoic acid (DHA), with the addition of folic acid (FA) as a surface adornment. They examined the potential antitumor effects of these PTX/DHA-FA-LNs in the context of breast cancer. The PTX/DHA-FA-lipid NPs (LNs) demonstrated a correlation between dosage and cytotoxicity, as well as a heightened incidence of apoptosis in MCF-7 cells. The results of the cellular uptake investigation showed that the MCF-7 cells and M2 macrophages rapidly internalized these LNs in an in vitro setting. Furthermore, the specific impact of PTX/DHA-FA-LNs was facilitated by the process of FA receptor-mediated endocytosis, and the level of its toxicity to cells was directly related to its effectiveness in being taken up by the cells. In 2022, Song and his group introduced a new type of NPs called core-shell lipid-gelatin-epigallocatechin-3-gallate hybrid NPs (LGE-N). These NPs aimed to enhance the stability and effectiveness of epigallocatechin-3-gallate (EGCG) in combating tumors. The cytotoxicity of the EGCG solution (EGCG-S) and LGE-N was evaluated using a conventional 3-(4,5-dimethylthiazol-2-yl)-2, 5-diphenyltetrazolium bromide test. The results of in vitro cytotoxicity experiments demonstrated that LGE-N displayed a reduced half maximum inhibitory concentration in comparison to EGCG-S when tested on the MCF-7 breast cancer cell line.

4.3 Carbon-Based Nanotherapy

Researchers have discovered that NPs such as carbon nanotubes or carbon nanofibers (CNTs/CNFs) could be used as carriers for direct drug delivery, especially in conjunction with or together with other chemotherapy drugs. CNTs/CNFs are tubes made of carbon which has nanometer dimensions. The dimensions of CNTs expressed in various ratios with 10 15 nm diameters. Most of these carriers are multi-wall carbon nanotubes (MWCNT) generally utilized for cancerous cell targeting.

4.3.1 Carbon Nanotubes/Carbon Nanofibers

A study in 2023 was aimed to evaluate the antiproliferative effects of CNTs in combination with anticancer drugs against human colorectal carcinoma. A variety of carbon nanotube (CNT) samples were synthesized, including distinct functional groups (-O, -N, -S) and exhibiting diverse textural characteristics. The samples that exhibited the highest degree of interaction with anticancer drugs and showed enhanced efficacy in the treatment of HT-29 cells were the samples that were modified with nitrogen and sulfur groups (Abreu et al. 2023).

The research was conducted to evaluate the impact of carbon nanofibers (CNFs) and multi-walled carbon nanotubes (MWCNTs) on cellular functionality, cellular absorption, and their capacity to augment the responsiveness of urological cancer cells to the standard chemotherapeutic agent's cisplatin and carboplatin. The cellular function was shown to be somewhat impaired by both carbon nanofibers (CNFs) and carbon nanotubes (CNTs) at concentrations ranging from 1 to 200 mg/mL. Notably, CNFs exhibited a greater detrimental effect compared to CNTs. The combined use of carbon nanofibers (CNFs) and carbon nanotubes (CNTs) significantly improved the efficacy of cisplatin and carboplatin in terms of cellular survival. This enhancement ranged from 1.2- to 2.8-fold in prostate, bladder, and cisplatin-resistant prostate cancer cells, surpassing the effects observed when these drugs were used individually.

A significant observation was made in the study conducted by Ringel et al. 2014, about the potential role of albumin in facilitating the internalization of carbon nanomaterials into cells. The co-localization of albumin and carbon nanomaterials on the cellular membrane was substantiated by the utilization of fluorescence microscopy. The use of transmission electron microscopy revealed that cancer cells internalized carbon nanomaterials, leading to a heightened probability of elevated concentrations of chemotherapeutic drugs. In summary, the utilization of carbon nanomaterial-based applications presents a promising approach for addressing chemoresistance by enhancing the susceptibility of cancer cells to traditional chemotherapeutic agents (Ringel et al. 2014).

In 2017, Erdmann et al. conducted a study to examine the chemosensitizing capabilities of carbon nanofibers (CNFs) and carbon nanotubes (CNTs) with the conventional medicines docetaxel (DTX) and mitomycin C (MMC). The DU-145 prostate cancer cells were subjected to treatment with DTX and MMC either alone or in conjunction with the carbon nanomaterials. The vitality of prostate cancer (PCa) cells was reduced to 94% and 68% when treated with DTX and MMC alone, respectively. However, when CNFs were coupled with the therapy, the cellular viability decreased to just 30%. Therefore, the concurrent administration of traditional drugs with carbon nanomaterials exhibits a greater capacity to decrease the dosage of chemotherapy treatments, thereby restricting the occurrence of systemic toxicities caused by the pharmaceuticals.

4.3.2 Carbon Dots

Carbon dots, often known as C-dots, are a novel nanomaterial that exhibits remarkable adaptability. Their first discovery dates back to the year 2004, as seen in Fig. 4.2. C-dots have gained significant prominence as a versatile platform for the conjugation of receptors and chemotherapeutic medicines, primarily owing to their abundant surface functional groups, such as carboxylic and amino groups (Li et al. 2015).

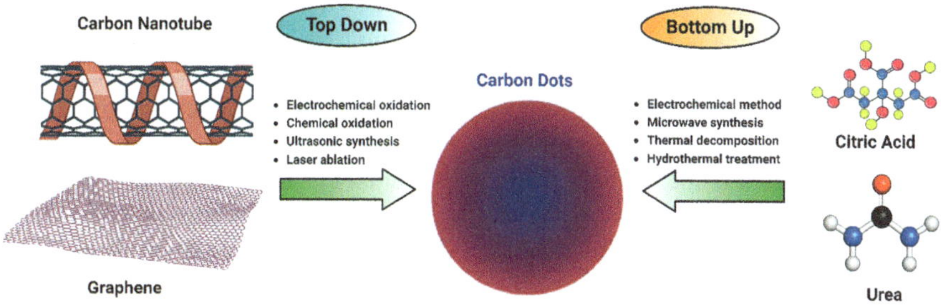

Fig. 4.2 Top-down and bottom-up approaches for the synthesis of C-dots. (Adopted from Ross et al. 2020)

Cationic carbon dots (CCDs) have been identified as promising NPs for therapeutic applications due to their ability to internalize into cells and traverse the nuclear membrane. The research performed by Chen et al. (2023) involved the preparation of cationic carbon dots (Y15-CDs) produced from 1, 2, 4, and 5-benzene tetramine, which is known as Y15 and acts as a focal adhesion kinase inhibitor. The results of cytotoxicity experiments conducted on human foreskin fibroblasts (HFF), prostate cancer (PC3), and osteosarcoma cells (U2OS) demonstrated the nuclear localization properties and anticancer efficacy of Y15-CDs. One significant observation is that Y15-CDs demonstrated enhanced cytotoxicity toward cancer cells (PC3 and U2OS) compared to normal cells (HFF), suggesting that Y15-CDs may hold promise as a component of cancer chemotherapeutic treatments. Another study was conducted to develop a new type of hybrid liposome nanocarrier consisting of poly(vinylidene fluoride-tetrafluoroethylene), graphene quantum dots (GQDs), and silibinin, which were activated by ultrasound. These NPs were designed to have piezoelectric properties. The cytotoxicity experiments conducted on MCF-7, 4 T1, and NIH3T3 cell lines showed that the presence of NPs did not have a significant impact on cell viability at concentrations up to 125 μg/mL (Ramedani et al. 2022).

In a study, monodisperse and spherical mesoporous silica NPs (MSNs) were first synthesized and loaded with hydrothermally synthesized anticancer carbon dots (CDs). Using anti-MUC1 aptamer, the formulation was made targeted toward cancer cells. Potent and selective anticancer activity was obtained against MCF-7 and MDA-MB-231 cancer cells. The apoptotic-mediated cell death was recognized as the primary cytotoxic mechanism of the MSNs-CDs (Kajani et al. 2023).

Photodynamic therapy (PDT) has considerable potential as an innovative approach to cancer treatment. Nonetheless, the clinical use of photosensitizers is hindered by the constraints imposed by hydrophobic interactions and inadequate water solubility. In a research endeavor undertaken in the year 2022, the authors undertook the integration of carbon dots with various porphyrins to enhance the inherent characteristics of porphyrins. Subsequently, the researchers proceeded to assess the effectiveness of these modified porphyrins as photodynamic therapy

(PDT) drugs. The biocompatibility was improved by the enhancement of covalent bonds between cyclodextrins (CDs) and porphyrins. The produced conjugates also exhibited pH sensitivity, resulting in a reduction in the hemolysis ratio. This characteristic makes them a promising option for photodynamic therapy (PDT). The biocompatibility of porphyrins was enhanced by the integration of carbon dots, resulting in a reduction in their toxicity (Sajjad et al. 2022).

According to reports, Tian and colleagues from the group of 2023 have synthesized three pentacyclic triterpenoids (PTs), namely glycyrrhetinic acid (GA), ursolic acid (UA), and oleanolic acid (OA). These compounds possess anticancer properties but suffer from hydrophilicity when used as raw materials. To overcome this limitation, the researchers have employed cyclodextrins (CDs) to enhance the properties of the PTs. The carboxyl edges of the examined samples showed remarkable water solubility and a notable ability to selectively target mitochondria in tumor cells. This selectivity can be attributed to the increased susceptibility of tumor cell mitochondria to damage caused by reactive oxygen species (ROS) and their elevated levels of internal oxidative stress. The induction of cell death in cancer cells by PT-CDs occurs through three distinct mechanisms, including apoptosis, ferroptosis, and autophagy. Significantly, the functionalized PT-CDs demonstrated remarkably low systemic toxicity.

4.4 Dendrimers Based Nanotherapy

Dendrimers are another type of polymeric structure that is used to synthesize drug-conjugated nanoformulations. The branching steps represent the generations in the dendrimeric structure, and they can be combined with any form of specific moieties or compounds to enhance the therapeutic efficacy of the bioactive (Fig. 4.3) (Choudhary et al. 2017).

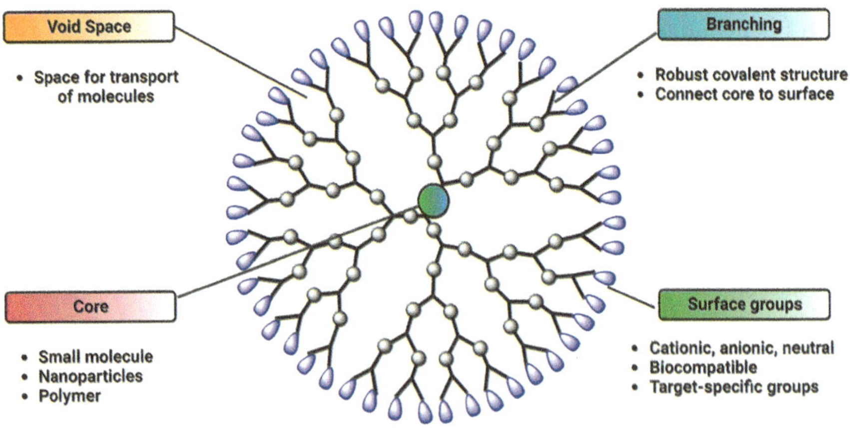

Fig. 4.3 A basic structure of a dendrimer (adopted from Choudhary et al. 2017)

The synthesis of amine-terminated nanodiamonds (NP) was achieved by the covalent modification of carboxylic acid functionalized nanodiamonds (ND-COOH) with poly(amidoamine) dendrimer (PAMAM) in a conducted study by Patil et al. 2022. This research focuses on the physical encapsulation of cabazitaxel (CTX), a drug with low hydrophilicity, using NP-formed ND-PAMAM-CTX (NPC) as the encapsulating agent. Significant internalization of NPC was found in both the cytoplasm and nucleus of U87 cells. The study demonstrated an increase in cell mortality and a decrease in the expression of tubulin protein in cells treated with NPC. Therefore, this work has brought attention to the potential of a recently developed nanocarrier and can offer valuable insights for the treatment of glioblastoma multiforme (GBM) in the future (Patil et al. 2022).

Another investigation was undertaken to examine the effects of the covalent conjugation of PAMAM G3 dendrimer with two recognized anticancer drugs, namely celecoxib and simvastatin. The study involved in vitro testing of these conjugates on human squamous carcinoma (SCC-15-15) and glioblastoma (U-118 MG) cells, as well as normal human fibroblasts (BJ). The conjugates that were acquired were further modified with biotin and R-glycidol to enhance their binding affinity toward cancer cells. The conjugates, consisting of two celecoxib and four simvastatin residues, exhibited remarkable efficacy and substantially reduced the viability of SCC-15 and U-118 MG cells even at extremely low doses. The potential of biotinylated and R-glycosylated PAMAM G3 dendrimers, which have been modified with both celecoxib and simvastatin, shows that they are effective drugs for the treatment of cancer. (Wróbel et al. 2022).

A research group developed an innovative class of dendrimers known as enzyme-triggerable cell-penetrating peptides (ETCPP) that incorporate a camptothecin (CPT) warhead. Among the recently synthesized dendrimer conjugates of ETCPP, the high-generation dendrimer known as BL_Oc-SS-CPT had the most pronounced cytotoxic action against a range of cancer cells, surpassing the effectiveness of CPT by a factor of 3–10. Significantly, BL_Oc-SS-CPT demonstrated exceptional in vivo tumor targeting capacity, together with notable antitumor effectiveness and minimal safety profiles (Mai et al. 2022).

In a study conducted to examine the impact of functionalizing poly(amidoamine) (PAMAM) dendrimer G3 with L, F, or both L and F on breast cancer cells exhibiting distinct receptor statuses (ER-positive MCF-7, triple-negative MDA-MB-231, and HER2-positive SK-BR-3 cells). The conjugation of L- and F-PAMAM resulted in an enhanced cytostatic and cytotoxic effect compared to the individual administration of L and F. This enhanced effect was followed by an increase in autophagy levels. The formulation also resulted in the eradication of doxorubicin-induced senescent breast cancer cells by apoptosis-mediated analysis, irrespective of receptor status. Therefore, these innovative nanoplatforms have significant potential in the field of breast cancer care (Lewińska et al. 2022).

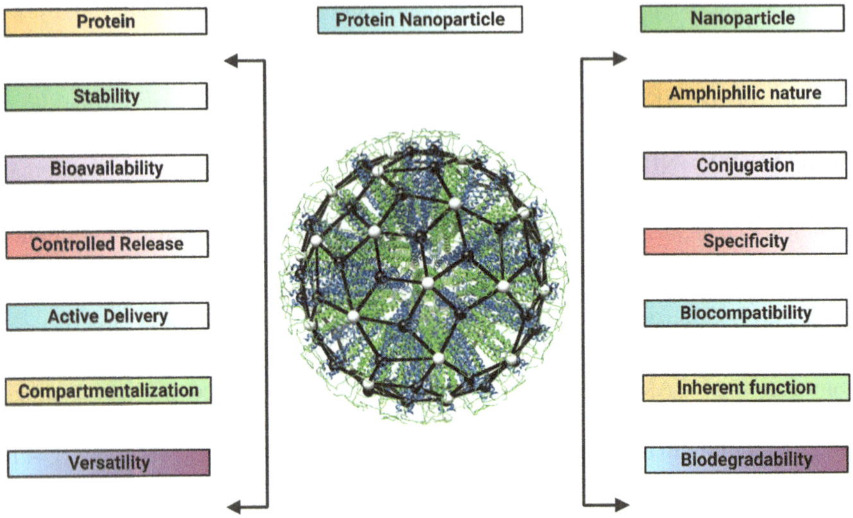

Fig. 4.4 Advantages and versatility offered by protein-based nanoparticles (adopted from Habibi et al. 2022)

4.5 Protein-Based Nanotherapy

Protein-based nanosystems are biocompatible and biodegradable, with a long half-life and high purity than any other nanomaterial-based drug delivery carriers (Fig. 4.4). These targeted protein-based nanosystems can be designed by utilizing ligand-receptor interactions owing to their high-precision specific binding ability, and such interactions as avidin-biotin binding.

In 2022, Khella et al. performed an investigation to assess the efficacy of polymeric NPs loaded with CA in enhancing the anticancer effects on MCF-7 breast cancer cells and Caco-2 colon cancer cells. The compound CA was enclosed within NPs composed of bovine serum albumin (BSA), chitosan (CH), and cellulose (CL). The NPs loaded with CA-BSA (CA-BSA-NPs) exhibited a highly favorable formulation with superior loading capacity. The anticancer efficacy of the CA-BSA-NPs was assessed by the quantification of cell viability, apoptosis rate, and the expression of GCLC, COX-2, and BCL-2 genes in MCF-7 and Caco-2 cell lines. The results of the cytotoxicity experiment, namely the MTT assay, indicated that CA-BSA-NPs exhibited increased anticancer activity against MCF-7 and Caco-2 cell lines in comparison to free CA and BSA-NPs. Additionally, the findings obtained from the apoptosis test indicated that the Caco-2 cells were arrested at the G2/M phase, while the MCF-7 cells were also arrested at the G2/M phase when treated with CA-BSA-NPs. The gene expression study conducted using RT-PCR demonstrated an observed increase in the expression of the GCLC gene, while a decrease in expression was seen for the BCL-2 and COX-2 genes in cells subjected to treatment with CA-BSA-NPs. In summary, CA-BSA-NPs have been identified as a potentially effective NP formulation for the treatment of breast and colorectal cancer.

In another study, the utilization of pH-responsive albumin NPs containing doxorubicin was applied. The drug release exhibited limited efficacy under physiological settings; however, these NPs demonstrated full release of their chemotherapeutic payload at a little lower pH, about one unit below the physiological pH, mimicking the conditions seen in the tumor microenvironment. The results of in vitro experiments revealed that albumin NPs containing DOX had much greater apoptotic activity against human lung cancer cells, with a four-fold increase compared to other tests. The capacity of albumin to inhibit the aggregation of partly hydrophobic DOX and thereafter release it over an extended duration, without any immediate release, elucidated its potential to enhance the efficacy of DOX against lung cancer cells (Aziz et al. 2022).

The efficacy of NPs containing albumin visnagin (AV) has also been assessed in triple-negative breast cancer cell lines. To evaluate the effectiveness of NPs in inducing cell death in MDA-MB-468 cell lines, many approaches were employed. These techniques included the MTT test, dual AO/EB staining, DCFH-DA assay, Annexin-V-FITC/PI staining, propidium iodide staining, analysis of apoptotic proteins, assessment of antioxidant enzymes, and study of the PI3K/Akt/mTOR signaling pathway. The introduction of NPs resulted in a reduction in cell viability accompanied by an elevation in reactive oxygen species (ROS) levels. Additionally, NPs disrupted membrane integrity, a drop in antioxidant enzyme levels, induction of cell cycle arrest, and activation of the PI3K/Akt/mTOR signaling cascade, finally culminating in cellular demise. According to Alsrhani et al. (2023), AV NPs have significant potential as a robust therapeutic alternative for combating cancer. In an additional intriguing investigation, the researchers created NPs loaded with DTX, namely DTX@HSA-(PLA)1–3 NPs. The murine and human oral cancer cell lines, MOC2 and FaDu, were subjected to testing. Monolayers and multicellular spheroids were utilized to evaluate the cellular uptake, amount of cell viability, and induction of apoptosis after treatment with NPs. The treatment with DTX@ HSA-(PLA)2 NPs resulted in increased expression of apoptotic markers, cell cycle arrest in the G2/M phase, DNA damage, and mitochondrial depolarization compared to the administration of free DTX and DTX@HSA NPs in both cell lines (Kumbham et al. 2022).

Chitosan as a polymer has been actively evaluated as a drug delivery system. A novel NP system based on modified chitosan, which exhibits pH-responsive behavior, has been produced to actively load doxorubicin (DOX). The cytotoxicity of the NPs was evaluated on the MCF-7 BrCA cell line at several pH levels. The results obtained from the in vitro experiments demonstrated that the cytotoxicity of DOX-loaded NPs against MCF-7 BrCA cells was much greater in comparison to free DOX (Matalqah et al. 2022). Another research group endeavored to selectively attack breast and prostate cancer cells by employing chitosan-based NPs that were loaded with very effective steroidal chemicals with anticancer properties. The formulations showed efficacy against estrogen receptor-negative breast cancer cells (MDA-MB-231) and androgen receptor-negative prostate cancer cell lines (PC-3). The chitosan NPs, which have been loaded with steroids, exhibit significant cytotoxic effects against MDA-MB-231 and PC-3 cell lines. Importantly, these NPs do not demonstrate any hormone-related action, suggesting that they are both safe and effective in their intended use (Kuzminac et al. 2022).

Zein is a prolamin protein variant obtained from corn that has received recognition from the United States Food and Drug Administration (FDA) as a biologically safe substance. The plant protein in question has garnered significant attention in academic studies due to its notable characteristics, including low immunogenicity, edibility, amphipathic, biodegradability, biocompatibility, and resilience to gastrointestinal conditions. The benefits render zein-based research particularly advantageous in terms of enhancing oral bioavailability and attaining prolonged and specific drug release for anticancer agents. Maytansine, often known as DM1, exhibits strong antineoplastic properties. According to reports, Yu and his group 2020 created zein NPs (ZNPs) loaded with DM1 to achieve targeted delivery to tumors and minimize the hazardous side effects associated with DM1. The researchers discovered that A549 cells exhibited a significant decrease in activity when exposed to DM1-loaded ZNPs. This result was attributed to the ZNPs' capacity to facilitate cellular internalization. Additionally, the researchers have shown that ZNPs loaded with DM1 have superior effectiveness in inhibiting tumor growth compared to DM1 alone. Therefore, it has been demonstrated that DM1-loaded ZNPs are a favorable choice for chemotherapy in the treatment of non-small cell lung cancer. Dong et al. (2016) demonstrated effective encapsulation of DOX within zein, resulting in the formation of spherical NPs. The observed release behavior of the system was pH-responsive, as evidenced by the accelerated release of DOX in acidic buffer solutions (pH 5.0 and 6.5) compared to neutral solutions (pH 7.4). The impact of nano-encapsulation on the inhibition of HeLa cell growth was also investigated. The findings suggest that the utilization of DOX-loaded zein NPs (DOX-zein-NPs) resulted in enhanced efficacy in eliminating cancer cells at lower concentrations of DOX, in comparison to the use of free DOX.

Curcumin in ***casein*** NPs (CasNPs) have been developed to evaluate their anticancer efficacy. Cur-CasNPs exhibit pH-dependent release properties, exhibiting increased drug release in mildly acidic conditions. Furthermore, they have exhibited augmented cytotoxicity toward cancer cells (MCF-7) in comparison to normal cells (CHO). Furthermore, it should be noted that the CasNPs are non-toxic to the biological system. Additionally, these NPs have demonstrated significant cellular internalization and have shown an increased level of toxicity toward MCF-7 cells when compared to pure drugs (Barick et al. 2021). In 2023, Khatun and the group reported the encapsulation of 10-hydroxycamptothecin in biocompatible casein NPs. The biocompatibility of the synthesized CS and CCS NPs was found to be excellent when tested on fibroblast cell lines L929, NIH-3 T3, and zebrafish embryos. The researchers observed an increased internalization of CCS NPs in 4 T1 cells, resulting in a cancer cell toxicity of around 80–85%. Observation of NPs exhibiting sub-cellular mitochondrial localization, together with modifications in mitochondrial membrane potential, lysosomal localization, and formation of reactive oxygen species leading to cancer cell death, has been reported. In addition, the study revealed growth inhibition of 4 T1 cell colonies and the presence of antimetastatic activity. It is due to an elevation in the levels of the CHOP protein, which acts as a reliable indicator for cell death resulting from endoplasmic reticulum stress. The three-dimensional model of 4 T1 cells demonstrated extensive infiltration into the tumor and showed notable therapeutic effectiveness for CCS NPs. The findings of this study have emphasized the potential of utilizing casein-based nanoformulation as a viable and cost-effective approach for cancer treatment in triple-negative breast cancer (TNBC) patients.

4.6 Synthetic Polymer-Based Nanotherapy

Apart from naturally occurring polymers, nanotherapy based on synthetic polymers has been greatly explored. These include PLGA [poly (lactide-co-glycolide)], PLA [polylactic acid], PBCA [Poly(butyl cyanoacrylate)], etc. They are synthetically produced and are then developed into various nanosystems.

4.6.1 PLGA Nanoparticles

PLGA is a biodegradable copolymer ester of two α-hydroxyacids (lactic and glycolic acids) and has been approved by the FDA for drug delivery. There are several physicochemical properties of PLGA that impact the release behavior of drugs from its nanocarrier form. In 2023, Hu et al. synthesized hyaluronic acid-coated Olaparib-loaded PEI-PLGA NPs, which are referred to as HA-Ola-PPNPs. The objective of this study was to investigate the potential anticancer efficacy of these nanoparticles against triple-negative breast cancer. The overexpression of Bax, Cytochrome C, and Caspase 3 demonstrated the induction of cell death by HA-Ola-PPNPs (Hu et al. 2022). According to Simitcioglu et al. (2022), a study was conducted to examine the impact of various molecular weights (Resomer 502 and 504) and terminal groups (Resomer 502H and 504H) of polylactic-co-glycolic acid NPs on their characteristics, as well as their effects on prostate cancer cells and healthy cells. The researchers aimed to investigate the influence of these Resomers on the properties of the NPs and their interactions with both cancerous and non-cancerous cells. The researchers provided evidence that the groups exhibiting large molecular weight had a discernible preference for cytotoxicity.

To enhance the systemic administration of FZ, a group of researchers undertook the task of transforming fat-soluble medicines into hydrophilic forms. As part of their investigation, they developed FZ-encapsulated poly(D, L-lactide-co-glycolide) acid (PLGA) NPs, referred to as FZ-PLGA-NPs. The study conducted by Chang et al. (2023) showed that the use of FZ-PLGA-NPs resulted in a considerable reduction in cellular proliferation inside the EOC cell line. In their study conducted in 2023, Kumari et al. developed a formulation consisting of poly (lactide-co-glycolide) (PLGA) NPs encapsulating docetaxel (DTX), with the monoclonal antibody Cetuximab (Cet) attached to the surface. This research aimed to explore the potential of this targeted therapeutic approach for the treatment of lung cancer. The cellular uptake in lung cancer cells that overexpressed EGFR (specifically A549 and NCI-H23) was boosted. The NPs exhibited enhanced therapeutic efficacy against non-small cell lung cancer (NSCLC) cell lines as well.

Another research group successfully synthesized nano-encapsulated hydroxytyrosol using a copolymer known as poly-lactide-co-glycolide-co-polyacrylic acid (PLGA-PAA). The estimation of the antiproliferative potential was conducted using an MTT test. The assessment of gene expression levels for CDKN1A, CDKN1B, and CCND1 was also conducted. The results obtained from the MTT experiment indicate that the nano-capsulated hydroxytyrosol exhibited more cytotoxicity toward the colorectal cancer cell line compared to free hydroxytyrosol. The study demonstrated an increase in the expression of CDKN1A and CDKN1B genes, as well as a decrease in the expression of the CCND1 gene, within colorectal cancer cells (Sani et al. 2022).

4.6.2 PLA and PBCA Nanoparticles

Poly(lactic acid) (PLA), poly(butyl cyanoacrylate) (PBCA), and their copolymers are among the most frequently used biodegradable polymers for encapsulated drugs. PLA is widely used for many biological applications due to its biocompatible and biodegradable properties. PLA and PBCA have been widely used in different biomedical applications like tissue engineering, fracture repairs, orthopedics, drug delivery carriers, and antimicrobial agents. In 2022 Wadhawan and their research group successfully produced polymeric NPs composed of polylactic acid-poly (ethylene glycol) (PLA-PEG) copolymers, encapsulated with biosurfactants. This was achieved by the utilization of the double emulsion solvent evaporation technique, with folic acid (FA) serving as the ligand for targeted delivery. The researchers observed that the presence of FA resulted in an increase in both the absorption and cytotoxicity of NPs loaded with biosurfactant in MDA-MB-231 cancer cells, as compared to NPs loaded with plain biosurfactant. Apoptosis induction was furthermore seen in cancer cells that were subjected to NP treatment (Wadhawan et al. 2022).

In 2022, Zhang et al. synthesized Gemcitabine-loaded nanospheres by emulsion polymerization (GEM-PBCA-NP), and then, anti-MUC1 and anti-human CD199 monoclonal antibodies were coupled separately. Both NPs demonstrated significantly higher cytotoxicity as well apoptosis rates against human pancreatic cancer cell lines. Even the tumor inhibition rate was higher in tumor-bearing mice for the nanoformulations (Zhang et al. 2022). Another study reported the development of an orthogonal experimental design (OED) for the preparation of PBCA NPs. The results of the in vitro investigation utilizing SH-SY5Y neuroblastoma cells indicated that osthol-loaded NPs exhibited a notable increase in the intracellular absorption process and cytotoxicity when compared to free osthol. Therefore, the osthol-PBCA NPs have promising potential for the delivery of hydrophobic drugs (Zheng et al. 2022).

4.7 Metal-Based Nanotherapy

Metals, including iron, zinc, copper, gold, silver, and others, alongside mesoporous silica NPs (MSNs), exhibit considerable potential as drug delivery materials owing to their diminutive dimensions, inert characteristics, and remarkably expansive specific surfaces that can be modified with therapeutic agents and targeting entities.

4.7.1 Mesoporous Silica Nanoparticles (MSNs)

Mesoporous silica nanoparticles (MSNs) are promising nanomaterials for use in biomedical applications. The functionalization properties of MSNs make them capable of co-delivery, controlled delivery, and stimuli-responsive delivery. Multi-drug resistance is a common reason behind the failure of chemotherapy. Therefore, Wang and group 2014 designed mesoporous silica NPs (MSNs). They tested the efficacy of DOX in the breast cancer cell line MCF-7/MDR1. The authors showed that the drug resistance in the human breast cancer cell line MCF-7/MDR1 could be overcome by treatment with DOX encapsulated within mesoporous silica NPs.

Research conducted in 2022 reported a novel method for synthesizing mesoporous silica NPs (MSNs) characterized by a distinctive core-shell structure resembling an egg yolk. The study conducted by Tiburcius et al. (2022) assessed the efficacy of Dox-laden mesoporous silica NPs (MSNs) in inhibiting the growth of prostate cancer cells. Two distinct prostate cancer cell lines, namely PC-3 and LNCaP, were utilized for this evaluation. The results indicated that the optimized MSN samples loaded with the drug exhibited more cytotoxicity compared to other formulations. Another research group conducted an intriguing experiment in which they examined the effects of fluorescent silicon NPs (DOX-NPs) on the MCF-7 breast cancer cell line. The researchers assessed the cytotoxicity of DOX-NP, as well as its impact on the cell cycle and the production of CD44 antigen. CD44 is a molecule that plays a role in cellular adhesion and the invasion of tumors. The evaluation includes a comparison between DOX-NP, free DOX, and SiNPs used independently. A particular capacity to secrete a small number of extracellular vesicles (EVs) expressing CD44+ was identified in both CD81-negative and CD81-positive populations. The modulation of CD44 levels on the cell surface in cancer cells has significant relevance in altering signaling networks that promote tumor growth (Sola et al. 2022).

4.7.2 Iron Oxide Nanoparticles

Among the wide amount of drug nanocarriers, magnetic iron oxide nanoparticles (IONs) stand out from the crowd and constitute robust nanoplatforms since they can achieve high drug loading as well as targeting abilities stemming from their remarkable properties (magnetic and biological properties). Iron oxide nanoparticles have been the most extensively studied inorganic nanocarrier systems for drug delivery.

These are biodegradable, biocompatible, and efficiently cleared from the human body via the iron metabolism pathways. In a study, Rostami et al. (2022) presented a confirmation of the anticancer properties of magnetic Fe_3O_4 NPs coated with L-lysine (referred to as F-Lys NPs) and loaded with Tamoxifen (TMX), specifically targeting breast cancer cells. To assess the cytotoxicity of the NPs that were specifically engineered, a combination of MTT and apoptosis tests, real-time PCR, and cell cycle analysis were utilized. An increase in the concentration of NPs and the duration of treatment resulted in greater suppression of cell growth and activation of apoptosis in cells treated with F-Lys-TMX NPs. The study revealed a decrease in the gene expression levels of ERBB2, cyclin D1, and cyclin E, whereas an increase in the expression levels of caspase-3 and caspase-9 was seen. The drug release mechanism has been the subject of research, which has consistently shown a pH-dependent release pattern of TMX. The findings of the cell cycle investigation revealed that the application of F-Lys-TMX NPs led to the cessation of cellular progression in the G0/G1 phase, therefore, advocating for the integration of metal NPs into the cancer therapy protocol. In another study, iron oxide (Fe_3O_4) NPs were coated with hyperbranched polyglycerol (HPG) and folic acid (FA). The FA@HPG@Fe_3O_4 NPs were seen to be internalized to a greater extent in both HeLa cells and murine L929 fibroblasts compared to the HPG@Fe_3O_4 NPs. Furthermore, the proportion of MRI signal enhancement was reduced with incubation of HeLa cells with NPs, with the most significant modification found following incubation with FA@HPG@Fe_3O_4 NPs (Ramezani Farani et al. 2022).

4.7.3 Zinc Oxide Nanoparticles

Zinc oxide (ZnO) NPs have emerged as a versatile nanocarrier to deliver the therapeutic moiety effectively to the targeted site. Specifically, when the drug alone is not capable of entering the targeted cells, the drug-loaded NPs penetrate into the cells through various intracellular endocytic pathways. ZnO NPs have become one of the most popular metal oxide nanoparticles in biological applications due to their excellent biocompatibility and low toxicity. Thomas et al. (2022) conducted a synthesis of ZnO NPs using Solanum nigrum and observed considerable anticancer effects against HeLa cell lines via the apoptotic mechanism. The evaluation of the cytotoxicity of zinc oxide (ZnO) NPs was conducted using several experimental approaches, such as the MTT test, wound healing assay, DAPI staining, and acridine orange and ethidium bromide double staining. The utilization of the reverse transcriptase-PCR methodology was performed to examine the patterns of expression of β-catenin, p53, caspase-3, and caspase-9. The results of the study indicate that ZnO NPs obtained from Solanum nigrum exhibit cytotoxic effects on the HeLa cell line, with the extent of toxicity depending on the dosage administered. The outcome is attained by the inhibition of β-catenin and the augmentation of p53, caspase-3, and caspase-9 levels.

In 2022, George et al. fabricated zinc oxide NPs (ZnO NPs) by a green synthesis approach employing *R. fairholmianus* root extract. Subsequently, the impact of these

NPs on MCF-7 cells was assessed. The treatment with *R. fairholmianus* capped zinc oxide NPs (RFZnO NPs) resulted the decreased in cellular proliferation, accompanied by an increase in cellular cytotoxicity and levels of reactive oxygen species (ROS). An increase in programmed cell death, known as apoptosis, was seen, as shown by elevated levels of cytoplasmic cytochrome c and caspase 3/7. These biomarkers are commonly used to assess the occurrence of apoptosis. Furthermore, an increase in the expression of pro-apoptotic proteins, namely p53 and Bax, was seen, whereas a decrease in the expression of the anti-apoptotic protein Bcl-2 was noted. In summary, the results of this work indicated that RFZnO NPs can induce apoptosis in breast cancer cells via activating a mitochondria-mediated caspase-dependent apoptotic pathway (George et al. 2022a).

4.7.4 Inert Metals Nanoparticles

Among all types of noble metal-based nanoparticles, gold nanoparticles (Au NPs) have shown a great capacity for use as potential drug delivery carriers. The therapeutic efficacy of gold NPs (AuNPs), gold NPs loaded with Galangin (AuNPs-GAL), and free Galangin as an antiangiogenic agent against breast cancer cells was investigated in a reported study. Various physicochemical and molecular methodologies have been employed to investigate the anticancerous properties of AuNPs-GAL on MCF-7 and MDA-MB-231 cells. These methodologies encompass the characterization of the treated cancer cells, assessment of their cytotoxicity, evaluation of their scratch wound healing capabilities, and examination of the gene expression levels of VEGF and ERKI. The findings derived from the MTT assay demonstrated a decrease in cellular proliferation that was contingent upon both the duration of exposure and the concentration of the treatment. Moreover, the data suggested a synergistic effect (Qaddoori and Al-Shmgani 2023).

So far, no previous research has provided empirical data on the in vitro anti-human pancreatic cancer capabilities of the newly formulated chemotherapeutic drug including silver NPs and *Berberis thunbergii* leaf (AgNPs). The objective of this study was to examine the potential anti-human pancreatic cancer characteristics of silver nitrate ($AgNO_3$), aqueous extract from *B. thunbergii* leaves, and silver NPs (AgNPs). To assess these properties, the researchers employed the 3-(4,5-dimethylthiazol-2-yl)-2,5-diphenyl-2H-tetrazolium bromide (MTT) test on commonly used human pancreatic cancer cell lines. The treatment with AgNPs exhibited low cell viability and was shown to be dependent on the dosage when tested against PANC-1, AsPC-1, and MIA PaCa-2 cell lines. Therefore, the research provides evidence in favor of employing silver AgNPs as a viable therapeutic approach for combating pancreatic cancer, as suggested by Guo et al. (2022). In a separate investigation, the MTT test was employed to evaluate the cytotoxic effects of CP, green AgNPs, and CP-AgNPs on PC-3, a cell line derived from human prostate cancer. The cytotoxic impact of CP-AgNPs on PC-3 cells was shown to be much higher compared to CP and AgNPs alone. The IC50 values for CP-AgNPs, CP, and AgNPs were determined to be 27.05 μg/mL, 57.64 μg/mL, and 125.4 μg/

mL, respectively. Therefore, our results indicate the potential use of CP-AgNPs as a more effective anticancer drug in comparison to free CP (El-Sheikh et al. 2022).

The magnetic $CuFe_2O_4$ material has an intrinsic protease activity that enables the hydrolysis of bovine serum albumin (BSA) and casein molecules under physiological circumstances. Protease is known to have a significant role in several pathophysiological mechanisms and is involved in every phase of tumor advancement. In a conducted study, it was observed that $CuFe_2O_4$ demonstrated an inhibitory effect on the proliferation of A549, SKOV3, HT-29, BABL-3 T3, and HUVEC cells. Notably, $CuFe_2O_4$ exhibited heightened sensitivity toward SKOV3 ovarian cancer cells, surpassing the effectiveness of conventional chemotherapy drugs oxaliplatin and artesunate by approximately 30-fold in inhibiting the growth of SKOV3 cells. Furthermore, $CuFe_2O_4$ was shown to modulate the intracellular protein expression of MMP-2, MMP-9, F-actin, and NF-kB, which are commonly linked to overall protein degradation. The compound $CuFe_2O_4$ also demonstrates the ability to impede cell motility. The $CuFe_2O_4$ NPs are very desirable for their uses in cancer treatment due to their exceptional magnetic characteristics, proficient protease-mimicking capacity, and notable antitumor efficiency (Chen et al. 2022).

4.8 Combinatorial Nanoparticles

Combination therapy of multiple drugs through a single system exhibits high therapeutic effects. Combination therapies of small-molecule drugs are common in clinical practice and it is anticipated that packaging into single macromolecular carriers will enable drug release in precisely balanced ratios and rates, and in selectively targeted tissues and cells. This vast level of pharmacological control is intriguing, especially from the perspective of tailoring personalized treatments with maximized therapeutic synergy for individual patients. In 2022, Esmaeili et al. conducted a significant study in which they developed a cancer nano theranostics system using a mesoporous silica@chitosan@gold (MCM@CS@Au) nanosystem. This nanosystem was specifically designed to target MUC-1 positive tumor cells, and aptamers were employed for this purpose. Moreover, the nanosystem was utilized for the loading of curcumin. The researchers discovered that the energy transfer occurring between the base pairs inside the hairpin structure of double-stranded DNA aptamers serves as a quenching mechanism for MCM@CS@Au fluorescence. This ultimately leads to the development of an optical biosensor with a binary "on/off" response. In contrast, the inclusion of chitosan NPs that are responsive to changes in pH offers an intelligent nanosystem for enhanced delivery of curcumin to specific cellular targets. The drug release and fluorescence recovery ("On" state) occur as a result of aptamer binding to the MUC-1 receptor in a low pH environment. The observed toxicity outcomes indicate that the nanosystem exhibited a higher degree of toxicity toward MUC-1-positive tumor cells compared to MUC-1-negative cells, hence demonstrating its capacity for selective targeting. Hence, our nanosystem possesses the potential to serve as an intelligent nanotheranostic system for the detection and monitoring of specific biomarkers (such as MUC-1), non-invasive fluorescence imaging, and precise administration of curcumin to selected sites.

In a recent study, researchers reported on the use of mesoporous silica NPs (MSNs) that were coated with a dual film consisting of calcium carbonate ($CaCO_3$) and a lipid bilayer. These NPs, referred to as MSNs@$CaCO_3$@liposomes, demonstrated the ability to provide sustained drug release specifically targeted to the tumor microenvironment. Additionally, this novel approach resulted in increased biocompatibility. The calcium carbonate coating, which is sensitive to changes in pH, served to seal the pore channels of mesoporous silica NPs (MSNs). This enabled the release of drugs to be triggered by pH variations when the NPs were internalized by cancer cells. The DOX-loaded MSNs@$CaCO_3$@liposomes (referred to as DOX/MSNs@$CaCO_3$@liposomes) exhibited non-specific protein adsorption, and the hemolysis test indicated improved biocompatibility. The in vitro drug release investigation showed that the DOX/MSNs@$CaCO_3$@liposomes exhibited a delayed release of DOX specifically at pH 5.0, whereas no release was seen at pH 7.4. The assessment of the anticancer efficacy of DOX/MSNs@$CaCO_3$@liposomes showed a favorable capacity for tumor regression. Hence, it is plausible to consider dual-film-coated mesoporous silica NPs (MSNs) as a promising option for the development of a drug delivery system targeting tumors (Wang et al. 2022).

The interaction between proteins and NPs is of utmost importance in facilitating drug binding and maintaining its stability, hence improving the effectiveness of targeted drug delivery. The present work aimed to investigate the relationship between whey protein β-lactoglobulin (BLG) conjugated with gold NPs (AuNPs) and two specific compounds, curcumin (CUR) and gemcitabine (GEM). The cytotoxicity experiments were conducted with breast cancer cell lines known as MCF-7. The examination of these conjugates may be deemed relevant in the context of targeted drug delivery investigations within the field of cancer research (Waghmare et al. 2022).

In a study, Jaiswal and group (2022) reported the synthesis of two bi-functional nanoconjugates, namely AuNR@EGFP-BSAFA and AuNR@RNaseA-BSAFA, which consisted of gold nanorods (AuNR) conjugated with proteins. These nanoconjugates were coated with folic acid-modified bovine serum albumin (BSAFA), which served as a biomimetic protein corona. The purpose of this study was to demonstrate the targeted delivery of enhanced green fluorescent protein (EGFP) and therapeutic ribonuclease A enzyme (RNase A) to the cytosol while preserving their functional properties. The use of Folate receptor-targeting BSAFA on the surface of AuNRs promotes the process of receptor-mediated internalization, subsequently leading to the liberation of EGFP and RNase A within HT29 cells. The presence of green fluorescence distributed throughout the cytoplasm of the cell suggests that the AuNR@EGFP-BSAFA complex has effectively delivered EGFP to the cytosol. The therapeutic administration of RNase A using AuNR@RNaseA-BSAFA in multicellular 3D spheroids of HT29 cells results in a significant decrease in cellular RNA fluorescence intensity, reducing it to 38%. This drop indicates the degradation of RNA and eventual cell death. The current AuNR-protein nanoconjugates demonstrate promise as a possible platform for cancer therapy due to their variable nanoformulation strategy, which encompasses anisotropic particle shape, diverse protein types, and targeted delivery capabilities in their functional form.

The research team has successfully developed an acidity-responsive nanovehicle as a delivery mechanism for doxorubicin (DOX). The aforementioned outcome was attained by the collaborative assembly of amphiphilic PEGylated chitosan20k and hydrophobic poly(lactic-co-glycolic acid) (PLGA) segments at a pH condition of 8.5. The findings from the in vitro trials investigating cellular uptake and cytotoxicity indicate that the DOX-PC20kPNs were successfully internalized by prostate TRAMP-C1 cancer cells. Furthermore, the free drug was promptly dispersed across various cellular compartments. This phenomenon facilitated the accumulation of the drug in the nucleus, hence augmenting its anticancer efficacy. In addition, it was shown that the presence of hydrated polyethylene glycol (PEG) shells on DOX-PC20kPNs led to a significant reduction in their internalization by macrophage-like RAW264.7 cells (Huang et al. 2022).

Another study endeavor was done to examine the enhanced cytotoxic impacts of polymer-coated gold NPs (GNPs) containing 5-fluorouracil (5-FU) in the treatment of lung adenocarcinoma. Based on the findings obtained from the MTT assay, it was shown that the cytotoxic impact of PFGNPs on the A549 cells was 82.89%, which was higher in comparison to the 74.91% cytotoxic effect exhibited by the 5-FU solution. The present study investigated the impact of PFNPs, FGNPs, PFGNPs, and 5-FU on the expression of EGFR and KRAS genes. Our findings revealed that PFNPs exhibited the highest level of efficacy among the tested NPs and chemotherapeutic agent, 5-FU. In conclusion, this strategy has promise as a chemotherapeutic alternative that offers a reduction in toxicity and an enhancement in efficacy (Gupta et al. 2022).

4.9 Conclusion

The previous decade has seen tremendous upgradation and innovation in the field of nanomedicine against cancer. A plethora of novel nanoplatforms or nanosystems has been developed to cater to the treatment of various carcinoma. Many biological phenomena specific to tumorigenesis and tumor microenvironment have been studied to exploit them for the enhancement of nanoformulations' chemotherapeutic efficacy. The NPs have not only been developed from a variety of naturally occurring as well synthetically produced materials but also have been purposefully optimized and modified to expand the therapeutic window of conventional chemotherapeutic drugs such as enhancing their bioavailability, targetability as well as minimizing their systemic toxicities The incorporation of different shapes of NPs, utilization of inert metals and/or elements, magnetic metals, naturally occurring proteins as well as synthetic polymers have been tested in permutations and combinations to formulate the finest nanosystems to be utilized either in combination or as an adjunct therapeutic modality in cancer management regimen.

4.10 Future Perspectives

Nanomedicine is an extremely dynamic field of research and entrepreneurship. It has time and again proven its utility in cancer treatment. Scientists have been actively stretching the limitations as well as the possibilities of the nanomaterials to formulate highly specific and versatile nanosystems as in the case of theranostics. Various molecules such as DNA, RNA, inert metals, and even carbon are being evaluated extensively to support cancer management and improvement in patient compliance. Additionally, it is a ripe time to develop new and sensitive methodologies for the synthesis and testing of novel and combinatorial NPs as this area grows and caters to the intersection of various disciplines as well as pathologies. This field has a huge potential to contribute not only to human welfare but also to strengthen industry and academic collaboration generating advancement in science and human resource.

References

Abreu S, Vale N, Soares OSGP (2023) Combination of CNTs with classical drugs for treatment in human colorectal adenocarcinoma (HT-29) cell line. Nanomaterials (Basel) 13(13):1933. https://doi.org/10.3390/nano13131933; PMID: 37446448; PMCID: PMC10343277

Ahmed MM, Fatima F, Anwer MK, Aldawsari MF, Alsaidan YSM, Alfaiz SA, Haque A, Az A, Alhazzani K (2020) Development and characterization of Brigatinib loaded solid lipid nanoparticles: in-vitro cytotoxicity against human carcinoma A549 lung cell lines. Chem Phys Lipids 233:105003. https://doi.org/10.1016/j.chemphyslip.2020.105003; Epub 2020 Oct 20

Alajami HN, Fouad EA, Ashour AE, Kumar A, Yassin AEB (2022) Celecoxib-loaded solid lipid nanoparticles for colon delivery: formulation optimization and in vitro assessment of anti-cancer activity. Pharmaceutics 14(1):131. https://doi.org/10.3390/pharmaceutics14010131; PMID: 35057027; PMCID: PMC8780659

Alsrhani A, Elderdery AY, Alzahrani B, Alzerwi NAN, Althobiti MM, Rayzah M, Idrees B, Elkhalifa AME, Subbiah SK, Mok PL (2023) Therapeutic potential of albumin nanoparticles encapsulated Visnagin in MDA-MB-468 triple-negative breast cancer cells. Molecules 28(7): 3228. https://doi.org/10.3390/molecules28073228; PMID: 37049991; PMCID: PMC10096807

Aziz A, Sefidbakht Y, Rezaei S, Kouchakzadeh H, Uskoković V (2022) Doxorubicin-loaded, pH-sensitive albumin nanoparticles for lung cancer cell targeting. J Pharm Sci 111(4): 1187–1196. https://doi.org/10.1016/j.xphs.2021.12.006; Epub 2021 Dec 9

Barick KC, Tripathi A, Dutta B, Shelar SB, Hassan PA (2021) Curcumin encapsulated casein nanoparticles: enhanced bioavailability and anticancer efficacy. J Pharm Sci 110(5):2114–2120. https://doi.org/10.1016/j.xphs.2020.12.011; Epub 2020 Dec 15

Chang CS, Ryu JY, Choi JK, Cho YJ, Choi JJ, Hwang JR, Choi JY, Noh JJ, Lee CM, Won JE, Han HD, Lee JW (2023) Anti-cancer effect of fenbendazole-incorporated PLGA nanoparticles in ovarian cancer. J Gynecol Oncol 34:e58. https://doi.org/10.3802/jgo.2023.34.e58; Epub ahead of print

Chen D, Jiang L, Lei T, Xiao G, Wang Y, Zuo X, Li B, Li L, Wang J (2022) Magnetic $CuFe_2O_4$ with intrinsic protease-like activity inhibited cancer cell proliferation and migration through mediating intracellular proteins. Biomater Biosyst 5:100038. https://doi.org/10.1016/j.bbiosy.2021.100038; PMID: 36825110; PMCID: PMC9934488

Chen J, Li F, Gu J, Zhang X, Bartoli M, Domena JB, Zhou Y, Zhang W, Paulino V, Ferreira B, Michael Brejcha N, Luo L, Arduino C, Verde F, Zhang F, Zhang F, Tagliaferro A, Olivier JH, Zhang Y, Leblanc RM (2023) Cancer cells inhibition by cationic carbon dots targeting the cellular nucleus. J Colloid Interface Sci 637:193–206. https://doi.org/10.1016/j.jcis.2023.01.086; Epub 2023 Jan 21. PMID: 36701865; PMCID: PMC9957951

Choudhary S, Gupta L, Rani S, Dave K, Gupta U (2017) Impact of dendrimers on solubility of hydrophobic drug molecules. Front Pharmacol 8:261. https://doi.org/10.3389/fphar.2017.00261; PMID: 28559844; PMCID: PMC5432624

Dong F, Dong X, Zhou L, Xiao H, Ho PY, Wong MS, Wang Y (2016) Doxorubicin-loaded biodegradable self-assembly zein nanoparticle and its anti-cancer effect: preparation, in vitro evaluation, and cellular uptake. Colloids Surf B Biointerfaces 140:324–331. https://doi.org/10.1016/j.colsurfb.2015.12.048; Epub 2015 Dec 29

El-Sheikh SMA, Edrees N, El-Sayed H, Khamis T, Arisha AH, Metwally MMM, Eleiwa NZ, Galal AAA (2022) Could cisplatin loading on biosynthesized silver nanoparticles improve its therapeutic efficacy on human prostate cancer cell line and reduce its in vivo nephrotoxic effects? Biol Trace Elem Res 200(2):582–590. https://doi.org/10.1007/s12011-021-02677-3; Epub 2021 Mar 23

Erdmann K, Ringel J, Hampel S, Wirth MP, Fuessel S (2017) Carbon nanomaterials sensitize prostate cancer cells to docetaxel and mitomycin C via induction of apoptosis and inhibition of proliferation. Beilstein J Nanotechnol 8:1307–1317. https://doi.org/10.3762/bjnano.8.132; PMID: 28690966; PMCID: PMC5496539

Esmaeili Y, Khavani M, Bigham A, Sanati A, Bidram E, Shariati L, Zarrabi A, Jolfaie NA, Rafienia M (2022) Mesoporous silica@chitosan@gold nanoparticles as "on/off" optical biosensor and pH-sensitive theranostic platform against cancer. Int J Biol Macromol 202:241–255. https://doi.org/10.1016/j.ijbiomac.2022.01.063; Epub 2022 Jan 15

Gao M, Xu Y, Qiu L (2017) Sensitization of multidrug-resistant malignant cells by liposomes co-encapsulating doxorubicin and chloroquine through autophagic inhibition. J Liposome Res 27(2):151–160. https://doi.org/10.1080/08982104.2016.1185731; Epub 2016 Jun 2

George BP, Rajendran NK, Houreld NN, Abrahamse H (2022a) *Rubus* capped zinc oxide nanoparticles induce apoptosis in MCF-7 breast cancer cells. Molecules 27(20):6862. https://doi.org/10.3390/molecules27206862; PMID: 36296460; PMCID: PMC9611499

George TA, Chen MM, Czosseck A, Chen HP, Huang HS, Lundy DJ (2022b) Liposome-encapsulated anthraquinone improves efficacy and safety in triple negative breast cancer. J Control Release 342:31–43. https://doi.org/10.1016/j.jconrel.2021.12.001; Epub 2021 Dec 8

Guo J, Li Y, Yu Z, Chen L, Chinnathambi A, Almoallim HS, Alharbi SA, Liu L (2022) Novel green synthesis and characterization of a chemotherapeutic supplement by silver nanoparticles containing Berberis thunbergii leaf for the treatment of human pancreatic cancer. Biotechnol Appl Biochem 69(3):887–897. https://doi.org/10.1002/bab.2160; Epub 2021 Apr 17

Gupta R, Vishwakarma L, Guleri SK, Kumar G (2022) 5-fluorouracil-impregnated PLGA coated gold nanoparticles for augmented delivery to lung cancer: in vitro investigations. Anti Cancer Agents Med Chem 22(12):2292–2302. https://doi.org/10.2174/1871520622666211224103110

Habibi N, Mauser A, Ko Y, Lahann J (2022) Protein nanoparticles: uniting the power of proteins with engineering design approaches. Adv Sci (Weinh) 9(8):e2104012. https://doi.org/10.1002/advs.202104012; Epub 2022 Jan 25. PMID: 35077010; PMCID: PMC8922121

Hu H, Zhang Y, Ji W, Mei H, Wu T, He Z, Wang K, Shi C (2022) Hyaluronic acid-coated and olaparib-loaded PEI - PLGA nanoparticles for the targeted therapy of triple negative breast cancer. J Microencapsul 39(1):25–36. https://doi.org/10.1080/02652048.2021.2014586; Epub 2021 Dec 20

Huang SJ, Wang TH, Chou YH, Wang HD, Hsu TC, Yow JL, Tzang BS, Chiang WH (2022) Hybrid PEGylated chitosan/PLGA nanoparticles designed as pH-responsive vehicles to promote intracellular drug delivery and cancer chemotherapy. Int J Biol Macromol 210:565–578. https://doi.org/10.1016/j.ijbiomac.2022.04.209; Epub 2022 May 2

Jain P, Pandey V, Soni V (2022) Bioconjugate-loaded solid lipid nanoparticles for enhanced anticancer drug delivery to brain cancer cells: an *in vitro* evaluation. Indian J Med Res 156(1):139–148. https://doi.org/10.4103/ijmr.IJMR_514_19; PMID: 36510906; PMCID: PMC9903388

Jaiswal N, Halder S, Mahata N, Chanda N (2022) Bi-functional gold nanorod-protein conjugates with biomimetic BSA@folic acid corona for improved tumor targeting and intracellular delivery of therapeutic proteins in colon cancer 3D spheroids. ACS Appl Bio Mater 5(4):1476–1488. https://doi.org/10.1021/acsabm.1c01216; Epub 2022 Mar 14

Kajani AA, Rafiee L, Javanmard SH, Dana N, Jandaghian S (2023) Carbon dot incorporated mesoporous silica nanoparticles for targeted cancer therapy and fluorescence imaging. RSC Adv 13(14):9491–9500. https://doi.org/10.1039/d3ra00768e; PMID: 36968033; PMCID: PMC10034601

Khatun S, Pebam M, Putta CL, Rengan AK (2023) Camptothecin loaded casein nanosystem for tuning the therapeutic efficacy against highly metastatic triple-negative breast cancer cells. Biomater Sci 11(7):2518–2530. https://doi.org/10.1039/d2bm01814d

Khella KF, Abd El Maksoud AI, Hassan A, Abdel-Ghany SE, Elsanhoty RM, Aladhadh MA, Abdel-Hakeem MA (2022) Carnosic acid encapsulated in albumin nanoparticles induces apoptosis in breast and colorectal cancer cells. Molecules 27(13):4102. https://doi.org/10.3390/molecules27134102; PMID: 35807348; PMCID: PMC9268188

Kumari L, Ehsan I, Mondal A, Al Hoque A, Mukherjee B, Choudhury P, Sengupta A, Sen R, Ghosh P (2023) Cetuximab-conjugated PLGA nanoparticles as a prospective targeting therapeutics for non-small cell lung cancer. J Drug Target 31(5):521–536. https://doi.org/10.1080/1061186X.2023.2199350; Epub 2023 Apr 14

Kumbham S, Ghosh A, Ghosh B, Biswas S (2022) Human serum albumin-poly(Lactide)-conjugated self-assembly NPs for targeted docetaxel delivery and improved therapeutic efficacy in oral cancer. Int J Biol Macromol 222(Pt A):1287–1303. https://doi.org/10.1016/j.ijbiomac.2022.09.250; Epub 2022 Oct 1

Kuzminac IZ, Ćelić AS, Bekić SS, Kojić V, Savić MP, Ignjatović NL (2022) Hormone receptor binding, selectivity and cytotoxicity of steroid D-homo lactone loaded chitosan nanoparticles for the treatment of breast and prostate cancer cells. Colloids Surf B Biointerfaces 216:112597. https://doi.org/10.1016/j.colsurfb.2022.112597; Epub 2022 May 25

Lewińska A, Wróbel K, Błoniarz D, Adamczyk-Grochala J, Wołowiec S, Wnuk M (2022) Lapatinib- and fulvestrant-PAMAM dendrimer conjugates promote apoptosis in chemotherapy-induced senescent breast cancer cells with different receptor status. Biomater Adv 140:213047. https://doi.org/10.1016/j.bioadv.2022.213047; Epub 2022 Jul 26

Li S, Peng Z, Leblanc RM (2015) Method to determine protein concentration in the protein-nanoparticle conjugates aqueous solution using circular dichroism spectroscopy. Anal Chem 87(13):6455–6459. https://doi.org/10.1021/acs.analchem.5b01451; Epub 2015 Jun 16

Li H, Qu X, Qian W, Song Y, Wang C, Liu W (2022) Andrographolide-loaded solid lipid nanoparticles enhance anti-cancer activity against head and neck cancer and precancerous cells. Oral Dis 28(1):142–149. https://doi.org/10.1111/odi.13751; Epub 2020 Dec 20

Mai R, Deng B, Zhao H, Li L, Fang Y, Li S, Deng X, Chen J (2022) Design, synthesis, and bioevaluation of novel enzyme-triggerable cell penetrating peptide-based dendrimers for targeted delivery of camptothecin and cancer therapy. J Med Chem 65(7):5850–5865. https://doi.org/10.1021/acs.jmedchem.2c00287; Epub 2022 Apr 5

Matalqah SM, Aiedeh K, Mhaidat NM, Alzoubi KH, Al-Husein BA (2022) Preparation of modified chitosan-based nanoparticles for efficient delivery of doxorubicin and/or cisplatin to breast cancer cells. Curr Cancer Drug Targets 22(2):133–141. https://doi.org/10.2174/1568009622666220126100532

Patil S, Mishra VS, Yadav N, Reddy PC, Lochab B (2022) Dendrimer-functionalized nanodiamonds as safe and efficient drug carriers for cancer therapy: nucleus penetrating nanoparticles. ACS Appl Bio Mater 5(7):3438–3451. https://doi.org/10.1021/acsabm.2c00373; Epub 2022 Jun 27

Qaddoori MH, Al-Shmgani HS (2023) Galangin-loaded gold nanoparticles: molecular mechanisms of antiangiogenesis properties in breast cancer. Int J Breast Cancer 2023:3251211. https://doi.org/10.1155/2023/3251211; PMID: 36844680; PMCID: PMC9950320

Ramedani A, Sabzevari O, Simchi A (2022) Hybrid ultrasound-activated nanoparticles based on graphene quantum dots for cancer treatment. Int J Pharm 629:122373. https://doi.org/10.1016/j.ijpharm.2022.122373; Epub 2022 Nov 8

Ramezani Farani M, Azarian M, Heydari Sheikh Hossein H, Abdolvahabi Z, Mohammadi Abgarmi Z, Moradi A, Mousavi SM, Ashrafizadeh M, Makvandi P, Saeb MR, Rabiee N (2022) Folic acid-adorned curcumin-loaded iron oxide nanoparticles for cervical cancer. ACS Appl Bio Mater 5(3):1305–1318. https://doi.org/10.1021/acsabm.1c01311; Epub 2022 Feb 24. Erratum in: ACS Appl Bio Mater. 2023 Feb 20;6(2):932. PMID: 35201760; PMCID: PMC8941513

Ringel J, Erdmann K, Hampel S, Kraemer K, Maier D, Arlt M, Kunze D, Wirth MP, Fuessel S (2014) Carbon nanofibers and carbon nanotubes sensitize prostate and bladder cancer cells to platinum-based chemotherapeutics. J Biomed Nanotechnol 10(3):463–477. https://doi.org/10.1166/jbn.2014.1758

Ross S, Wu RS, Wei SC, Ross GM, Chang HT (2020) The analytical and biomedical applications of carbon dots and their future theranostic potential: a review. J Food Drug Anal 28(4):677–695. https://doi.org/10.38212/2224-6614.1154; PMID: 35696139; PMCID: PMC9261808

Rostami S, Tafvizi F, Kheiri Manjili HR (2022) High efficacy of tamoxifen-loaded L-lysine coated magnetic iron oxide nanoparticles in cell cycle arrest and anti-cancer activity for breast cancer therapy. Bioimpacts 12(4):301–313. https://doi.org/10.34172/bi.2021.23337; Epub 2021 Dec 1. PMID: 35975200; PMCID: PMC9376161

Sajjad F, Han Y, Bao L, Yan Y, Shea DO, Wang L, Chen Z (2022) The improvement of biocompatibility by incorporating porphyrins into carbon dots with photodynamic effects and pH sensitivities. J Biomater Appl 36(8):1378–1389. https://doi.org/10.1177/08853282211050449; Epub 2021 Dec 30

Sani NS, Onsori H, Akrami S, Rahmati M (2022) A comparison of the anti-cancer effects of free and PLGA-PAA encapsulated hydroxytyrosol on the HT-29 colorectal cancer cell line. Anti Cancer Agents Med Chem 22(2):390–394. https://doi.org/10.2174/1871520621666210308095712

Satari N, Taymouri S, Varshosaz J, Rostami M, Mirian M (2020) Preparation and evaluation of inhalable dry powder containing glucosamine-conjugated gefitinib SLNs for lung cancer therapy. Drug Dev Ind Pharm 46(8):1265–1277. https://doi.org/10.1080/03639045.2020.1788063; Epub 2020 Jul 13

Simitcioglu B, Karagoz ID, Ilbasmis-Tamer S, Tamer U (2022) Effect of different molecular weight and terminal group PLGA on docetaxel nanoparticles: characterization and cytotoxic activity of castration-resistant prostate cancer cells. Pharm Dev Technol 27(7):794–804. https://doi.org/10.1080/10837450.2022.2120004; Epub 2022 Sep 12

Sola F, Montanari M, Fiorani M, Barattini C, Ciacci C, Burattini S, Lopez D, Ventola A, Zamai L, Ortolani C, Papa S, Canonico B (2022) Fluorescent silica nanoparticles targeting mitochondria: trafficking in myeloid cells and application as doxorubicin delivery system in breast cancer cells. Int J Mol Sci 23(6):3069. https://doi.org/10.3390/ijms23063069; PMID: 35328491; PMCID: PMC8954043

Song J, Yu X, Lv C, Wang D (2022) Single-step assembly of lipid-gelatin-epigallocatechin-3-gallate hybrid nanoparticles for cancer therapy. Anticancer Drugs. https://doi.org/10.1097/CAD.0000000000001484; Epub ahead of print, Publish Ahead of Print

Tekade RK et al (2017) Solid lipid nanoparticles for targeting and delivery of drugs and genes. In: Nanotechnology-based approaches for targeting and delivery of drugs and genes. Academic Press, Amsterdam, pp 256–286

Thomas S, Gunasangkaran G, Arumugam VA, Muthukrishnan S (2022) Synthesis and characterization of zinc oxide nanoparticles of Solanum nigrum and its anticancer activity via the induction of apoptosis in cervical cancer. Biol Trace Elem Res 200(6):2684–2697. https://doi.org/10.1007/s12011-021-02898-6; Epub 2021 Aug 27

Tiburcius S, Krishnan K, Jose L, Patel V, Ghosh A, Sathish CI, Weidenhofer J, Yang JH, Verrills NM, Karakoti A, Vinu A (2022) Egg-yolk core-shell mesoporous silica nanoparticles for high doxorubicin loading and delivery to prostate cancer cells. Nanoscale 14(18):6830–6845. https://doi.org/10.1039/d2nr00783e

Wadhawan A, Singh J, Sharma H, Handa S, Singh G, Kumar R, Barnwal RP, Pal Kaur I, Chatterjee M (2022) Anticancer biosurfactant-loaded PLA-PEG nanoparticles induce apoptosis in human MDA-MB-231 breast cancer cells. ACS Omega 7(6):5231–5241. https://doi.org/10.1021/acsomega.1c06338; PMID: 35187338; PMCID: PMC8851644

Waghmare MN, Qureshi TS, Krishna CM, Pansare K, Gadewal N, Hole A, Dongre PM (2022) β-Lactoglobulin-gold nanoparticles interface and its interaction with some anticancer drugs—an approach for targeted drug delivery. J Biomol Struct Dyn 40(13):6193–6210. https://doi.org/10.1080/07391102.2021.1879270; Epub 2021 Jan 29

Wang X, Teng Z, Wang H, Wang C, Liu Y, Tang Y, Wu J, Sun J, Wang H, Wang J, Lu G (2014) Increasing the cytotoxicity of doxorubicin in breast cancer MCF-7 cells with multidrug resistance using a mesoporous silica nanoparticle drug delivery system. Int J Clin Exp Pathol 7(4):1337–1347; PMID: 24817930; PMCID: PMC4014214

Wang Y, Zhao K, Xie L, Li K, Zhang W, Xi Z, Wang X, Xia M, Xu L (2022) Construction of calcium carbonate-liposome dual-film coated mesoporous silica as a delayed drug release system for antitumor therapy. Colloids Surf B Biointerfaces 212:112357. https://doi.org/10.1016/j.colsurfb.2022.112357; Epub 2022 Jan 22

Wróbel K, Wołowiec S, Markowicz J, Wałajtys-Rode E, Uram Ł (2022) Synthesis of biotinylated PAMAM G3 dendrimers substituted with *R*-glycidol and celecoxib/simvastatin as repurposed drugs and evaluation of their increased additive cytotoxicity for cancer cell lines. Cancers (Basel) 14(3):714. https://doi.org/10.3390/cancers14030714; PMID: 35158983; PMCID: PMC8833738

Xu SY, Su H, Zhu XY, Li XY, Li J, Chen X, Wang Q, Hao RY, Yan XY (2022) Long-circulating doxorubicin and schizandrin a liposome with drug-resistant liver cancer activity: preparation, characterization, and pharmacokinetic. J Liposome Res 32(2):107–118. https://doi.org/10.1080/08982104.2021.1884093; Epub 2021 Sep 13

Yu X, Wu H, Hu H, Dong Z, Dang Y, Qi Q, Wang Y, Du S, Lu Y (2020) Zein nanoparticles as nontoxic delivery system for maytansine in the treatment of non-small cell lung cancer. Drug Deliv 27(1):100–109. https://doi.org/10.1080/10717544.2019.1704942; PMID: 31870183; PMCID: PMC6968508

Zhang L, Hou Y, Li C, Liu H, Wang Y (2022) Comparative study on the antitumor effects of gemcitabine polybutylcyanoacrylate nanoparticles coupled with anti-human MUC1 and CA199 monoclonal antibodies on pancreatic cancer in vitro and in vivo. Arab J Gastroenterol 23(4):263–269. https://doi.org/10.1016/j.ajg.2022.06.005; Epub 2022 Aug 1

Zheng L, Shen L, Li Z, Zhang X, Wu M, Zhang Y, Liu J (2022) Design, preparation, and evaluation of Osthol poly-butyl-cyanoacrylate nanoparticles with improved in vitro anticancer activity in neuroblastoma treatment. Molecules 27(20):6908. https://doi.org/10.3390/molecules27206908; PMID: 36296500; PMCID: PMC9609737

Zhu Y, Wang A, Zhang S, Kim J, Xia J, Zhang F, Wang D, Wang Q, Wang J (2023) Paclitaxel-loaded ginsenoside Rg3 liposomes for drug-resistant cancer therapy by dual targeting of the tumor microenvironment and cancer cells. J Adv Res 49:159–173. https://doi.org/10.1016/j.jare.2022.09.007; Epub 2022 Sep 24. PMID: 36167294; PMCID: PMC10334248

Testing of Nanocarriers Loaded with Chemotherapeutic Agents in the Animal Models of Cancers: Preclinical Trials for Efficacy and Safety Profiling of Nanocarriers

5

Laiba Iftikhar, Mishaal Fareed, Naeem Ullah Khan, and Farah Rehan

Abstract

Researchers are focusing on nanotechnology use in cancer treatment for decades. The primary processes used in nanotechnology involve the segregation, combining, and modification of materials by a single atom or molecule. It offers a distinctive strategy and sophisticated technology for treating cancers through early detection, forecasting, prevention, individualized therapy, and medical care. Chemotherapeutic medicines that are hard to penetrate blood vessels are encased in nanoparticles between 50 and 400 nm, their small nano size making it hard for them to cross a healthy person's blood vessel walls and cells. Polymers, liposomes, micelles, dendrimers, and carbon nanotubes are among various nano-delivery carriers being employed against cancer. These nanocarriers can be delivered through multiple routes of drug delivery, such as oral, injection, intraocular, and parenteral, and later tested in animal models for their preclinical testing. This chapter focused on the use of various animal models been used so far for the testing of these nanocarriers to assess their efficacy in preclinical setup. Animal models are classified according to the method by which cancer is induced in the animal. Spontaneous tumor models are the most primitive models. The study of carcinogenesis has benefited greatly from the use of mouse models of human cancer. The zebrafish model is a low-cost, high-throughput animal model that can be used to study the mechanisms behind tumor formation, metastasis, and therapeutic response as well as forecast the efficacy of personalized cancer treatment. The hu-PBL model is now being used in numerous areas of cancer research, including investigations into malignancies of the thyroid, thyroid gland, breast, nasopharynx, and other cancers. However, the creation of secure and

L. Iftikhar · M. Fareed · N. U. Khan · F. Rehan (✉)
Department of Pharmacy, Forman Christian College University, Lahore, Pakistan
e-mail: farahrehan@fccollege.edu.pk

F. A. Khan (ed.), *Nano Drug Delivery for Cancer Therapy*,
https://doi.org/10.1007/978-981-99-6940-1_5

effective transport vectors is a key obstacle to the success of gene therapy using nanoparticles. There is at least a theoretical possibility that NPs could cause protein unfurling, crosslinking, or unfortunate enzymatic movement, leading to unfavorable and unexpected outcomes. Safety profiling of nanoparticles ensures that the treatment is effective and safe for patients. Without proper safety profiling, there is a risk of harmful side effects and potential damage to healthy cells. Before testing on humans, in vivo, safety profiling technique is used to assess the possible harmful effects of nanoparticles on healthy cells in case they occurred.

Keywords

Cancer treatment · Nanotechnology · Nanoparticles · Chemotherapeutic drugs · Animal models · Zebrafish model · Mouse model · Hu-PBL model · Safety profiling

5.1 Cancer

5.1.1 Historical Turning Points in Cancer

Although cancer is as ancient as people, pale pathologic research demonstrates that tumors were present in animals long before humans set foot on Earth. In his Materia Medica, Pliny the Roman scientist (AD 23–79) compiled a list of cancer therapies. His most effective therapy consisted of a boiling concoction made up of egg white, honey, powdered falcon feces and sea crab ash (Hajdu 2012). Aretaeus (AD 81–138), an Egyptian physician who lived and worked in Alexandria, provided the earliest comprehensive explanation of the causes, symptoms, and treatments of uterine cancer. There are two different sorts of cancer, according to what he noted in his notes: one is hard and ulcerated to the touch, while the other is odorous and ulcerated (Wang and Thanou 2010). Hippocrates used the word "Carcinos" to denote both ulcerative and non-ulcerative tumors. Greek language uses the word "cancer" to refer to a crab, which makes sense given that the disease has crab-like spreading projections that resemble fingers. The Latin word for crab, cancer, was later used to translate the Greek phrase by the Roman physician Celsus (25 BC–50 AD). Galen (130–200 AD), a different Greek physician, called tumors oncos (Greek for swelling). Although a comparison of Hippocrates and Celsus' crab is still used to explain dangerous cancers (Cheong et al. 2011).

5.1.2 Treatment of Cancer

Cancer treatment depends on how enormous it has been spread, plus understanding how we can overcome it. We can go through various modalities, just as cancer drugs, chemotherapy, surgery for cancer, radiotherapy, hormone therapy, bone marrow, and stem cell transplants, immunotherapy, and individualized and complementary therapies.

5.1.3 Introduction to Nanotechnology

The definition of nanotechnology is "technology at the nanoscale." It is the twenty-first century's most promising technology. The definition of "nano" is one billionth of a meter (10^{-9} m). It includes the creation of nanoscale materials, their characterization, engineering, and size-controlled application. Numerous biological applications of nanotechnology make use of the physical and chemical features of nanoparticles. The creation and use of chemical, physical, and biological systems at scales ranging from single molecules to submicron dimensions, as well as the integration of the resulting nanomaterials into more intricate systems, are all included in this (Nasrollahzadeh et al. 2019). Synthetic approaches for the fabrication of nanoengineered materials can be classified into two groups: "top-down" and "bottom-up" approaches. In the top-down method, externally controlled and larger devices are used to create nanoscale devices to direct their assembly. To cut, grind, and shape materials into the appropriate arrangement or shape when using a top-down approach, traditional workshop or microfabrication techniques are frequently used. In the bottom-up approach, devices, and molecular components which do not require any external agent assembly are used to construct devices and materials. They chemically assemble by recognizing molecules of their own type. This approach starts with the collection and combination of atoms and molecules to build complex structures (Silva 2004).

5.2 Applications of Nanotechnology

Nanotechnology application is a young field with many possible applications. Recently, the use of nanotechnology in the treatment of cancer has received a lot of attention. It offers a distinctive strategy and sophisticated technology for treating cancers through early detection, forecasting, prevention, individualized therapy, and medical care (Misra et al. 2010). Chemotherapy, radiation therapy, and surgery are the only cancer treatments that are commonly used, but these are not always the best options due to their numerous drawbacks. Delineating tumor margins, diagnosing recurrent tumor cells and micrometastase, and determining if a tumor has been completely removed therefore are urgent needs and have significant potential. Nanotechnology can be used in many fields related to health care services including cancer therapy.

- Nanotechnology can be used for better cancer diagnosis.
- Nanotechnology can provide better drug delivery to tumor/cancer cells.
- It can provide molecular targeted cancer therapy.
- This technology works better with dyes which allows better tumor cells detection.

5.2.1 Nanomaterials Used for Cancer Therapy

There are various nanomaterials that are broadly used for the treatment of cancer. Some of these nanomaterials are;

1. Polymeric nanoparticles.
2. Lipid-based nanomaterials.
3. Nano-emulsions.
4. Carbon nanomaterials.

5.2.1.1 Polymeric Nanoparticles

These are colloidal macromolecules of 10–1000 nm size. Polymeric nanoparticles are the drug carriers used for the sustained release of chemotherapeutic agents to target cancerous sites (Society 2018). To prevent inflammation and toxicity, polymeric nanoparticles made of polystyrene, PMMA, polyacrylamide, and polyacrylates are utilized. Chemotherapeutic chemicals can enter malignant regions thanks to leaky blood vessels in tumor-initiated angiogenesis. In comparison to conventional particles, polymeric nanoparticles, and their associated chemical medicines can more readily collect at target areas because of reduced lymphatic drainages. Due to PNPs' high surface-to-volume ratio, attaching targeting polymers to the surface is not difficult. Polymers' bioavailability can be raised by encasing them in polysorbates, according to research.

5.2.1.2 Lipid-Based Nanomaterials

The three primary types of lipid-based nanoparticles are liposomes, solid lipid nanoparticles (SLNs), and nanostructured lipid carriers (NLCs). The typical liposome size range is 20 nm to more than 1 μm (Batist et al. 2009). The main uses of liposomes are co-delivery and sustained release. Advanced solid tumors respond better to new liposomes containing chemotherapy drugs loaded with both floxuridine and irinotecan (Batist et al. 2009). Because of their low toxicity, biocompatibility, and non-immunogenicity, liposomes shield loaded pharmaceuticals from degradation. However, due to their shorter half-life, unsatisfactory stability, poor potency, and quick elimination by MPS, their use is restricted. SLNs are nanocarriers with particles as small as 1–100 nm (Fig. 5.1). Drugs are enclosed in SLNs' micelles-like structure, which has a non-aqueous core (Kraft et al. 2014). SLNs outperform liposomes in terms of stability and sustained release. SLNs and liposomes are less effective than NLCs. They can be given via oral, parenteral, inhalational, and ocular routes among others.

5.2.1.3 Nanoemulsions

Oil, emulsifying agents, and an aqueous phase combine to generate nanoemulsions (NE). The size of nanoemulsions can vary from 10 to 1000 nm. It is common practice to use drug nanocarriers known as nanoemulsions, which are typically solid spheres with an amorphous, lipophilic surface and a negative charge. The three most prevalent types of nanoemulsions that can be produced are bi-continuous

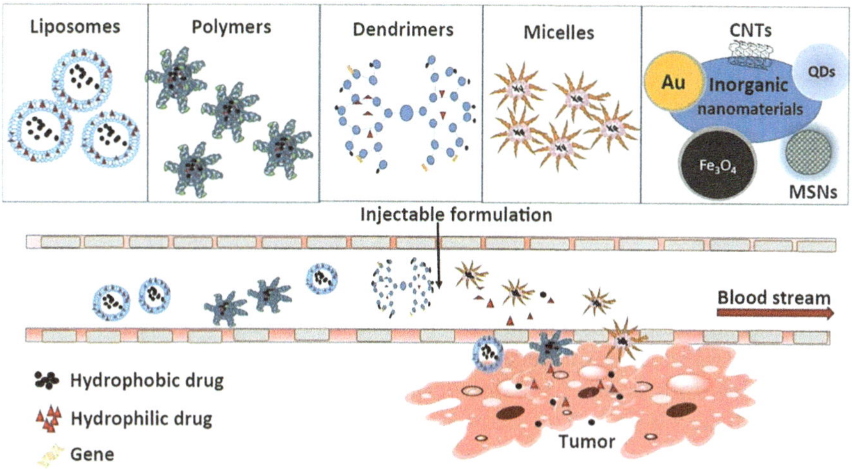

Fig. 5.1 Nanomaterials employed for drug delivery for cancer treatment (Credit: Lôbo et al. 2021)

nanoemulsions, water in oil nanoemulsion systems, and oil in water nanoemulsion system (Jaiswal et al. 2015). Nanoemulsions provide a variety of benefits over the majority of lipid-based nanomaterials and nanoparticles, including optical clarity, thermodynamic stability, a sizable surface area, ease of manufacturing, biodegradability, and the ideal drug release profile. The NE drug carrier system loaded with spirulina polysaccharides and PTX indicated that it may enhance the anti-tumor activity of PTX by regulating immunity via Toll-like receptor 4/nuclear factor kappa B (TLR4/NF-B) signaling pathways (Du et al. 2020).

5.2.1.4 Carbon Nanoparticles

"Carbon nanomaterials" (CNMs) are materials with various categories based on the element of carbon. In comparison to metal-based nanomaterials, CNMs are considered to be safer and more biocompatible (Saleem et al. 2018). Due to their inherent hydrophobicity, CNMs can load chemical medicines by "stacking" or hydrophobic interactions, making them effective drug delivery platforms (Ou et al. 2016). Graphene, fullerenes, carbon nanotubes, carbon nanohorns, carbon quantum dots, and graphyne (GDY) are among the carbon nanomaterials that have undergone extensive research in the area of cancer therapy. Graphene is a great substrate for drug administration, PDT, and PTT because it has a high surface-to-volume ratio and a high no. of branches that contain oxygen. Studies carried out on animals showed that CNTs could cause fibrosis, granuloma development, persistent inflammation, and pathophysiology similar to mesothelioma.

5.2.2 Use of Nanocarriers in Cancer Therapy

Chemotherapy drugs are toxic to carcinomas. These medications are hazardous to all cells, even healthy ones, due to their high toxicity and lack of selectivity. All medications that are too tiny to fit inside molecules cause this poisonous response. Almost all parts of the organism are endothelial. With careful management, both the target area and other unaffected areas can be reached. a sickness that therefore results in a lot of drug-related side effects. A colloidal nanoparticle system is a potential tactic that could enhance the therapeutic efficacy of chemotherapeutic medicines while lowering agents and their negative effects.

Since these medications are unable to traverse the blood vessel walls of a healthy person's system when they are enclosed in nanoparticles of 50–800 nm (the distance between these cells is just 15–30 nm). This is distinct from what usually occurs. Instead of the resulting healthy zones, areas that are inflamed or even have tumors will have endothelial cells that are less closely packed together. They further kill the tumor cells owing to their enhanced penetration and retention effect (EPR) resulting release of cancer therapeutics inside the tumor (Prabhakar et al. 2013). Allows for the precise targeting of nanocarriers to particular cancer cells with an active mode of action based on the expressed molecule on the tumor surface. Drug distribution to the site of the disease in a precise and effective way is one of the pillars of nanomedicine. In general, there are numerous ways to accomplish this, such as transdermal, nasal, and oral administration methods. Most importantly intravenously. However, the encapsulation or association of a certain type of nanoparticles can frequently boost the efficacy of medications while somewhat reducing their effectiveness.

Here, we provide an overview of nanoparticles for cancer imaging and treatment that have been tried and tested in clinical settings. Many various kinds of nanoparticles are presently being researched for cancer, according to both clinical and preclinical investigations. Other kinds of nanoparticles are now in the early stages of development and could eventually enter the preclinical stage for cancer imaging and therapy. We want to draw attention to the potential value of diagnostic nanoparticles among all of these other nanoparticles.

5.2.3 Opportunities and Challenges of Nanoparticles for Cancer Therapy

Nanomedicine is a great alternative to traditional cancer treatment, but it doesn't overcome all of the limitations of oncology. Nanoparticles used in cancer therapy can serve as carriers for drugs, nucleic acids, contrast media or enhance the radiation sensitivity of cells. Significant improvements in drug delivery efficiency and direct targeting to tumors have been achieved by using nanoparticles as drug carriers. However, nanoparticles alone do not significantly improve the effectiveness of chemotherapeutic agents. Most nanomedicines can reduce the toxicity of chemotherapy but do not impair the effectiveness of the therapy. Tumors exhibit unique pathophysiological features that differ from normal tissues, such as: B. extensive angiogenesis, damage to vascular structures and lymphatic drainage. Nanoparticles exploit these anatomically impaired properties of the tumor and result in significant leakage of the nanoparticles into the tumor tissue. Due to slow vascular reflux of into cancerous tissue and lymphatic clearance, nanoparticles are stored in the tumor. This is known as improved permeability and retention. Similarly, specific targeting of can be achieved by targeting tissues in the vicinity of the tumor (e.g., Kupffer cells in the liver).

Nanoparticles can be delivered through multiple routes of drug delivery, such as oral, injection, intraocular, parenteral, etc. It has been noted that nanoparticles are better option than microparticles (>1 μm) as carriers of drugs with greater cellular uptake demonstrated. These features allow finer control of drug targeting and delivery to malignant tissues (Awasthi et al. 2018). Improving nanoparticle's capacity to use the immune system to deliver chemotherapy to sensitize lymph nodes, which will then deliver chemotherapy to certain metastatic locations, is an utmost need. Hence, the most cutting-edge approach to treating cancer, however, is gene therapy, which eliminates the illness's origin by changing the genetic makeup (Pedziwiatr-Werbicka et al. 2021). Delivery of these experts via a nanometer-scale carrier will increase their power in a few ways, including by protecting them from enzymatic contamination and delaying elimination (Nabil et al. 2019).

5.3 Preclinial Testing of Nanocarriers

Preclinical in vivo model systems (animal model) are used to investigate safety and efficacy underpin advancements in cancer therapy. Animal models include animal experimental objects established in medical research and similar materials that can simulate the human body (Li et al. 2021). Workers at laboratories take care of the animals, promote top-notch research, and meet other particular requirements. Studies of the tumor microenvironment generally reveal similarities between oncology treatment models and the genetic landscape of clinical cancer. For biomedical research, the following animal model is ideal:

- Mimics human disease at the molecular level.
- Is reliable and predictable.
- Makes significant differences in survival, and Allows for accurate treatment assessment.
- Easily visualized.
- Setting like human disease appearing in the background (Obeid et al. 2018).

Animal models are classified according to the method by which cancer is induced in the animal. The simplest models are those that represent spontaneous tumors. Although these models mimic natural human diseases, they have not kept pace with the development of experimental treatment programs. It was therefore necessary to move away from clinical concerns and take a further step towards artificiality in search of a satisfactory test method. Cancer can be reproducibly induced in animals by exposing them to various drugs and now by transplanting tumor cells or tissues (Phillips and Roth 2019).

5.3.1 Uses of Animal Model

- Animal models are an excellent tool to explore the pathophysiology of human diseases, the principles of disease prevention and therapy, and they are a constant source of new ideas for cancer research.
- Humans and animal models differ from one another in terms of physiology, genetics, and immunity.
- As a result, it is critically important for the advancement of biomedicine to design animal models that can take the place of humans in preclinical research and more accurately represent human biological traits.
- It is impossible to accurately portray the biological processes of human tumor tissue using the cell line-derived xenograft (CDX) model and traditional in vitro culture, which reduces the effectiveness of cancer research.
- PDX model is high consistency rate of treatment response allows it to more accurately replicate the progression of tumors in people.
- The humanized animal model was created on the basis of this in order to copy the physiological or pathological condition of humans.
- Making a range of cancer animal models is now possible thanks to the ongoing development and advancement of gene-editing technologies, which has significantly aided the study of cancer.
- An improved model that is more like the human genome is being developed.
- Gene-editing technology will someday allow us to mix several building methods to produce a model that is better suited for cancer research.
- One future strategy for using zebra fish as a model organism of tumor growth and progression is to better mimic the tumor environment of the human body (Li et al. 2021).

5.3.2 Various Animal Models Employed in Cancer Models

5.3.2.1 Canine Cancer Models

Customers' dogs provide a special chance to research naturally occurring tumors in a setting that is advantageous to both animals and people. The great genetic variety throughout the canine population and the many physiologic similarities to humans make these animals useful as comparative cancer models. Osteosarcoma, soft tissue sarcoma (STS), lung cancer, oral melanoma, breast cancer, oral squamous cell carcinoma, nasal tumors, and malignant non-Hodgkin's lymphoma are among the most prevalent tumors in dogs. Given the striking similarities, this is perhaps the finest cancer model that dogs can provide.

The canine cancer model is distinctive because it displays tumors with various traits that are similar to those found in humans on its own (for example, large dog osteosarcoma). The quick study of DNA damage and epigenetic alterations that build over time and result in carcinogenesis is made possible by the spontaneous onset and history of cancer, especially given the high degree of genome similarity observed between the canine and human genomes. Enable it to happen. These benefits have made it possible for researchers to use canine cancer models to find pertinent genetic changes and cancer-causing factors that are present in human tumors. Canine participants also skip the clinical trial phase, hastening the process of developing new drugs. Utilizing canine cancer models, numerous medicines have undergone preclinical research. A medication called resiniferatoxin works as an agonist for the pain brought on by bone cancer since dogs have a very strong tendency to injure themselves when they are in agony. There are a number of problems with using canines to simulate human cancer. Canine cancer models often incorporate lymphoid and sarcoma tumor types rather than carcinomas. Studies on the development of anticancer drugs in dogs are not necessarily applicable to people because humans and dogs react to drugs differently (Wu et al. 2023).

5.3.2.2 Patient-Derived Xenograft (PDX) Model

Greek word xenos, which meaning foreign, is where the phrase "xenograft" originates. It is taken out of one organism and inserted into another. Organs, tissues, or living cells are used in these transplants, which are often carried out on mice with immunocompetent immune systems. Xenografts are employed in the field of cancer research to tackle significant studies that rely on using animal models that closely mirror the tumor growth of human patients. In order to preserve the original or primary tumor properties, xenograft models with primary carcinoma tissue are constructed using very few transit numbers, for instance, fewer than 10 passes discharged from human patients. Cell heterogeneity, clinical biomolecular markers, malignant genotypes and phenotypes, tumor structure, and vasculature are a few of these characteristics. In order to better understand the molecular cancer biology pertinent to human cancer and how patients respond to cancer therapy, the PDX models show enhanced preclinical investigations. Depends on the assumptions you make. PDX models have proven useful for.

- Studies of cancer metastasis and drug inhibition.
- Individualized dosing and therapy.
- Preclinical studies and the identification of new anticancer medication candidates are growing (Liu et al. 2023).

Surgery or biopsy techniques are used to collect and preserve primary or metastatic tumors as tissue structures. Using patient-derived tumor tissue xenograft models to study anticancer drugs and treatment options can result from this progressive development of tumor specimens in immune-deficient mice.

Primary or metastatic tumors are gathered and kept as tissue structures by surgical or biopsy techniques. The investigation of anticancer medicines and therapies may switch to using xenograft models of tumor tissue from patients after resection of this tumor specimen in immunocompromised mice. The subcutaneous region (dorsal region) is the most well-known location for tumor implantation in mice. Other locations include the pancreas, brain, oral cavity, ovary, breast, etc. In order to boost the rate of engraftment, numerous methods and initiatives have been used to implant tumor in the renal capsular site (Liu et al. 2023).

PDX models are used to identify possible therapeutic efficacy biomarkers as well as to organize potential clinical symptoms. Cetuximab does not alter Kirsten rat sarcoma (KRAS)-mutant PDX models of colorectal cancer, according to numerous testings. This is shown in the ovarian malignant tumor in particular, where exposure to cisplatin causes the beginning of cisplatin resistance in a platinum-sensitive animal, exactly as was shown in a clinical situation. Breast cancer PDXs models serve as a useful model for doing translational research since they mimic various facets of the biology of the tumor (Rivera et al. 2021). However, due to their unique tumor microenvironment and inability to be genetically modified or to integrate the immune system because they are created in immunodeficient mice, patient-derived xenografts have limitations. They are not appropriate for low-throughput drug screening and immunomodulatory assays for cancer prevention, initiation, and progression of cancer gene modeling.

5.3.2.3 Mouse Models

The study of carcinogenesis and the provision of experimental insights that other systems are unable to supply have greatly benefited from the use of human cancer mouse models. Even while there are differences between mice and humans, current models, for instance, can more accurately simulate sporadic human malignancies by carefully controlling the timing and location of mutations, even within single cell. Therefore, mice offer a model system that can be used in experiments and have a variety of research tools established to understand the fundamental causes of cancer. Because tumors form from scratch in the context of a healthy immune system and co-evolve with the surrounding stroma, many of these problems can be addressed using mice that have been genetically altered to have cell-specific human cancer-associated defects. De novo mouse cancer models are being used to provide preclinical direction for human therapeutic trials. In these methods, immune-compromised mice at the orthotopic site are transplanted with human critical tumors, as well as mice that have been "humanized" by receiving transplanted human hematopoietic and other organ frameworks. By creating constitutive or conditional erasures within the quality, "knockout" approaches are most frequently used in loss-of-function investigations to remove the quality of interest. The primary few exons that encode the start codon are typically sufficient to stop translation or interpretation into a useful protein item for traits that span large genomic areas.

In order to generate systems where genetic processes can be precisely controlled both geographically and temporally, complex conditional genetic engineering technology has been developed to go beyond conventional knockout constraints. Site-specific recombinases, such as yeast FLP and bacterial Cre, catalyze accurate recombination between specified 34 bploxP and FRT sites, respectively. Therefore, in the presence of Cre or FLP protein synthesis, homologous recombination between loxP or FRT sites that flank the gene of interest and are oriented in the same direction is activated, recombining out the flanked genetic sequence and deleting the gene of interest. Then, it is possible to control the deletion of the target gene by controlling the expression of the recombinase throughout time and space, overcoming interference from lethality and developmental defects. To conditionally knockout the target gene in a specific tissue or cell type, such as progenitor cells, or at specific times during development, mice carrying the target gene flanked by loxP or FRT sites are crossed with mice carrying the Cre or the FLP recombinase.

Using genetic techniques adapted from Drosophila systems, sporadic single-cell knockout mouse models have been created. Recombination-enabling enzymes. Under tissue-specific and temporal conditions, can be used to promote uncommon G2 mitotic interchromosomal recombination events between FRT or loxP sites on homologous mouse chromosomes, resulting in the production of a wild-type and a mutant daughter cell within a population of heterozygous cells. In the cell cycle's G2, homologous chromosomes line up to prepare for mitosis. The Cre or FLP recombinase can recombine homologues when they are aligned to make certain chromosomal regions homozygous in the daughter cells. Gain-of-function assays are often used to investigate oncogenes in mouse models. To learn how an oncogene promotes cancer in vivo, transgenic or knockout animals that are constitutively overexpress the oncogene can be employed.

The Rosa26 region is widely used as an insertion site for knockin mice since it is devoid of any essential genes and provides for the best transgene expression. Transgenic animals may have many insertions of the transgene, whereas knocking animals only have one copy of the transgene. Knockin approaches can also be used to replace a normal gene copy with a mutant version of the gene in order to examine specific mutational events in the context of the normal control of the gene (Proetzel et al. 2014).

5.3.2.4 Transgenic Mouse Models (of Breast Cancer): The MMTV Paradigm

Taking advantage of the common preference of tumor-inducing infections known to produce distinct types of cancer easily in mice was one way to achieve tissue-specific regulation of oncogene expression. The mouse mammary tumor infection (MMTV), a retrovirus known as the "milk agent" that is transmitted from mother to pup and is responsible for developing mammary tumors in susceptible strains of innate mice, is a crucial example of such an infection. Years of MMTV research led to the discovery that the virus contains a regulatory DNA structure called the long terminal rehash (LTR) that promotes steroid-hormone-inducible transcription. When the MMTV provirus is inserted into the murine genome with a proto-oncogene, it imparts overexpression on mammary epithelial cells that respond to hormonal stimulation during pregnancy and lactation, leading to mammary tumors in female mice with MMTV infection. Using the provirus coordinates as a "tag," the analysis of MMTV inclusion destinations led to the disclosure of new oncogenes. However, the tissue-specificity of MMTV-driven tumorigenesis, the hormone-inducible nature of oncogene expression in these tumors, and the characterization of the transcriptional enhancer responsible for this additionally displayed a tremendous opportunity for the design of novel transgenic mouse models of cancer.

After discovering the potential value of the MMTV-LTR for breast cancer research, Phil Leder made an attempt to employ this viral enhancer to create the first transgenic mouse models of this illness. By this time, it was known that the injection of a MMTV-H-Ras vector could trigger the production of hormone-induced oncogenes in cultured cells. After that, Leder and Timothy Stewart, a postdoctoral fellow in his lab, used pronuclear microinjection to produce transgenic mice that expressed the oncogene Myc under the control of the MMTV-LTR. These MMTV-Myc animals underwent careful inspection, and it was discovered that female mice from two of the founder lines underwent Myc-driven mammary tumor development following numerous nursing and pregnancy cycles that induced the expression of the MMTV-LTR. This was defining once more.

The MMTV-Myc and MMTV-Ras mice were crossed in this work, leading to an acceleration of carcinogenesis and the first in vivo evidence of oncogene collaboration. In 1988, Harold Varmus' group released the MMTV-Int1 (now referred to as MMTV-Wnt1) model, establishing this proviral insertion site as a legitimate proto-oncogene capable of causing breast cancers. This extraordinary fusion of molecular oncology, MMTV virology, and novel transgenic technologies. In addition, the MMTV-LTR was used to show the effectiveness of the oncogene ErbB2, which is amplified and overexpressed in 20% of human breast tumors (Slamon et al. 1987), to transform the mouse mammary epithelium. For mechanical studies of mammary gland transformation by this crucial oncogene and therapeutic target, transgenic mouse models of ErbB2-driven breast cancer have proven invaluable. These models have grown in complexity and their capacity to accurately model the human disease over time. The MMTV-LTR is still an essential tool for the in vivo validation of possible breast cancer onco-genes, and many of the initial MMTV-based breast cancer mice models are still in use. To develop transgenic mice, endogenous promoters that are specific to the mouse mammary epithelium have also been characterized as a result of research from other. One example of a promoter in this group is the whey acidic protein (Wap) gene. A mild phenotype with alveolar hyperplasias was produced when oncogenic H- Ras was produced from this hormone-dependent promoter, in contrast to the MMTV-driven Ras model, which normally proceeds to adenocarcinoma. The shape and phenotype of these cancers differ dramatically despite the nearly complete penetrance of breast tumors brought on by Wap-promoted My expression during numerous cycles of pregnancy and lactation.

5.3.2.5 Zebrafish Model

As a new model system for cancer research, tumors have been implanted in zebrafish. Both laboratory cell lines and actual cancer cells extracted from patients were injected. Traditional patient-derived xenografts (PDXs) are created from cancer cells or masses that have been surgically removed from patients following biopsy or excision. These PDXs are then implanted into a model animal, like immunodeficient mice (Keller and Murtha 2004; Patton et al. 2005; Zon and Peterson 2005; Barros et al. 2008; Marques et al. 2009; Smith et al. 2010; Eguiara et al. 2011; Nguyen et al. 2012; Dai et al. 2014; Tang et al. 2014; Mandelbaum et al. 2018; Choo et al. 2019). The zebrafish model is a low-cost, high-throughput animal model that can be used to study the mechanisms behind tumor formation, metastasis, and therapeutic response as well as forecast the efficacy of personalized cancer treatment (Chen et al. 2021).

One of the best models at the moment is the zebrafish cancer model, a vertebrate model that has gained popularity recently. Zebrafish genomes are conservative and similar to those of humans, providing a strong basis for research into the origins of various cancers. The zebrafish model offers various distinct benefits over the most used mouse models for cancer research, including:

- Rapid reproduction, cheap cost, and tiny size.
- Clear embryos.
- Zebrafish that are transgenic or immunodeficient can stay transparent even after they reach adulthood. Real-time visualization and tracking of cancer cell growth, spread, and metastasis are useful.
- Because zebrafish is fertilized in vitro, the transgenic animal model can be made fast and the gene operation is not too difficult. With the aid of genetic engineering, gene editing, xenotransplantation, drug-induced toxic damage, and other methods, numerous zebrafish cancer models have currently been created (Fig. 5.2) (Li et al. 2021; Veldman and Lin 2008). Nanoparticles-Based Oligonucleotides Delivery in Cancer: Role of Zebrafish as Animal Model Oligonucleotide (ON) therapies are chemically produced DNA or RNA molecules that can be used as molecular target agents to suppress protein or gene expression. The capacity of ON compounds to cross the plasma membrane may depend on their chemical makeup. Zebrafish (ZF) models could increase the range of possibilities for evaluating the efficacy of the proposed therapeutic techniques in useful in vivo

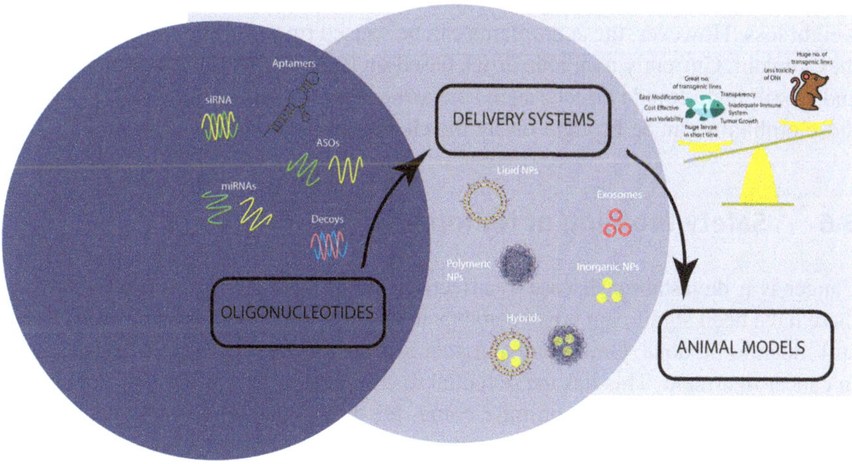

Fig. 5.2 Animal Model for Genetic Manipulation (Credit: Proetzel et al. 2014)

models, which is still a crucial need. The main characteristics of ZF models and the advantages and disadvantages of using them in the creation of ON-based therapy strategies are also covered.

5.4 Hu-PBL Model

Immunodeficient mice were injected with human peripheral blood cells to create the Hu-PBL model. It is currently the most simple and economical humanized mouse model available. The hu-PBL model is currently being used in numerous areas of cancer research, including investigations into malignancies of the thyroid, thyroid gland, breast, nasopharynx, and other cancers. Because it can reproduce numerous T cells, as opposed to the hu-HSC model, it is the ideal paradigm for the study of mature effectors T cells. This model is prone to graft-versus-host disease (GVHD), which shortens research time and diminishes mouse endurance due to the immune system's rejection of human T cells and mouse cells (Li et al. 2021).

5.5 Toxicities of Nanocarriers

Magnesium oxide nanoparticles (MgO NPs) are becoming more widely known for their uses in cancer treatment methods such nano-cryotherapy and hyperthermia. The current study showed how normal lung fibroblast cells and several malignant cell types responded to magnesium oxide nanoparticles (MgO NPs) in terms of cytotoxicity (van Zandwijk and Frank 2019). Inflammation of lungs and cardiac problems may be side effects of inhaled nanoparticles. Regardless of their level of oxidation, copper- and zinc-based nanomaterials exhibit the highest levels of toxicity among any nanoparticles examined till now. Current drugs based on nanotechnology, such as Abraxane and Doxil, include side effects, such as diarrhea, nausea, and weight loss. However, these problems can be caused by the chemotherapeutic drugs they contain. Currently available drugs based on nanotechnology, such as Abraxane and Doxil, include side effects, such as diarrhea, nausea, and weight loss. However, these problems can be brought on by the chemotherapeutic drugs they contain.

5.6 Safety Profiling of Nanoparticles

Cancer is a devastating disease k, affecting millions of people worldwide. While there have been significant advancements in cancer treatment over the years, there is still much to be done. One promising area of research is nanoparticles and their use in cancer treatment. This has the potential to completely change how we handle this illness. However, with great promise comes great responsibility. The safety of these nanoparticles must be thoroughly studied and understood before they can be used in clinical settings. This is where safety profiling comes into play. By understanding the potential risks associated with nanoparticles, we can develop safer and more effective treatments for cancer. Nanoparticles are advantageous in cancer therapy because they can specifically target cancer cells while sparing healthy cells. The possibility of toxicity and the difficulty of transporting nanoparticles to particular parts of the body are two possible disadvantages, though.

5.6.1 Importance of Safety Profiling

Safety profiling of nanoparticles is critical in cancer therapy because it ensures that the treatment is effective and safe for patients. Without proper safety profiling, there is a risk of harmful side effects and potential damage to healthy cells. For example, in a recent study, researchers found that certain types of nanoparticles used in cancer therapy had toxic effects on human liver cells. This highlights the need for rigorous safety profiling to prevent harm to patients. Additionally, statistics show that adverse reactions to nanoparticles are on the rise, making safety profiling even more crucial.

5.6.2 Current Methods of Safety Profiling

Currently, there are several methods used for safety profiling of nanoparticles in cancer therapy. One common method is in vitro testing, which involves testing the toxicity of nanoparticles on cell cultures. This method allows researchers to know about the toxic effects of nanoparticles on healthy cells before they are tested on humans. Another method is in vivo testing, which involves testing the toxicity of nanoparticles on living organisms. This method provides fair representation of the toxic effects of nanoparticles on humans. To see how nanoparticles and cells interact, researchers also employ imaging methods including atomic force microscopy (AFM) and transmission electron microscopy (TEM). These methods help researchers better understand the behavior of nanoparticles in biological systems and the possible hazardous impacts they may have.

5.6.3 Challenges and Limitations

One of the major challenges in safety profiling of nanoparticles in cancer therapy is their potential toxicity. While nanoparticles have shown promise in delivering drugs directly to cancer cells, they can also cause harm to healthy cells if not properly targeted. This is why it's important to develop methods that can accurately predict the toxicity of nanoparticles before they are used in clinical trials. The use of computer models, which can assist researchers in identifying potential hazardous effects of nanoparticles without the requirement for animal testing, is one approach to this problem.

Another challenge is the lack of standardized methods for safety profiling. Currently, different labs may use different techniques to evaluate the safety of nanoparticles, making it difficult to compare results across studies. To address this challenge, there is a need for more collaboration between researchers and regulatory agencies to establish standardized protocols for safety profiling. This will ensure that the results of safety studies are reliable and can be used to inform clinical decision-making.

5.6.4 Future Directions

As we look toward the future of safety profiling for nanoparticles in cancer therapy, there are several exciting possibilities on the horizon. The use of machine learning techniques to analyze massive datasets and spot potential safety issues before they become a problem is one of the most promising areas of research. By leveraging the power of artificial intelligence, we can streamline the safety profiling process and make it more efficient than ever before. Another area of research that shows great promise is the development of new imaging techniques that allow us to visualize nanoparticles inside the body in real-time. This will enable us to better understand how nanoparticles interact with different tissues and organs, and could lead to more targeted and effective cancer therapies in the future.

References

Awasthi R et al (2018) Nanoparticles in cancer treatment: opportunities and obstacles. Curr Drug Targets 19(14):1696–1709

Barros TP, Alderton WK, Reynolds H, Roach AG, Berghmans S (2008) Zebrafish: an emerging technology forin vivopharmacological assessment to identify potential safety liabilities in early drug discovery. Br J Pharmacol 154(7):1400–1413. https://doi.org/10.1038/bjp.2008.249

Batist G et al (2009) Safety, pharmacokinetics, and efficacy of CPX-1 liposome injection in patients with advanced solid tumors. Clin Cancer Res 15(2):692–700

Chen X et al (2021) Benefits of zebrafish xenograft models in cancer research. Front Cell Dev Biol 9:9

Cheong S et al (2011) Simple synthesis and functionalization of iron nanoparticles for magnetic resonance imaging. Angew Chem Int Ed Engl 50(18):4206–4209

Choo BKM, Kundap UP, bin Johan Arief MF, Kumari Y, Yap JL, Wong CP, Othman I, Shaikh MF (2019) Effect of newer anti-epileptic drugs (AEDs) on the cognitive status in pentylenetetrazol induced seizures in a zebrafish model. Prog Neuro-Psychopharmacol Biol Psychiatry 92:483–493

Dai Y-J et al (2014) Zebrafish as a model system to study toxicology. Environ Toxicol Chem 33 (1):11–17

Du M et al (2020) Design and development of spirulina polysaccharide-loaded nanoemulsions with improved the antitumor effects of paclitaxel. J Microencapsul 37(6):403–412

Eguiara A et al (2011) Xenografts in zebrafish embryos as a rapid functional assay for breast cancer stem-like cell identification. Cell Cycle 10(21):3751–3757

Hajdu SI (2012) A note from history: landmarks in history of cancer, part 4. Cancer 118(20): 4914–4928

Jaiswal M et al (2015) Nanoemulsion: an advanced mode of drug delivery system. 3 Biotech 5(2): 123–127

Keller ET, Murtha JM (2004) The use of mature zebrafish (Danio rerio) as a model for human aging and disease. Comp Biochem Physiol C Toxicol Pharmacol 138(3):335–341. https://doi.org/10.1016/j.cca.2004.04.001

Kraft JC et al (2014) Emerging research and clinical development trends of liposome and lipid nanoparticle drug delivery systems. J Pharm Sci 103(1):29–52

Li Z et al (2021) Application of animal models in cancer research: recent progress and future prospects. Cancer Manag Res 13:2455–2475

Liu Y et al (2023) Patient-derived xenograft models in cancer therapy: technologies and applications. Signal Transduct Target Ther 8(1):160

Lôbo GCNB et al (2021) Nanocarriers used in drug delivery to enhance immune system in cancer therapy. Pharmaceutics 13(8):1167

Mandelbaum J, Shestopalov IA, Henderson RE, Chau NG, Knoechel B, Wick MJ, Zon LI (2018) Zebrafish blastomere screen identifies retinoic acid suppression of MYB in adenoid cystic carcinoma. J Exp Med 215(10):2673–2685

Marques IJ, Weiss FU, Vlecken DH et al (2009) Metastatic behaviour of primary human tumours in a zebrafish xenotransplantation model. BMC Cancer 9:128. https://doi.org/10.1186/1471-2407-9-128

Misra R et al (2010) Cancer nanotechnology: application of nanotechnology in cancer therapy. Drug Discov Today 15(19):842–850

Nabil G et al (2019) Nano-engineered delivery systems for cancer imaging and therapy: recent advances, future direction and patent evaluation. Drug Discov Today 24(2):462–491

Nasrollahzadeh M et al (2019) Chapter 1—an Introduction to nanotechnology. In: Nasrollahzadeh M, Sajadi SM, Sajjadi M, Issaabadi Z, Atarod M (eds) Interface science and technology, vol 28. Elsevier, Amsterdam, pp 1–27

Nguyen AT et al (2012) An inducible krasV12 transgenic zebrafish model for liver tumorigenesis and chemical drug screening. Dis Model Mech 5(1):63–72

Obeid M et al (2018) Translational animal models for liver cancer. Am J Interv Radiol 2:1–7

Ou L et al (2016) Toxicity of graphene-family nanoparticles: a general review of the origins and mechanisms. Part Fibre Toxicol 13(1):57

Patton EE, Widlund HR, Kutok JL, Kopani KR, Amatruda JF, Murphey RD, Berghmans S, Mayhall EA, Traver D, Fletcher CD, Aster JC, Granter SR, Look AT, Lee C, Fisher DE, Zon LI (2005) BRAF mutations are sufficient to promote nevi formation and cooperate with p53 in the genesis of melanoma. Curr Biol 15(3):249–254. https://doi.org/10.1016/j.cub.2005.01.031

Pedziwiatr-Werbicka E et al (2021) Nanoparticles in combating cancer: opportunities and limitations. A brief review. Curr Med Chem 28(2):346–359

Phillips NLH, Roth TL (2019) Animal models and their contribution to our understanding of the relationship between environments, epigenetic modifications, and behavior. Genes (Basel) 10(1):47

Prabhakar U et al (2013) Challenges and key considerations of the enhanced permeability and retention effect for nanomedicine drug delivery in oncology. Cancer Res 73(8):2412–2417

Proetzel G et al (2014) Genetically engineered humanized mouse models for preclinical antibody studies. BioDrugs 28(2):171–180

Rivera M et al (2021) Patient-derived xenograft (PDX) models of colorectal carcinoma (CRC) as a platform for chemosensitivity and biomarker analysis in personalized medicine. Neoplasia 23(1):21–35

Saleem J et al (2018) Carbon-based nanomaterials for cancer therapy via targeting tumor microenvironment. Adv Healthc Mater 7(20):1800525

Silva GA (2004) Introduction to nanotechnology and its applications to medicine. Surg Neurol 61(3):216–220

Slamon DJ, Clark GM, Wong SG, Levin WJ, Ullrich A, McGuire WL (1987) Human breast cancer: correlation of relapse and survival with amplification of the HER-2/neu oncogene. Science 235 (4785):177–182. https://doi.org/10.1126/science.3798106

Smith ACH, Raimondi AR, Salthouse CD, Ignatius MS, Blackburn JS, Mizgirev IV, Storer NY, de Jong JLO, Chen AT, Zhou Y, Revskoy S, Zon LI, Langenau DM (2010) High-throughput cell transplantation establishes that tumor-initiating cells are abundant in zebrafish T-cell acute lymphoblastic leukemia. Blood 115(16):3296–3303. https://doi.org/10.1182/blood-2009-10-246488

Society AC (2018) Understanding what cancer is: ancient times to present. Retrieved 4/10/2023, from https://www.cancer.org/cancer/understanding-cancer/history-of-cancer/what-is-cancer.html

Tang Q et al (2014) Optimized cell transplantation using adult rag2 mutant zebrafish. Nat Methods 11(8):821–824

van Zandwijk N, Frank AL (2019) Awareness: potential toxicities of carbon nanotubes. Transl Lung Cancer Res 8(Suppl 4):S471–s472

Veldman MB, Lin S (2008) Zebrafish as a developmental model organism for pediatric research. Pediatr Res 64(5):470–476

Wang M, Thanou M (2010) Targeting nanoparticles to cancer. Pharm Res 62(2):90–99

Wu K et al (2023) Analyses of canine cancer mutations and treatment outcomes using real-world clinico-genomics data of 2119 dogs. NPJ Precis Oncol 7(1):8

Zon LI, Peterson RT (2005) In vivo drug discovery in the zebrafish. Nat Rev Drug Discov 4(1):35–44. https://doi.org/10.1038/nrd1606

Testing of Nanocarriers Loaded with Chemotherapeutic Agents: Cancer Patients and Clinical Trials

6

Firdos Alam Khan

Abstract

The materials which possess biocompatibility, better tumor targeting, and lesser toxicity are the most desirable nanomaterials for clinical applications. Some of the best-studied nanomaterials are metallic nanomaterials, polymeric nanomaterials or other organic and inorganic nanomaterials which showed promising applications in cancer diagnosis and cancer therapy. As per the report which was published in the 2016, there are 25 different types of nanomaterials-based carriers that are approved by the US Food and Drug Administration (FDA) or European Medicines Agency (EMA) for the clinical applications. In addition, more than 45 other nanoparticle-based nanocarriers that were approved by the USFDA and EMA but were presently being assessed in different phases of clinical trials. The nano-based drug carriers are tested to examine how anti-cancer drugs can be effectively targeted cancer tissues. There are four different phases (I, II, III, and IV) of clinical trials of testing nano-based carriers. In these past 3 years, several nanomedicine-based products are approved by US-FDA and EMA for their application in cancer diagnosis and cancer therapy. It is estimated that more than 75 new trials are under progress in different phases of clinical trials.

Keywords

Nanocarriers · Nanocarriers-loaded chemotherapeutic agents · Cancer patients · Clinical trials

F. A. Khan (✉)
Department of Stem Cell Research, Institute for Research and Medical Consultations (IRMC), Imam Abdulrahman Bin Faisal University, Dammam, Saudi Arabia
e-mail: fakhan@iau.edu.sa

F. A. Khan (ed.), *Nano Drug Delivery for Cancer Therapy*,
https://doi.org/10.1007/978-981-99-6940-1_6

6.1 Clinical Trials and Different Phases

There are four different phases (I, II, III, and IV) of clinical trials of testing nano-based carriers (Fig. 6.1). The application of nanomaterials in biomedical applications has attracted many researchers to develop special nanomaterials for clinical applications (Khan 2020a). One of the main attributes of the nanomaterials to be used as drug carriers is safety and better tumor targeting. The materials which possess biocompatibility, better tumor targeting, and lesser toxicity are the most desirable nanomaterials for clinical applications (Khan 2020b). Some of the best-studied nanomaterials are liposomes, polymeric, micelles, albumin-bound nanoparticles, and inorganic nanoparticles (Anselmo and Mitragotri 2019). There are about 45 different non-approved nanoparticles used in 80 different trials in various patients associated with radiation exposure, arthritis, pneumonia, amyloidosis, hepatitis, and fibrosis (Anselmo and Mitragotri 2019). It has been reported that out of 80 clinical trials, 28 trials have been successfully completed and 12 trials were terminated early due to failure (Anselmo and Mitragotri 2019). There are various application of different types of nanomaterials in clinical condition for example, these trials were used for tumor targeting. The main objective of the clinical trials is to examine whether prepared nanomaterials or nanocarriers possess desirable capabilities for cancer treatments in the different types of patients and at the same time also examine any side-effects or harmful effects associated with nanocarriers. It is very necessary to choose those nanomaterials which possess potential anti-cancer capabilities but also produce less harmful effects on human subjects. In order to start

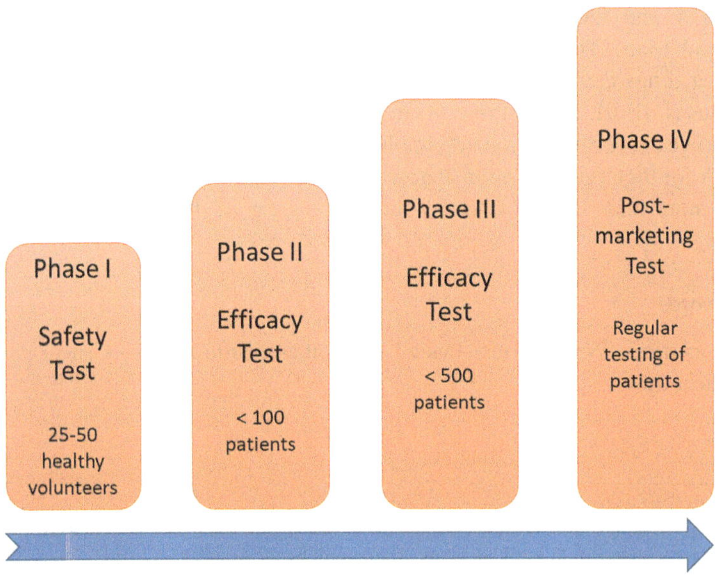

Fig. 6.1 Illustration of different phases of clinical trials

any clinical trials, it is necessary to evaluate the anticancer properties of nanocarriers in in vitro as well as in in vivo (animal) studies. Once in vitro and animal studies showed very good results with effective anticancer properties and less cytotoxic effects, then nanomaterials or nanocarriers may be taken for the clinical trials. Another important aspects of clinical trials is that prepared nanomaterials or nanocarriers should be produced as per the Good Laboratory Practice (GLP) guidelines for the pre-clinical trials (animal studies) and for the clinical trials nanomaterials or nanocarriers should be prepared as per the Good Manufacturing Practice (GMP) guidelines. To produce GLP and GMP grade nanomaterials, the facilities and staff who are involved in the production should be trained and approved by the respective country government agencies such as FDA is the organization which responsible GLP and GMP guidelines in the USA.

6.2 Application of Nanocarriers in Anticancer Drug Delivery

The application of different shapes, sizes, and types of nanomaterials (Fig. 6.2) have been involved in the production of nanocarriers and based on the published papers which are available in PubMed, NIH clinical trial sites, we have gathered the information and have described the individual type of nanomaterials for their application as the nanocarriers delivery the anti-cancer agents or drug or molecules for both diagnostic and therapy purposes.

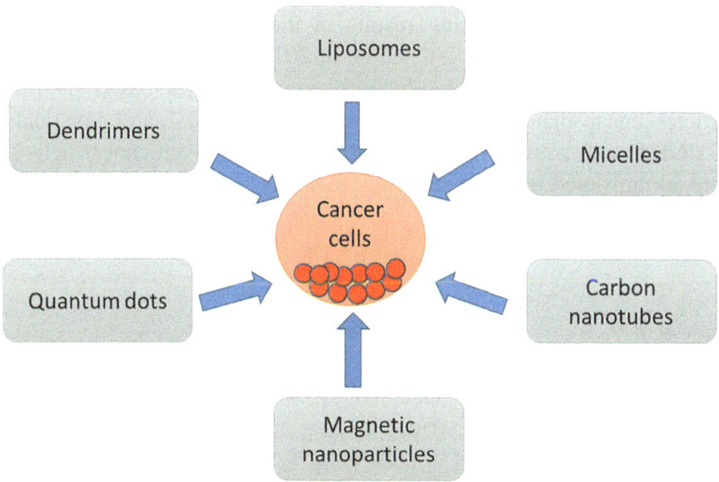

Fig. 6.2 Application of different types of nanomaterials in cancer treatment

6.2.1 Poly (Lactic-Co-Glycolic Acid) (PLGA) Nanocarriers

The application of Poly(lactic-co-glycolic acid) (PLGA)-nanoparticles in the development of nanocarriers has attracted many researchers as they possessed many advantages such better biocompatibility and better drug conjugation capabilities (Fig. 6.3) and with low cytotoxicity (Creemers et al. 2021; Kumari et al. 2010). It has been reported that nanoparticle which encapsulated antigens are taken up more efficiently by dendritic cells (Foged et al. 2002). In addition, it has been observed that PLGA-encapsulated peptides produced impact on major histocompatibility complex classes I and II molecules that produced functional and tumor-antigen specific ($CD8^{+)}$ and ($CD4^{+}$) T-cell responses as found in the pre-clinical studies (Waeckerle-Men and Groettrup 2005). The PLGA can be easily and rapidly internalized in the myeloid cells after systemic injection. The administration of PLGA nanocarriers contribute to disturb macrophage divergence for successful anticancer action (Reichel et al. 2019; Feng et al. 2019). The PLGA has also been used in the clinical conditions, where PLGA-encapsulated component which target the immune response, is the called as cancer-testis antigen NY-ESO-1 have been many advanced cancers lung, melanoma, bladder, prostate, ovarian, esophageal, breast, and gastric (Gnjatic et al. 2006; Caballero and Chen 2009; Sharma et al. 2003; Sugita et al. 2004; Vaughan et al. 2004; Iura et al. 2017; Gure et al. 2005; Tessari et al. 2018). The application of nanoparticles in drug encapsulation studied using long and short peptides (85–111, 117–143 and 157–165), which combined cover more than 80% of the European population for both class I and class II Human Leukocyte Antigen (HLA) alleles (IEDB 2019).

The application of NY-ESO-1-specific peptides as produced as per GMP guidelines have shown promising results in the cancer patients (Baumgaertner et al. 2016; Sabbatini et al. 2012; Chen et al. 2015; Odunsi et al. 2012). In one of the first human trials, immunomodulatory nanomedicines based on the PLGA nanoparticles encapsulated with tumor-antigen NY-ESO-1 and invariant natural killer T (iNKT) cell activator IMM60 showed profound anticancer activity when they were administered intravenously (Creemers et al. 2021).

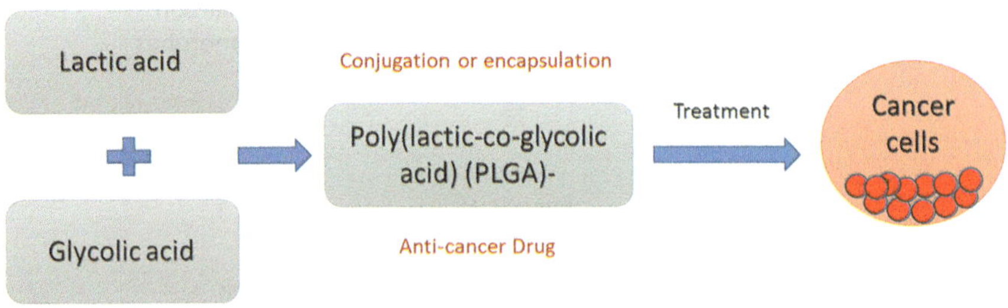

Fig. 6.3 Anti-cancer drug delivery by liposome conjugated nanoparticles

6.2.2 Magnetic Nanoparticle-Based Nanocarriers

Magnetic nanoparticles such as super-para-magnetic iron oxide nanoparticles (SPIONs) are one of the most extensively studied targeted nanomaterials for cancer treatment. The SPIONs have been used in drug delivery (Schleich et al. 2014; Wallyn et al. 2019), magnetic resonance imaging (MRI) (Hour et al. 2020; Du et al. 2020), and cellular-specific targeting (Ni et al. 2020), and with prolonged blood circulation (Ryu et al. 2014). It has been reported that SPIONs store in the tumor site through higher permeability under an external magnetic field (Zou et al. 2010). The application of SPIONs have been improved by the addition of folic acid (Montazerabadi et al. 2019; Shaghaghi et al. 2019), transferrin (Wang et al. 2020) to improve their targeting performance. SPIONs have been progressively studied for their excellent superparamagnetic, magnetic heating properties, and for enhanced magnetic resonance imaging (MRI). To deliver the drugs or anti-cancer agents to the cancer tissues, SPIONs can be conjugated with anti-cancer drugs and this nano delivery system has several benefits such as better magnetic targeted functionalization, cancer imaging, magnetic thermotherapy, and combined delivery of anti-cancer agents. There are few magnetic nanomaterial-based nanocarriers loaded with drugs were assessed in patients and did not produce convincing results, leading to termination of clinical trials of Phase II and III (Zhi et al. 2020).

6.2.3 Plant-Based Nanocarriers

There are numerous plant-based anticancer nanoformulations have been approved by the US-FDA, and there about 12 formulations are in different stages of clinical trials (Dhupal and Chowdhury 2020). Plant-based molecules are conjugated with anti-neoplastic drugs that can easily diffuse through the cell membrane escaping immune defense mechanism is known as passive targeting. It has been reported that plant-based-nanocomposites circulates through the cancer cell cytoplasm, diffused into intracellular organelles inducing oxidative stress and loss of mitochondrial membrane potential that led to apoptosis in cancer cells (Dhupal and Chowdhury 2020). The application of nanoparticles-based drug carrier having anti-neoplastic agents plus phyto-molecules have transformed conventional cancer therapy greatly advancing clinical practice. We have summarized the various plant-based molecules which are conjugated with anticancer drug to be used for the clinical applications as approved by US-FDA (www.ClinicalTrials.gov) (Table 6.1).

We have listed clinical trials using in cancer patients with nano-formulations as approved by US-FDA (www.ClinicalTrials.gov) (Table 6.2).

Table 6.1 List of phytochemical-based nano-formulation approved by the US FDA

Phytochemicals	Nano-formulation	Cancer type
Paclitaxel	Albumin-stabilized nanoparticle formulation	Breast cancer, non-small cell lung cancer, pancreatic cancer
Paclitaxel	Paclitaxel liposome injection of cholesterol and lecithin	Breast cancer, non-small cell lung cancer, ovarian cancer
Paclitaxel	Paclitaxel-loaded polymeric micelle	Breast cancer, gastric cancer, non-small cell lung cancer, ovarian cancer, bladder cancer, pancreatic cancer
Paclitaxel	Paclitaxel-incorporating micellar nanoparticle	Breast cancer, solid tumors
Irinotecan	Irinotecan hydrochloride liposome	Pancreatic cancer, refractory pediatric malignancies, NSCLC, advanced breast cancer
Vinorelbine tartrate	Liposomal vinorelbine tartrate	Non-small cell lung cancer, breast cancer, ovarian cancer
Vincristine	Vincristine sulfate liposome	Acute lymphoblastic leukemia (ALL)
Camptothecin	Liposomal 9-nitrocamptothecin	Endometrial cancer, Ewing's sarcoma
Camptothecin	Polymer nanoparticle formulation of camptothecin	Rectal cancer, advanced non-small cell lung cancer, metastatic gastroesophageal cancer
Curcumin	Liposomal curcumin	Solid tumors
Curcumin	Curcumin/doxorubicin-encapsulating nanoparticle	Advanced solid tumors
Docetaxel	Liposomal docetaxel	Solid tumors
Docetaxel	Docetaxel polymeric micelle	Head and neck squamous cell carcinoma, bladder canceresophageal carcinoma
Docetaxel	Docetaxel containing CriPec nanoparticles	Solid tumorsovarian cancer
Docetaxel	Docetaxel- polymeric nanoparticles (DOC-PNP)	Advanced solid cancer
Docetaxel	Docetaxel-loaded nanopharmaceuticals	Solid tumor

6.3 Clinical Trials of Nano-Based Formulation for Cancer Diagnosis and Therapy-Completed Trials

Till March 2023, so far 26 clinical trials have been completed as per the www.ClinicalTrials.gov website. Trials include both diagnostic assay of the different types of cancer and also treatment of various types of cancer. The list of all 26 trials with basic information is given in Table 6.3.

Table 6.2 List of clinical trials using in cancer patients with nano-formulations as approved by the

Name	Material description	Indication(s)
Adagen®/pegademase bovine (sigma-tau pharmaceuticals)	PEGylated adenosine deaminase enzyme	Severe combined immunodeficiency disease (SCID)
Cimzia®/certolizumab pegol (UCB)	PEGylated antibody fragment (Certolizumab)	Crohn's disease; Rheumatoid arthritis; Psoriatic arthritis; Ankylosing spondylitis
Copaxone®/Glatopa (Teva)	Random copolymer of L-glutamate, L alanine, L-lysine and L-tyrosine	Multiple sclerosis (MS)
Eligard® (Tolmar)	Leuprolide acetate and polymer (PLGH (poly (DL-lactide-coglycolide)))	Prostate cancer
Macugen®/Pegaptanib (Bausch & Lomb)	PEGylated anti-VEGF aptamer (vascular endothelial growth factor) aptamer	Macular degeneration, neovascular age-related
Mircera®/Methoxy polyethylene glycol-epoetin beta (Hoffman-La Roche)	Chemically synthesized ESA (erythropoiesis-stimulating agent)	Anemia associated with chronic kidney disease
Neulasta®/pegfilgrastim (Amgen)	PEGylated GCSF protein	Neutropenia, chemotherapy induced
Pegasys® (Genentech)	PEGylated IFN alpha-2a protein	Hepatitis B; hepatitis C
PegIntron® (Merck)	PEGylated IFN alpha-2a protein	Hepatitis C
Renagel®[sevelamer hydrochloride]/ Renagel®[sevelamer carbonate] (Sanofi)	Poly(allylamine hydrochloride)	Chronic kidney disease
Somavert®/pegvisomant (Pfizer)	PEGylated HGH receptor antagonist	Acromegaly
Oncaspar®/pegaspargase (Enzon pharmaceuticals)	Polymer–protein conjugate (PEGylated L-asparaginase)	Acute lymphoblastic leukemia
Krystexxa®/pegloticase (horizon)	Polymer–protein conjugate (PEGylated porcine-like uricase)	Chronic gout
Plegridy® (Biogen)	Polymer–protein conjugate (PEGylated IFN beta-1a)	Multiple sclerosis
ADYNOVATE (Baxalta)	Polymer–protein conjugate (PEGylated factor VIII)	Hemophilia
Zilretta	Triamcinolone acetonide with a poly lactic-*co*-glycolic acid (PLGA) matrix microspheres	Osteoarthritis (OA) of the knee

(continued)

Table 6.2 (continued)

Name	Material description	Indication(s)
Rebinyn	Coagulation fator IX (recombinant) GlycoPEGylated	Control and prevention of bleeding episodes and prevention of bleeding in the perioperative setting for hemophilia B patients
DaunoXome® (Galen)	Liposomal daunorubicin	Karposi's sarcoma
DepoCyt© (sigma-tau)	Liposomal cytarabine	Lymphomatous meningitis
Marqibo® (Onco TCS)	Liposomal vincristine	Acute lymphoblastic leukemia
Onivyde® (Merrimack)	Liposomal irinotecan	Pancreatic cancer
AmBisome® (Gilead Sciences)	Liposomal amphotericin B	Fungal/protozoal infections
DepoDur® (Pacira pharmaceuticals)	Liposomal morphine sulphate	Analgesia (post-operative)
Visudyne® (Bausch and Lomb)	Liposomal verteporfin	Macular degeneration, wet age-related; myopia; ocular histoplasmosis
Doxil®/Caelyx™ (Janssen)	Liposomal doxorubicin	Karposi's sarcoma; Ovarian cancer; Multiple myeloma
Abelcet® (sigma-tau)	Liposomal amphotericin B lipid complex	Fungal infections
Curosurf®/Poractant alpha (Chiesei farmaceutic)	Liposome–proteins SP-B and SP-C	Pulmonary surfactant for respiratory distress syndrome
Vyxeos (jazz pharma)	Lipossomal combination of daunorubicin and cytarabine	Acute myeloid leukemia (AML) or AMLA with myelodysplasia-related changes (AML-MRC)
Estrasorb™ (Novavax)	Micellar estradiol	Menopausal therapy
Abraxane®/ABI-007 (Celgene)	Albumin-bound paclitaxel nanoparticles	Breast cancer; NSCLC; Pancreatic cancer
Ontak® (Eisai Inc)	Engineered protein combining IL-2 and diphtheria toxin	Cutaneous T-cell Lymphoma
Emend® (Merck)	Aprepitant	Antiemetic
Tricor® (Lupin Atlantis)	Fenofibrate	Hyperlipidemia
Rapamune® (Wyeth pharmaceuticals)	Sirolimus	Immunosuppressant
Megace ES® (par pharmaceuticals)	Megestrol acetate	Anti-anorexic
Avinza® (Pfizer)	Morphine sulphate	Psychostimulant
Focalin XR® (Novartis)	Dexmethylphenidate HCl	Psychostimulant
Ritalin LA® (Novartis)	Methylphenidate HCl	Psychostimulant
Zanaflex® (Acorda)	Tizanidine HCl	Muscle relaxant
Vitoss® (Stryker)	Calcium phosphate	Bone substitute
Ostim® (Heraseus Kulzer)	Hydroxyapatite	Bone substitute

(continued)

Table 6.2 (continued)

Name	Material description	Indication(s)
OsSatura® (IsoTis Orthobiologics)	Hydroxyapatite	Bone substitute
NanOss® (Rti surgical)	Hydroxyapatite	Bone substitute
EquivaBone® (Zimmer Biomet)	Hydroxyapatite	Bone substitute
Invega® Sustenna® (Janssen pharms)	Paliperidone palmitate	Schizophrenia; Schizoaffective disorder
Ryanodex® (eagle pharmaceuticals)	Dantrolene sodium	Malignant hypothermia
Inorganic and metallic nanoparticles Nanotherm® (MagForce)	Iron oxide	Glioblastoma
Feraheme™/ferumoxytol (AMAG pharmaceuticals)	Ferumoxytol SPION with polyglucose sorbitol carboxymethylether	Deficiency anemia iron deficiency in chronic kidney disease (CKD)
Venofer® (Luitpold pharmaceuticals)	Iron sucrose	Iron deficiency in chronic kidney disease (CKD)
Ferrlecit® (Sanofi Avertis)	Sodium ferric gluconate	Iron deficiency in chronic kidney disease (CKD)
INFeD® (Sanofi Avertis)	Iron dextran (low MW)	Iron deficiency in chronic kidney disease (CKD)
DexIron®/Dexferrum® (Sanofi Avertis)	Iron dextran (low MW)	Iron deficiency in chronic kidney disease (CKD)
Feridex®/Endorem® (AMAG pharmaceuticals)	SPION coated with dextran	Imaging agent
GastroMARK™; umirem® (AMAG pharmaceuticals)	SPION coated with silicone	Imaging agent

6.4 FDA Approved Nano-Based Materials for Cancer

There are few nanomedicine-based products which are approved and regulated by the US-FDA (Patra et al. 2018). In addition, there are also number of nano-based drug molecules which are under various phases of clinical trials for cancer diagnosis and cancer therapy. There 18 different types of nano-based drug delivery materials have been approved for chemotherapeutics applications; 15 were approved for antimicrobial agents and 28 were approved for different medical applications such as psychological diseases, autoimmune (Patra et al. 2018). The list of US-FDA approved nano-based materials is given in Table 6.2.

Table 6.3 List of completed clinical trials

Study 1	
Title	Measuring Urinary Nano Particles and Metallic Compounds in TCC Patients
Status	Completed
Study results	No results available
Conditions	Bladder cancer
Interventions	Other: Without intervention
Locations	Carmel Lady Davis medical center, Haifa, Israel
URL	https://ClinicalTrials.gov/show/NCT02811952
Study 2	
Title	Nano drug interventional therapy using digital subtraction angiography(DSA) for lung carcinoma
Status	Completed
Study results	No results available
Conditions	Lung cancer
Interventions	Procedure: Interventional therapy
Locations	Central laboratory in Fuda cancer hospital, Guangzhou, Guangdong, China
URL	https://ClinicalTrials.gov/show/NCT02449122
Study 3	
Title	Nano drug interventional therapy using digital subtraction angiography(DSA) for liver carcinoma
Status	Completed
Study results	No results available
Conditions	Liver cancer
Interventions	Procedure: Interventional therapy
Locations	Central laboratory in Fuda cancer hospital, Guangzhou, Guangdong, China
URL	https://ClinicalTrials.gov/show/NCT02449109
Study 4	
Title	Electroporation (NanoKnife) as treatment for advanced pancreatic cancer
Status	Completed
Study results	No results available
Conditions	Pancreatic cancer
Interventions	Procedure: Electroporation/ Nano knife treatment|device: NanoKnife
Locations	Surgical Department of Gastroenterology, Aalborg University Hospital, Aalborg, Nothern Jutland, Denmark
URL	https://ClinicalTrials.gov/show/NCT02079623
Study 5	
Title	Real time in vivo dosimetry for gynecologic brachytherapy
Status	Completed
Study results	No results available
Conditions	Cancer of the cervix, cancer of the endometrium
Interventions	Device: Novel nano-scintillator fiber-optic dosimeter (nanoFOD)
Locations	Duke cancer center, Durham, North Carolina, United States
URL	https://ClinicalTrials.gov/show/NCT02040155

(continued)

Table 6.3 (continued)

Study 6	
Title	A micro/Nano device for exhaled breath analysis
Status	Completed
Study results	No results available
Conditions	Lung cancer|pulmonary disease
Interventions	
Locations	James Graham Brown cancer center-University of Louisville, Louisville, Kentucky, United States
URL	https://ClinicalTrials.gov/show/NCT02123030
Study 7	
Title	Real-time in vivo body surface dosimetry for external beam radiation therapy
Status	Completed
Study results	No results available
Conditions	Cancer of the gastrointestinal, genitourinary or gynecologic systems
Interventions	Other: Novel nano-scintillatorfiber-optic dosimeter (nanoFOD)
Locations	Duke University Medica center, Durham, North Carolina, United States
URL	https://ClinicalTrials.gov/show/NCT02407977
Study 8	
Title	Percutaneous irreversible Electroportion for Unresectable breast cancer
Status	Completed
Study results	No results available
Conditions	Breast cancer
Interventions	Device: NanoKnife LEDC system
Locations	FUDA cancer hospital, Guangzhou, Guangdong, China
URL	https://ClinicalTrials.gov/show/NCT02340858
Study 9	
Title	Targeted atomic Nano-generators (Actinium-225-labeled humanized anti-CD33 monoclonal antibody HuM195) in patients with advanced myeloid malignancies
Status	Completed
Study results	No results available
Conditions	Leukemia|myelodysplastic syndrome
Interventions	Biological: ACTINIUM-225-LABELED HUMANIZED ANTI-CD33 MONOCLONAL ANTIBODY HuM195
Locations	Memorial Sloan Kettering cancer center, New York, New York, United States
URL	https://ClinicalTrials.gov/show/NCT00672165
Study 10	
Title	Assessment of stromal response to nab-paclitaxel in combination with gemcitabine in pancreatic cancer
Status	Completed
Study results	No results available
Conditions	Pancreatic cancer
Interventions	Drug: Gemcitabine plus nab-paclitaxel
Locations	Centro integral Oncologico Clara Campal, Madrid, Spain
URL	https://ClinicalTrials.gov/show/NCT01442974

(continued)

Table 6.3 (continued)

Study 11	
Title	Neoadjuvant Pembrolizumab(Pbr)/nab-paclitaxel followed by Pbr/Epirubicin/cyclophosphamide in TNBC
Status	Completed
Study results	No results available
Conditions	Malignant neoplasm of breast
Interventions	Drug: Pembrolizumab|drug: Nab-paclitaxel|drug: Epirubicin|drug: Cyclophosphamide
Locations	Department of Gynecology, Tübingen university hospital, Tübingen, Baden-Württemberg, Germany|Department of Gynecology and Obstetrics, Erlangen university hospital, Erlangen, Bavaria, Germany|Department for Gynecology and Obstetrics, Marienhospital Bottrop gGmbH, Bottrop, North Rhine-Westphalia, Germany|Department of Gynecology and Obstetrics, HELIOS hospital Berlin Buch GmbH, Berlin, Germany
URL	https://ClinicalTrials.gov/show/NCT03289819
Study 12	
Title	Nano-pulse stimulation (NPS) in seborrheic keratosis optimization study
Status	Completed
Study results	No results available
Conditions	Lesion Skin|seborrheic keratosis|Skin lesion|benign Skin tumor
Interventions	Device: Nano-pulse stimulation (NPS)
Locations	Clear Dermatology & Aesthetics Center/InvestigateMD, Scottsdale, Arizona, United States|mountain dermatology specialists, Edwards, Colorado, United States|SKIN Associates of South Florida, Coral Gables, Florida, United States|Chicago cosmetic surgery and dermatology, Chicago, Illinois, United States|skin care physicians, Chestnut Hill, Massachusetts, United States|Zel Skin & Laser Specialists, Edina, Minnesota, United States|dermatology, Laser & Vein Specialists of the Carolinas, Charlotte, North Carolina, United States
URL	https://ClinicalTrials.gov/show/NCT04249115
Study 13	
Title	Radiosensitization of multiple brain metastases using AGuIX gadolinium based nanoparticles
Status	Completed
Study results	No results available
Conditions	Brain metastases
Interventions	Drug: AGuIX|radiation: Whole brain radiation therapy
Locations	University hospital Grenoble, Grenoble, France
URL	https://ClinicalTrials.gov/show/NCT02820454
Study 14	
Title	Diagnostic and prognostic accuracy of gold nanoparticles in salivary gland Tumours
Status	Completed
Study results	No results available
Conditions	Carcinoma ex pleomorphic adenoma of salivary glands|pleomorphic adenoma of salivary glands

(continued)

Table 6.3 (continued)

Interventions	Diagnostic test: CD24-gold nanocomposite expression using real-time quantitative polymerase chain reaction|diagnostic test: Non-conjugated CD24 expression using real-time quantitative polymerase chain reaction
Locations	Faculty of Dentistry, October 6 university, Giza, Egypt
URL	https://ClinicalTrials.gov/show/NCT04907422
Study 15	
Title	Development of a nanosecond pulsed electric field system to treat skin cancer
Status	Completed
Study results	No results available
Conditions	Basal cell carcinomas
Interventions	Device: PulseCure pulse generator and Derm-pulse electrode
Locations	Children's hospital research center Oakland, Oakland, California, United States
URL	https://ClinicalTrials.gov/show/NCT01463709
Study 16	
Title	Safety and efficacy of vitamin C infusion in combination with local mEHT to treat non-small-cell lung cancer
Status	Completed
Study results	No results available
Conditions	Carcinoma, non-small-cell lung|stage IIIA non-small cell lung cancer|stage IIIB non-small cell lung cancer|stage IV non-small cell lung cancer|recurrent non-small cell lung cancer
Interventions	Drug: Vitamin C|device: Modulated electro-hyperthermia (mEHT)|other: Supportive care
Locations	Clifford Hospital, Guangzhou, China
URL	https://ClinicalTrials.gov/show/NCT02655913
Study 17	
Title	Nano-pulse stimulation (NPS) in seborrheic keratosis study
Status	Completed
Study results	No results available
Conditions	Lesion skin|seborrheic keratosis
Interventions	Device: Nano-pulse stimulation (NPS)
Locations	Premier plastic surgery, San Mateo, California, United States|skin care physicians, Chestnut Hill, Massachusetts, United States|Zel Skin & Laser Specialists, Edina, Minnesota, United States|Laser & Dermatologic Surgery Center, Chesterfield, Missouri, United States
URL	https://ClinicalTrials.gov/show/NCT03846531
Study 18	
Title	Atezolizumab in combination with carboplatin plus Pemetrexed in chemotherapy-naïve patients with asymptomatic brain metastasis
Status	Completed
Study results	No results available
Conditions	Non-small cell lung cancer stage IV
Interventions	Drug: Atezolizumab
Locations	Hospital general Universitario de Elche, Elche, Alicante, Spain|hospital Universitari Germans tries i Pujol, Badalona, Barcelona, Spain|hospital Universitari de Bellvitge, L'Hospitalet de Llobregat, Barcelona, Spain|Complejo

(continued)

Table 6.3 (continued)

	Hospitalario de la Coruña, La Coruña, Coruña, Spain|hospital insular de gran Canaria, Las Palmas de gran Canaria, Las Palmas, Spain|Complexo Hospitalario Universitario de Vigo, Vigo, Pontevedra, Spain|hospital general de Alicante, Alicante, Spain|H.U.Vall D'Hebrón, Barcelona, Spain|hospital de Santa Creu i Sant Pau, Barcelona, Spain|hospital Dr. Josep Trueta, Girona, Spain|hospital Fundación Jiménez Díaz, Madrid, Spain|H. La Paz, Madrid, Spain|hospital general Universitario de Valencia, Valencia, Spain|hospital La Fe, Valencia, Spain|hospital Clínico Universitario de Valladolid, Valladolid, Spain
URL	https://ClinicalTrials.gov/show/NCT03526900
Study 19	
Title	Breast cancer diet intervention study
Status	Completed
Study results	No results available
Conditions	Early-stage breast cancer|carbohydrate engorgement|proliferation|insulin resistance|estrogen receptor-positive breast cancer
Interventions	Dietary supplement: PreOP
Locations	
URL	https://ClinicalTrials.gov/show/NCT03886389
Study 20	
Title	Local mEHT + TCM versus intraperitoneal Chemoinfusion in treatment of malignant ascites: Phase II RCT
Status	Completed
Study results	Has results
Conditions	Peritoneal neoplasms|ascites|Yang deficiency|yin deficiency
Interventions	Device: Modulated electro-hyperthermia (mEHT)|dietary supplement: TCM herbal decoction (Shi pi)|drug: IPCI (CDDP+5FU)
Locations	Clifford Hospital, Guangzhou, Guangdong, China
URL	https://ClinicalTrials.gov/show/NCT02638051
Study 21	
Title	CellFX Treat & Resect low-Risk BCC feasibility study
Status	Completed
Study results	Has results
Conditions	BCC—basal cell carcinoma|BCC|excision margin
Interventions	Device: CellFX system
Locations	Surgical dermatology group, Vestavia Hills, Alabama, United States|Moy-Fincher-Chipps dermatology, Beverly Hills, California, United States|Palm Harbor dermatology, Clearwater, Florida, United States|SkinCare physicians, Chestnut Hill, Massachusetts, United States|clinical research Center of the Carolinas, Charleston, South Carolina, United States
URL	https://ClinicalTrials.gov/show/NCT04918381
Study 22	
Title	CellFX system for the treatment of cutaneous non-genital warts
Status	Completed
Study results	No results available
Conditions	Warts|warts hand|verruca
Interventions	Device: CellFX device

(continued)

Table 6.3 (continued)

Locations	Investigate MD, LLC, Scottsdale, Arizona, United States|Scripps clinic Carmel Valley, San Diego, California, United States|skin search of Rochester, Inc., Rochester, New York, United States|dermatology, Laser & Vein Specialists of the Carolinas, PLLC, Charlotte, North Carolina, United States|Austin Institute for Clinical Research, Inc., Pflugerville, Texas, United States
URL	https://ClinicalTrials.gov/show/NCT04554394
Study 23	
Title	A study of safety, efficacy and pharmacodynamics of Azacitidine in children and young adults with acute myeloid leukemia
Status	Completed
Study results	No results available
Conditions	Leukemia, myeloid, acute
Interventions	Drug: Azacitidine|other: Control arm
Locations	Rigshospitalet, Copenhagen, Denmark|Charite Berlin, Berlin, Germany|Universitatsklinikum Essen, Essen, Germany|Klinikum der Johann Wolfgang Goethe-Universität Frankfurt/Main, Frankfurt am Main, Germany|Universitatsklinik, Freiburg, Germany|medical School of Hannover, Hannover, Germany|VU University medical center, Amsterdam, Netherlands|Erasmus University medical center, Rotterdam, Netherlands
URL	https://ClinicalTrials.gov/show/NCT02450877
Study 24	
Title	CellFX comparison to cryosurgery in cutaneous non-genital common warts
Status	Completed
Study results	No results available
Conditions	Warts|warts hand|verruca
Interventions	Device: CellFX system device|device: Cryosurgery liquid nitrogen sprayer
Locations	Investigate MD, Scottsdale, Arizona, United States|Moy-Fincher-Chipps dermatology, Beverly Hills, California, United States|AboutSkin dermatology and DermSurgery, Greenwood Village, Colorado, United States|Palm Harbor dermatology, Clearwater, Florida, United States|oak dermatology, Joliet, Illinois, United States|Juva Skin & Laser Center, New York, New York, United States|dermatology, Laser & Vein Specialists of the Carolinas, Charlotte, North Carolina, United States|Dermatology & Laser Center of Charleston, Charleston, South Carolina, United States|Austin Institute for Clinical Research, Inc., Houston, Texas, United States|Austin Institute for Clinical Research, Inc., Pflugerville, Texas, United States
URL	https://ClinicalTrials.gov/show/NCT04738734
Study 25	
Title	Study of TG02 in elderly newly diagnosed or adult relapsed patients with anaplastic astrocytoma or glioblastoma
Status	Completed
Study results	No results available
Conditions	Astrocytoma, grade III|glioblastoma
Interventions	Drug: TG02|radiation: Radiation therapy|drug: Temozolomide
Locations	Universitaetskliniken der Uni Wien—Universitaetsklinikum Wien—AKH uniklinieken, Vienna, Austria|CHU de Lyon—CHU Lyon—Hopital neurologique Pierre Wertheimer, Bron, France|CHRU de Lille, Lille, France|

(continued)

Table 6.3 (continued)

	assistance Publique—Hopitaux de Marseille—Hôpital de La Timone, Marseille, France|Universitaetsklinikum Bonn, Bonn, Germany|Klinikum Der J.W. Goethe Universitaet-Klinik und Poliklinik fur Neurochirurgie, Frankfurt, Germany| Universitaetsklinikum Heidelberg—UniversitaetsKlinikum Heidelberg—head hospital, Heidelberg, Germany|Universitaetskliniken Regensburg—Universitaetsklinikum Regensburg, Regensburg, Germany|Erasmus MC cancer institute—location Daniel den hoed, Rotterdam, Netherlands|UniversitaetsSpital Zurich, Zürich, Switzerland
URL	https://ClinicalTrials.gov/show/NCT03224104
Study 26	
Title	Antisecretory factor in primary glioblastoma 1
Status	Completed
Study results	No results available
Conditions	Glioblastoma|cerebral edema|chemotherapy effect
Interventions	Dietary supplement: Salovum
Locations	Department of Neurosurgery, Lund, Sweden
URL	https://ClinicalTrials.gov/show/NCT04116138

6.4.1 Polymer Nanoparticles-Synthetic Polymer Particles Combined with Drugs or Biologics

Name	Material description	Nanoparticle advantage	Indication(s)	Year approved
Eligard® (Tolmar)	Leuprolide acetate and polymer (PLGH (poly (DL-lactide-coglycolide)))	Controlled delivery of payload with longer circulation time	Prostate cancer	2002
Neulasta®/ pegfilgrastim (Amgen)	PEGylated GCSF protein	Improved stability of protein through PEGylation	Neutropenia, chemotherapy induced	2002
Oncaspar®/ pegaspargase (Enzon pharmaceuticals)	Polymer–protein conjugate (PEGylated L-asparaginase)	Improved stability of protein through PEGylation	Acute lymphoblastic leukemia	1994
Krystexxa®/ pegloticase (horizon)	Polymer–protein conjugate (PEGylated porcine-like uricase)	Improved stability of protein through PEGylation; introduction of unique mammalian protein	Chronic gout	2010
ADYNOVATE (Baxalta)	Polymer–protein conjugate (PEGylated factor VIII)	Improved stability of protein through PEGylation	Hemophilia	2015

(continued)

Name	Material description	Nanoparticle advantage	Indication(s)	Year approved
Rebinyn	Coagulation fator IX (recombinant) GlycoPEGylated	Effective control in 95% of bleeding episodes; 98% of bleeds were treated with 1–2 infusions	Control and prevention of bleeding episodes and prevention of bleeding in the perioperative setting for hemophilia B patients	2017

6.4.2 Liposome Formulations Combined with Drugs or Biologics

Name	Material description	Nanoparticle advantage	Indication(s)	Year approved
DaunoXome® (Galen)	Liposomal daunorubicin	Increased delivery to tumor site; lower systemic toxicity arising from side-effects	Karposi's sarcoma	1995
DepoCyt© (sigma-tau)	Liposomal cytarabine	Increased delivery to tumor site; lower systemic toxicity arising from side-effects	Lymphomatous meningitis	1996
Marqibo® (Onco TCS)	Liposomal vincristine	Increased delivery to tumour site; lower systemic toxicity arising from side-effects	Acute lymphoblastic leukemia	2012
Onivyde® (Merrimack)	Liposomal irinotecan	Increased delivery to tumor site; lower systemic toxicity arising from side-effects	Pancreatic cancer	2015
Doxil®/Caelyx™ (Janssen)	Liposomal doxorubicin	Improved delivery to site of disease; decrease in systemic toxicity of free drug	Karposi's sarcoma;	1995
Vyxeos (jazz pharma)	Lipossomal combination of daunorubicin and cytarabine	Sustained release of the molecules and co-loaded two molecules with synergistic anti-tumor activity	Acute myeloid leukemia (AML) or AMLA with myelodysplasia-related changes (AML-MRC)	2017

6.4.3 Micellar Nanoparticles Combined with Drugs or Biologics

Name	Material description	Nanoparticle advantage	Indication(s)	Year approved
Estrasorb™ (Novavax)	Micellar estradiol	Controlled delivery of therapeutic	Menopausal therapy	2003
Abraxane®/ ABI-007 (Celgene)	Albumin-bound paclitaxel nanoparticles	Improved solubility; improved delivery to tumor	Breast cancer	2005
Ontak® (Eisai Inc)	Engineered protein combining IL-2 and diphtheria toxin	Targeted T cell specificity; lysosomal escape	Cutaneous T-cell	1999

6.4.4 Nanocrystals

Name	Material description	Nanoparticle advantage	Indication(s)	Year approved
Ryanodex® (eagle pharmaceuticals)	Dantrolene sodium	Faster administration at higher doses	Malignant hypothermia	2014
Inorganic and metallic nanoparticles Nanotherm® (MagForce)	Iron oxide	Allows cell uptake and introduces superparamagnetism	Glioblastoma	2010
Feridex®/Endorem® (AMAG pharmaceuticals)	SPION coated with dextran	Superparamagnetic character	Imaging agent	1996 (2008)
GastroMARK™; umirem® (AMAG pharmaceuticals)	SPION coated with silicone	Superparamagnetic character	Imaging agent	2001 (2009)

6.5 Conclusion

There are many useful applications of nanomaterials especially in drug delivery as the nanocarriers. One of the main advantages of the nano-based drug delivery system is better biodegradability, and better biocompatibility with low cell toxicity. Some of the best-studied nanomaterials are metallic nanomaterials, polymeric nanomaterials or other organic and inorganic nanomaterials which showed promising applications in cancer diagnosis, and cancer therapy. As per the report which was published in 2016, there are 25 different types of nanomaterials-based carriers that are approved by the Food and Drug Administration (FDA) or European Medicines Agency (EMA) for the clinical applications. In addition, more than 45 other nanoparticle-based nanocarriers that were approved by the FDA and EMA but were presently being assessed in different phases of clinical trials. The nano-based drug carriers are tested to examine how anti-cancer drugs can be effectively targeted cancer tissues. It is estimated that more than 75 new trials are under progress in different phases of clinical trials. There regulatory mechanisms for nanomedicine-based products are constantly improving where safety and toxicity issues are taken seriously for the development of nanomedicine products.

References

Anselmo AC, Mitragotri S (2019) Nanoparticles in the clinic: an update. Bioeng Transl Med 4(3): e10143. https://doi.org/10.1002/btm2.10143; PMID: 31572799; PMCID: PMC6764803

Baumgaertner P, Costa Nunes C, Cachot A et al (2016) Vaccination of stage III/IV melanoma patients with long NY-ESO-1 peptide and CpG-B elicits robust $CD8^+$ and $CD4^+$ T-cell responses with multiple specificities including a novel DR7-restricted epitope. Onco Targets Ther 5:e1216290. https://doi.org/10.1080/2162402X.2016.1216290

Caballero OL, Chen Y-T (2009) Cancer/testis (CT) antigens: potential targets for immunotherapy. Cancer Sci 100:2014–2021. https://doi.org/10.1111/j.1349-7006.2009.01303.x

Chen J-L, Dawoodji A, Tarlton A et al (2015) NY-ESO-1 specific antibody and cellular responses in melanoma patients primed with NY-ESO-1 protein in ISCOMATRIX and boosted with recombinant NY-ESO-1 fowlpox virus. Int J Cancer 136:E590–E601. https://doi.org/10.1002/ijc.29118

Creemers JHA, Pawlitzky I, Grosios K, Gileadi U, Middleton MR, Gerritsen WR, Mehra N, Rivoltini L, Walters I, Figdor CG, Ottevanger PB, de Vries IJM (2021) Assessing the safety, tolerability and efficacy of PLGA-based immunomodulatory nanoparticles in patients with advanced NY-ESO-1-positive cancers: a first-in-human phase I open-label dose-escalation study protocol. BMJ Open 11(11):e050725. https://doi.org/10.1136/bmjopen-2021-050725

Dhupal M, Chowdhury D (2020) Phytochemical-based nanomedicine for advanced cancer theranostics: perspectives on clinical trials to clinical use. Int J Nanomedicine 15:9125–9157. https://doi.org/10.2147/IJN.S259628; PMID: 33244231; PMCID: PMC7683832

Du C, Liu X, Hu H et al (2020) Dual-targeting and excretable ultrasmall SPIONs for T1-weighted positive MR imaging of intracranial glioblastoma cells by targeting the lipoprotein receptor-related protein. J Mater Chem B 8(11):2296–2306. https://doi.org/10.1039/c9tb02391g

Feng X, Xu W, Li Z et al (2019) Immunomodulatory nanosystems. Adv Sci 6:1900101. https://doi.org/10.1002/advs.201900101

Foged C, Sundblad A, Hovgaard L (2002) Targeting vaccines to dendritic cells. Pharm Res 19:229–238. https://doi.org/10.1023/A:1014474414097

Gnjatic S, Nishikawa H, Jungbluth AA et al (2006) Ny-Eso-1: review of an immunogenic tumor antigen. Adv Cancer Res 95:1–30. https://doi.org/10.1016/S0065-230X(06)95001-5

Gure AO, Chua R, Williamson B et al (2005) Cancer-testis genes are coordinately expressed and are markers of poor outcome in non-small cell lung cancer. Clin Cancer Res 11:8055–8062. https://doi.org/10.1158/1078-0432.CCR-05-1203

Hour FQ, Moghadam AJ, Shakeri-Zadeh A et al (2020) Magnetic targeted delivery of the SPIONs-labeled mesenchymal stem cells derived from human Wharton's jelly in Alzheimer's rat models. J Control Release 321:430–441. https://doi.org/10.1016/j.jconrel.2020.02.035

IEDB (2019) IEDB analysis resource: National Institute of allergy and infectious diseases; http://tools.iedb.org/population/

Iura K, Maekawa A, Kohashi K et al (2017) Cancer-testis antigen expression in synovial sarcoma: NY-ESO-1, PRAME, MAGEA4, and MAGEA1. Hum Pathol 61:130–139. https://doi.org/10.1016/j.humpath.2016.12.006

Khan FA (2020a) Applications of nanomaterials in human health, 1st edn. Springer, Singapore. https://doi.org/10.1007/978-981-15-4802-4

Khan FA (2020b) Synthesis of nanomaterials: methods & technology. In: Khan F (ed) Applications of nanomaterials in human health. Springer, Singapore. https://doi.org/10.1007/978-981-15-4802-4_2

Kumari A, Yadav SK, Yadav SC (2010) Biodegradable polymeric nanoparticles based drug delivery systems. Colloids Surf B Biointerfaces 75:1–18. https://doi.org/10.1016/j.colsurfb.2009.09.001

Montazerabadi A, Beik J, Irajirad R et al (2019) Folate-modified and curcumin-loaded dendritic magnetite nanocarriers for the targeted thermo-chemotherapy of cancer cells. Artif Cells Nanomed Biotechnol 47(1):330–340. https://doi.org/10.1080/21691401

Ni JS, Li Y, Yue W et al (2020) Nanoparticle-based cell trackers for biomedical applications. Theranostics 10(4):1923–1947. https://doi.org/10.7150/thno.39915

Odunsi K, Matsuzaki J, Karbach J et al (2012) Efficacy of vaccination with recombinant vaccinia and fowlpox vectors expressing NY-ESO-1 antigen in ovarian cancer and melanoma patients. Proc Natl Acad Sci U S A 109:5797–5802. https://doi.org/10.1073/pnas.1117208109

Patra JK, Das G, Fraceto LF et al (2018) Nano based drug delivery systems: recent developments and future prospects. J Nanobiotechnology 16:71. https://doi.org/10.1186/s12951-018-0392-8

Reichel D, Tripathi M, Perez JM (2019) Biological effects of nanoparticles on macrophage polarization in the tumor microenvironment. Nano 3:66–88. https://doi.org/10.7150/ntno.30052

Ryu JH, Lee S, Son S et al (2014) Theranostic nanoparticles for future personalized medicine. J Control Release 190:477–484. https://doi.org/10.1016/j.jconrel.2014.04.027

Sabbatini P, Tsuji T, Ferran L et al (2012) Phase I trial of overlapping long peptides from a tumor self-antigen and poly-ICLC shows rapid induction of integrated immune response in ovarian cancer patients. Clin Cancer Res 18:6497–6508. https://doi.org/10.1158/1078-0432.CCR-12-2189

Schleich N, Po C, Jacobs D et al (2014) Comparison of active, passive and magnetic targeting to tumors of multifunctional paclitaxel/SPIO-loaded nanoparticles for tumor imaging and therapy. J Control Release 194:82–91. https://doi.org/10.1016/j.jconrel.2014.07.059

Shaghaghi B, Khoee S, Bonakdar S (2019) Preparation of multifunctional janus nanoparticles on the basis of SPIONs as targeted drug delivery system. Int J Pharm 559:1–12. https://doi.org/10.1016/j.ijpharm.2019.01.020

Sharma P, Gnjatic S, Jungbluth AA et al (2003) Frequency of NY-ESO-1 and LAGE-1 expression in bladder cancer and evidence of a new NY-ESO-1 T-cell epitope in a patient with bladder cancer. Cancer Immun 3:19

Sugita Y, Wada H, Fujita S et al (2004) Ny-Eso-1 expression and immunogenicity in malignant and benign breast tumors. Cancer Res 64:2199–2204. https://doi.org/10.1158/0008-5472.CAN-03-3070

Tessari A, Pilla L, Silvia D et al (2018) Expression of NY-ESO-1, MAGE-A3, PRAME and WT1 in different subgroups of breast cancer: an indication to immunotherapy? Breast 42:68–73. https://doi.org/10.1016/j.breast.2018.08.106

Vaughan HA, Svobodova S, Macgregor D et al (2004) Immunohistochemical and molecular analysis of human melanomas for expression of the human cancer-testis antigens NY-ESO-1 and LAGE-1. Clin Cancer Res 10:8396–8404. https://doi.org/10.1158/1078-0432.CCR-04-0809

Waeckerle-Men Y, Groettrup M (2005) Plga microspheres for improved antigen delivery to dendritic cells as cellular vaccines. Adv Drug Deliv Rev 57:475–482. https://doi.org/10.1016/j.addr.2004.09.007

Wallyn J, Anton N, Vandamme TF (2019) Synthesis, principles, and properties of magnetite nanoparticles for in vivo imaging applications-a review. Pharmaceutics 11(11):601. https://doi.org/10.3390/pharmaceutics11110601

Wang J, Zhang B, Yang G et al (2020) Transferrin-conjugated superparamagnetic iron oxide nanoparticles as in vivo magnetic resonance imaging contrast agents. J Nanosci Nanotechnol 20(4):2018–2024. https://doi.org/10.1166/jnn.2020.17311

Zhi D, Yang T, Yang J et al (2020) Targeting strategies for superparamagnetic iron oxide nanoparticles in cancer therapy. Acta Biomater 102:13–34. https://doi.org/10.1016/j.actbio.2019.11.027

Zou P, Yu Y, Wang YA et al (2010) Superparamagnetic iron oxide nanotheranostics for targeted cancer cell imaging and pH-dependent intracellular drug release. Mol Pharm 7(6):1974–1984. https://doi.org/10.1021/mp100273t

Nanocarriers-Based Products in the Market, FDA Approval, Commercialization of Nanocarriers, and Global Market

7

Firdos Alam Khan

Abstract

There is an increasing interest in the nanomaterial-based products especially for their applications in the biomedical fields. These nanomaterials possess small size together with astonishing physicochemical properties which makes most suitable for the therapeutic applications. Different types of nanomaterials such as metallic, carbon-based, fluorescent, and polymer-based which have been used in the tissue engineering, drug delivery, and in other therapeutic applications. In spite of many useful applications, nanomaterials possess some harmful effects on human body, vegetation, and the environment. Nanomaterials can enter cells or tissues and organs because of their nano-size, large surface area, and better penetrability which may cause serious health issues (inhibition of cell proliferation, cell wall damage, modification of function and morphology of organelles, oxidative stress, genotoxicity, and cell death) in living organisms including humans. In this chapter we have aimed to provide the negative or harmful effects of the nanomaterials as per published research papers. The topics related to nanomaterial's toxicity, mode of action, and penetration of nanomaterials in the human body are discussed. We have also discussed the government regulations, ethical issues related to nanomaterials toxicity, and possible health hazards.

Keywords

Health and environmental hazards · Synthesis of nanocarriers · Nano-drug delivery · Respiratory diseases · Government regulations · Ethical issues

F. A. Khan (✉)
Department of Stem Cell Research, Institute for Research and Medical Consultations (IRMC), Imam Abdulrahman Bin Faisal University, Dammam, Saudi Arabia
e-mail: fakhan@iau.edu.sa

F. A. Khan (ed.), *Nano Drug Delivery for Cancer Therapy*,
https://doi.org/10.1007/978-981-99-6940-1_7

7.1 Commercialization of Nanomedicine Products

The commercialization is the process of transforming designs or ideas into products for human consumption especially for biomedical applications. The commercialization of any product demands the big investment and also needed demand for product in the market and hence it is very necessary to know the demand of the products for particular disease conditions in the particular area of country or at the global level (Craig et al. 2013; Sreeramana and Shubhrajyotsna 2016). The commercial value of the product is based on the novelty of the product and its requirements such as market size, and the current economic situation (Rogers 1995). The process of commercialization of nanomedicine products is shown in Fig. 7.1. The product development process has approximately 10 steps, which starts with generation of idea, and idea can be used for finding solution to the problem. Once solution is identified then research is started to generate data for proof of concept, once proof of concepts showed positive results, then reach to patent office of the organization to start the filing the disclosure of invention. The patent office will assess the invention and if they find the application fulfills the patenting requirements, they will file application to US patent office or patent office of the respective country. The US patent office will examine the patent application, if everything is satisfied, they will grant the patent. Once the patent is granted, the investors will start commercialization of the product by themselves provided they sufficient funding or licensing the patent to third party with the exchange of monitory benefits. The product development process start with development of Good Laboratory Practices (GLP) and Good Manufacturing Practices (GMP) facilities and regulations and these facilities should be approved by the USFDA or FDA of the respective country. Then preclinical and

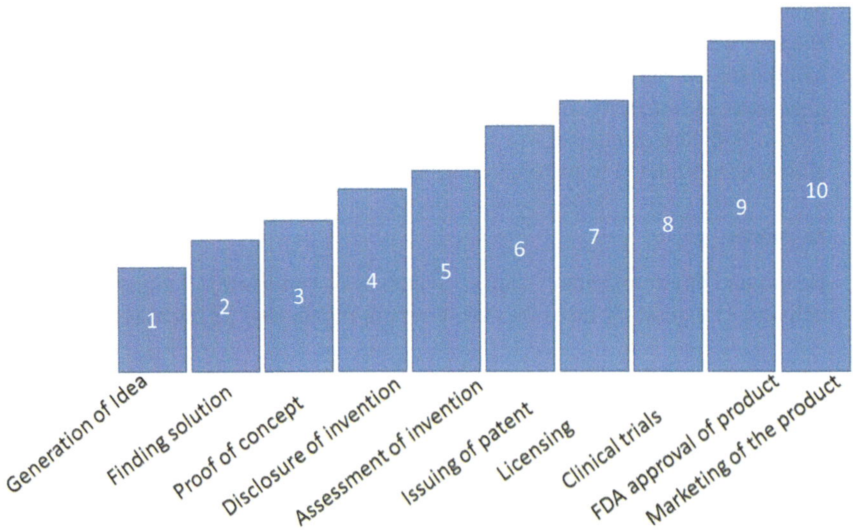

Fig. 7.1 The illustration showing the process of nanomedicine product development

clinical trials will start and once all trials are successful, the products will be approved by USFDA or FDA of the respective country for the large production and will be used for the human consumptions.

7.2 Process of Commercialization Nanomedicine Products

7.2.1 Identification of Problem

Any organization will start commercialization of the product when they have market survey data that show that there is sufficient and consistent demand for the product in the market. For example, in cancer therapy what are problems are associated with currently available treatments, and how these problems can be overcome by the innovative approach.

7.2.2 Finding Solution

Once problem is identified, then efforts should be directed toward finding the commercially viable solution, for example, the cancer can be better treated with nanomaterials because of their small size, better surface area, better bioavailability, and better tissue targeting capabilities. The organization may do technological feasibility study of producing such products.

7.2.3 Research and Invention

The research work will start to bring the data that support idea where new product possess many useful features what was initially expected to have. This is called proof of concept stage.

7.2.4 Disclosure of Invention

Once proof of concept is finalized, the organization will file a disclosure of invention to the organization's patent office.

7.2.5 Assessment of Invention

US Patent office will examine the invention and will examine the features of the products, reproducibility, and reliability and uniqueness of the products.

7.2.6 Issuing of Patent

If patent fulfills all the requirements, patent office will issue a patent which will protect the inventors for 20 years.

7.2.7 Licensing of Patented Technology

This is commercial decision to make by the organization where technology can be either transferred to the third part in return of money benefits or keep themselves for the product development process.

7.2.8 Preclinical and Clinical Trials

The nanomedicine product must be tested first in the animals which is called ad preclinical trial, and then tested in the humans in Phase I, Phase II, and Phase III for checking the effectiveness and safety of the nanomaterials.

7.2.9 Approval of the Product

Once all the trials are successful as per the USFDA or FDA of respective country, they will produce in large scale as per the market demand as per GMP requirements.

7.2.10 Marketing of Product

The nanomedicine product will be send to Hospitals for the human consumption with necessary packaging, transportation and storage condition as approved by USFDA or FDA of the respective country.

7.3 Nanomaterials as Commercial Products

The first commercialization of nanomaterials in the medical field has happened where metallic oxide nanomaterials, zinc oxide and titanium dioxide showed promising results, but the application of silver nanoparticles has been commercialized used in treating microbial infections many years ago (Hobson 2009). In addition, the application of silver nanomaterials has been effective in the treatment of chronic wound-care market where this product is used in wound dressing to combat microbial infections (Shahverdi et al. 2007; Vlachou et al. 2007). The commercial applications of nano-based products showing remarkable success as many big pharmaceutical companies are now catching up and started investing in the nano-based product development businesses especially in the development of product using nano-formulation in the form of liposomes, neo-somes, and nano-somes, respectively (Nohynek et al. 2007). The availability of nano-based products for both diagnostic and therapeutic applications showing continuous surge in the global markets as pharmaceutical and biotechnology companies are investing in the research and development of nano-based research. The application of nano-based product is not only applied to cancer diagnosis and therapy area but also used in the other medical conditions such as ophthalmology, gastroenterology, dermatology, dentistry, where nanomaterials have been applied.

The renowned Wall Street analysts, Andy Kessler has said that field of nanotechnology is already entering a significantly advanced state of product development (Hobson 2009). The ability of antibodies to link or conjugate with and other active biomolecules to make silicon microchips for nano-bio-electronic detectors outside of the body (Hobson 2009). The biobusiness of nano-based products for medical applications is complex affairs with lot of risk factors such as nanomedicine products may fail any point of clinical trials, where companies may face huge economical loss. With compared to other well-established pharmaceuticals companies, the field of nanomedicine is still emerging and face many challenges as the FDA regulations is not fully developed to safe guard for both patients and investors. We find considerable difference in the level of investment toward commercialization worldwide, as European Union lagging behind the US and Japan in terms of public funding and patents application. There is very low level about 3.5% of global venture capital is invested in the European countries, creating concerns for regional commercialization of nanomedicine products. A scarcity of funding or low investment is the principal cause for this difference of commercialization opportunities for nanomedicine products. The business pattern of the nanomedicine products is different than other business models and various constituents of business models of nanomedicine product is illustrated in Fig. 7.1.

The funds for nanomedicine product development may come either from with small investors or from big companies who have infrastructure and resources to develop the products. With compared to non-medical products, the nanomedicine products are complex affairs with unpredictable risk factors associated with them. The associated risk factors such as safety and efficacy of the nanomedicine product during the clinical trials may always pose serious risk, as nanomedicine may fail during the clinical trials, where companies may face huge financial loss. To avoid such a huge financial loss, the companies may seriously look into data obtained from non-clinical or animal trials and phase I of clinical trials and if data is showing profound results with regard to safety and efficacy of the products, then they can decide to carry the trial further or discontinue it. Another issue with nanomedicine product is approval of FDA which is complex and time consuming, because FDA will not approve any products, if products does not comply the guidelines of the FDA with sound proof that product is safe and effective for human consumption.

7.4 Global Market of Nanomedicine Products

The global market dynamics for the nanomedicine products has been increasing due to successful development of new and novel nanomedicine products. The nanomedicine market size (Fig. 7.2) was valued at $ 172.8 Billion in 2021 and is projected to reach $ 523.2 Billion by 2030, growing at a CAGR of 13.14% from 2022 to 2030 (Healthcare Nanotechnology (Nanomedicine) Market Size and Forecast 2023). There has been good demand for the nanomedicine products especially in the North America and European markets. The emergence of the nanomedicine market has impacted macroeconomic and microeconomic factors and due to positivity in the market, there has been surge of investments in the development and production of the new and navel nanomedicine products. As the market dynamics impact the supply and demand curves, companies and investors aim to examine the best way to use various financial tools to stem various policies for fast-moving the growth and lowering the risks.

7.4.1 Market Classification of Nanomedicine Products

The global nanomedicine market is classified based on the different factors which are shown below:

- **Indication**: The market is categorized into (1) cardiovascular diseases, (2) immunological diseases, (3) infectious diseases, (4) neurological diseases, (5) oncological diseases, (6) ophthalmological diseases, (7) orthopedic disorders, and (8) urological diseases (Fig. 7.3).

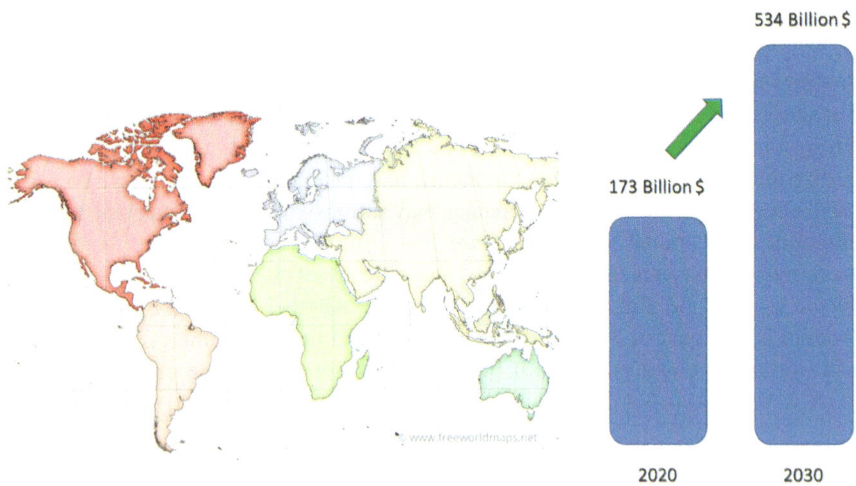

Fig. 7.2 The illustration showing projection of global market of nanomedicine products

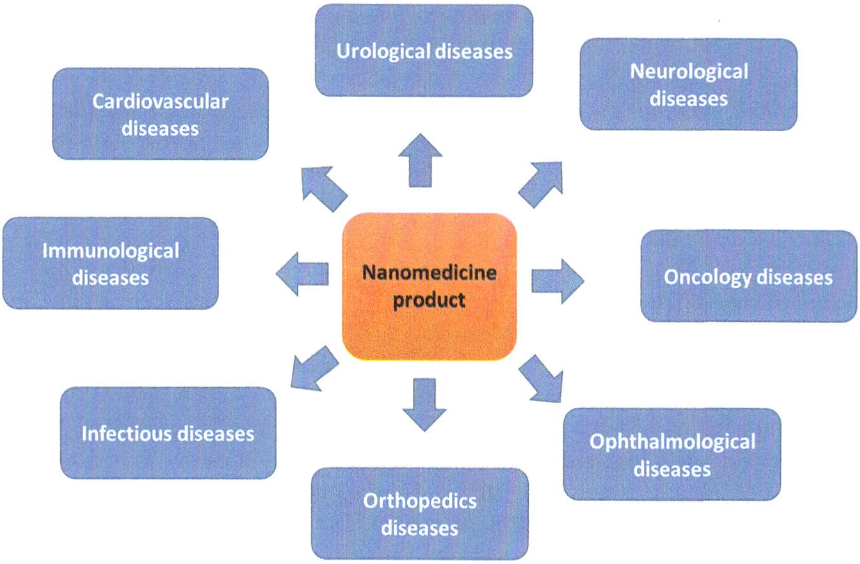

Fig. 7.3 Market demand of nanomedicine products for various disease conditions

- **Drug Delivery System**: The market is categorized into (1) carbon nanotube, (2) dendrimers, (3) exosomes, (4) graphene, (5) injectable nanoparticle generator, (6) liposomes, (7) nano-bubbles, (8) nano-clusters.
- **Product Type**: The market is classified into (1) in vitro diagnostics, (2) in vivo diagnostics, (3) regenerative medicine, (4) therapeutics, and (5) vaccines.
- **Disease Application**: the market is categorized into (1) anti-infectives, (2) anti-inflammatory, (3) cardiovascular, and (4) neurology.
- **Geographical location**: The global market is categorized into Americas, Europe, Middle-East and Africa and Asia-Pacific.

The global nanomedicine market is predictable to reach US$ 454.8 Billion by 2028 from US$ 177.1 Billion in 2021 (Anon (n.d.)). The tremendous increase in the nanomedicine market shows the positivity about the nanomedicine products at global level which in fact helping the companies to attract huge investments. It has been found that nanomedicine products especially for the diagnostic purposes are more reliable and give accurate results and most important require less time in diagnosis of diseases. Due to this, the nano-based diagnostic methods need lesser chemicals and reagents which is an attractive part for the investors. The nanomaterials also have other advantages like having larger surface to volume ratio compared to micro and macro sized materials which rises demand for nano-drug delivery. The better targeting is another important feature of the nanomaterials for effective delivery of the drug molecules in the targeted disease tissues with less side effects and less chance to effect the non-disease tissues or cells.

There has been increasing technological advancements in the nanomaterials for biomedical applications, and in addition of having promising regulatory framework, and surge research funding are some of the factors boosting growth of the market. Furthermore, growing government initiatives in the North America and European countries and with a high prevalence of chronic diseases such as diabetes, obesity, cancer, orthopedic diseases, and kidney disorder, growing geriatric population, and increasing preference for targeted nanomedicine is also likely to contribute to growth of the nanomedicine market.

There has been entry of new market players in the manufacturing of nanomedicine products which will help to reduce the gap between supply and demand. Currently, nanomedicine products that are available for the treatment of various cancers such as Depocyt, Doxil, Oncospar, Neulasta, and Abraxane, respectively. There has been increasing occurrence of cancer globally, hence, there is demand for the anti-cancer drug therapy including nanomedicine product. As per the National Cancer Institute, about 1.6 million new cases of cancer were diagnosed in the USA alone in 2016 (Anon (n.d.)). North America accounted for the largest market share in the global nanomedicine market. The market in Asia-Pacific is likely to see noteworthy growth due to growing geriatric population, growing healthcare awareness among the populace, and increasing disposable income in the region.

7.4.2 Leading Players in Nanomedicine Market

- CombiMatrix Corporation.
- Abbott Laboratories.
- GE Healthcare.
- Johnson & Johnson.
- Sigma-Tau Pharmaceuticals, Inc.
- Mallinckrodt plc, Nanosphere, Inc.
- Merck & Company, Inc.
- Celgene Corporation.
- Pfizer, Inc.
- UCB (Union chimique belge).
- Teva Pharmaceutical Industries Ltd.

7.5 Nanomedicine for Cancer Therapy

Nanomedicine products are expected to play a major role in the global pharmaceutical market and healthcare system. Since 1995, around 78 nanomedicine products have been approved by the US Food and Drug Administration (USFDA) and the European Medicines Agency (EMA) for clinical applications (Bulbake et al. 2017; Halwani 2022; Prasad et al. 2018; Anselmo and Mitragotri 2019). In addition, more than 140 nanomedicine products are under different stages of clinical trials. There has been growing market trends for the nanomedicine products as the more and more number of nanomedicine products are being approved and marketed. Many reports demonstrate that the number of medications based on nanotechnology are increasing annually. New nanomedicine products enter clinical trials to examine their improvement with regard to effectiveness compared with conventional preparations (Klein et al. 2019; Caster et al. 2017). This is due to the fast growth in research and the high market demand, which displays the prominence of nanomedicine products for the drug delivery area. Nevertheless, the majority of nanomedicine approved products have revealed a lower toxicity rather than better efficacy (Caster et al. 2017). The introduction of nanomedicine products in various countries for different the treatment of disease have shown remarkable improvement.

There are 78 different types of nanomedicine products have been approved by USFDA and EMA and have entered the global market for both diagnosis and treatment purposes. It has been reported that both the USFDA and the EMA have shared in the approval of 20 nanomedicine products globally, whereas other nanomedicine products have an approval from USFDA (43) and from EMA (12), respectively. Over the past 10 years, due to better healthcare benefits provided by the medical insurance companies, the focus on the development of nanomedicine products and the number of marketed nanomedicine products have significantly increased. Based on the types of nanomaterials used to produce nanomedicine products, the nanomedicine products are classified as nano-crystals, lipid-based nanomaterials, polymer-based nanomaterials, dendrimers-based nanomaterials, protein-based nanomaterials, or inorganic nanomaterials.

There has been growing demand for the inorganic nanomaterials for their application in drug delivery systems. The different types of inorganic nanomaterials such as metal, carbon nanotubes, calcium phosphate, iron oxide, silica, and quantum dot nanomaterials are very attractive for use in drug delivery due to their dual function as diagnostic area and also as therapeutic drug carriers (Anselmo and Mitragotri 2015). It is possible to modify the inorganic nanomaterials chemically or biologically to meet the perfect necessities for cellular drug delivery, and inorganic nanomaterials possess better biocompatibility, high charge density, and precise targeted drug delivery (Anselmo and Mitragotri 2015). It has been shown that inorganic nanomaterials display excellent stability over broad ranges of temperature variations and pH fluctuations and they are not subject to microbial attack. In spite of these different benefits, few inorganic nanomaterials have reached the patients due to slow dissolution and lack of biodegradation, which are the main shortcomings of inorganic nanomaterials (Anselmo and Mitragotri 2015). On the other hand, organic nanomaterial-based nanocarriers such as carbon nanomaterials that are usually characterized by their better biocompatibility, low toxicity, and improved drug-loading capacity. These organic-nanocarriers are mainly classified into three groups (1) lipid-based vectors, (2) polymer-based vectors, and (3) dendrimers. It has been found that these carriers provide special defense for therapeutic agents against environmental degradation. The surfaces of liposomes and micelles can also be functionalized to improve efficacy of the anti-cancer agents.

Cancer has 14 million new cases and more than 8.2 million cancer related deaths were reported in 2012 (Kumar et al. 2020). Consequently, new and effective anti-cancer molecules and drugs are immediately needed to control the high mortality. Anti-cancer nanomedicine products have made advancements in the past several decades for the cancer diagnosis and treatments. Several nanomaterials with different shapes and sizes have been developed for the effective delivery of therapeutic and diagnostic agents to tumors to increase the accumulation of chemotherapeutic agents within the tumor tissues (Peer et al. 2007; Schroeder et al. 2011; Farokhzad and Langer 2009). To target tumor, different strategies are used which can be classified as the passive targeting which is based on the enhanced permeability and retention (EPR) effect and active targeting focused by tumor specific moieties and stimuli-responsive tumor targeting (He et al. 2019). In the case of passive EPR targeting, a phenomenon that leads to better accumulation of nanomaterials in the tumor tissues as nanomaterials can easily pass through blood vessel of tumor (Prabhakar et al. 2013). Passive EPR targeting and nanomaterial-size control were the basic philosophies of first generation cancer nanomedicine and examples is Doxil, which is simple lipid vesicles protected with polyethylene glycol (PEG) to avoid opsonization by the host cells, henceforth preventing an immune response and prolonging circulation time. Doxil is a pegylated liposomal doxorubicin used to treat different kinds of cancer such as metastatic ovarian cancer and AIDS-related Kaposi's sarcoma with less side effects. Doxorubicin was developed by encapsulating an 80–90 nm size uni-lamellar liposome coated with PEG (Pillai 2014; Pillai and Ceballos-Coronel 2013). Doxorubicin is known to act on the cancer cells by two different mechanisms, in the first case, doxorubicin attacks cancer cells DNA and caused the disruption of topoisomerase II-mediated DNA repairing systems, and in the second case, doxorubicin caused the generation of free radicals, resulting in damage to cellular membranes, DNA, and proteins, respectively (Gordon et al. 2000). Nevertheless, doxorubicin has been known to cause severe heart problems, such as heart failure during the treatment or after treatment (Pillai 2014).

Another anti-cancer drug which is called as Abraxane which was approved by the USFDA for the treatment of metastatic breast cancer in 2005, lung cancer in 2012, and advanced pancreatic cancer in 2013, respectively. In this drug, human albumin is conjugated to paclitaxel in order to deliver it in the form of nanomaterials with 130 nm in diameter size. It has been reported that better biodegradability, low toxicity, and less immunogenicity make albumin a perfect carrier for drug delivery by uptake by cells. The bioavailability of albumin-paclitaxel combination in the tumor is enabled by the gp60 receptor-mediated pathway in the endothelial cell walls of tumors, which binds to albumin with a great affinity (Von Hoff et al. 2011; Desai et al. 2004, 2006). The accumulation of Abraxane at the tumor site is also aided by secreted protein, acidic and rich in cysteine, overexpressed in multiple types of tumors, including breast, prostate, gastric, lung, and kidney (Pillai 2014).

Another ground-breaking strategy for targeting cancer cells, which emphases on nanomaterials that can be activated to release their contents (drug molecules) on exposure to heat, light, or ultrasound method. ThermoDox is a stimuli-responsive targeting nanomedicine which was industrialized by Celsion Corporation in partnership with Duke University for the liver cancer treatment. ThermoDox is a temperature sensitive Doxorubicin-Pegylated Liposome that is able to release the anti-cancer drug at the targeted site by increasing the temperature from 39 to 42 ° Celsius through the application of radio-frequency (Pillai 2014). Furthermore, the accumulation of ThermoDox drug in tumor tissues upturns as the cancer blood vessels become leaky due to the local application of hyperthermia (Pillai 2014).

While more anti-cancer nanomedicine products are presently available on the market than any other drugs, many drugs are marketed for the treatment of autoimmune conditions, metabolic, ophthalmic, neurological, hematological, inflammatory diseases, respectively (Desai et al. 2006). The bulk of the developed nanomedicine products have used liposomes and polymers as incorporated nanomaterials (Desai et al. 2006). Additionally, there is a clear trend in the movement of nanomaterials from simple to complex multi-component drug delivery systems (Desai et al. 2004). These complex systems can include functions such as controlled release and active targeting to allow nanomaterials to improve drug efficacy, efficacy, and drug safety. Though most USFDA approved nanomedicine products are based on passive targeting through the EPR effect, some next generation nanomedicine products use active targeting methods in clinical trials (Bobo et al. 2016).

References

Anon (n.d.). https://www.coherentmarketinsights.com/insight/request-pdf/393

Anselmo A, Mitragotri S (2015) A review of clinical translation of inorganic nanoparticles. AAPS J 17:1041–1054. https://doi.org/10.1208/s12248-015-9780-2

Anselmo AC, Mitragotri S (2019) Nanoparticles in the clinic: an update. Bioeng Transl Med 4: e10143. https://doi.org/10.1002/btm2.10143

Bobo D, Robinson KJ, Islam J, Thurecht KJ, Corrie SR (2016) Nanoparticle-based medicines: a review of FDA-approved materials and clinical trials to date. Pharm Res 33:2373–2387. https://doi.org/10.1007/s11095-016-1958-5

Bulbake U, Doppalapudi S, Kommineni N, Khan W (2017) Liposomal formulations in clinical use: an updated review. Pharmaceutics 9:12. https://doi.org/10.3390/pharmaceutics9020012

Caster J, Patel AN, Zhang T, Wang A (2017) Investigational nanomedicines in 2016: a review of nanotherapeutics currently undergoing clinical trials. Wiley Interdiscip Rev Nanomed Nanobiotechnol 9:e1416. https://doi.org/10.1002/wnan.1416

Craig B, Richard M, Grant E, Darren M (2013) Success factors to a nanotech-enabled whole product. In: Tsuzuki T (ed) Nanotechnology commercialisation. Pan Stanford Publishing Pvt. Ltd, Singapore, p 171

Desai N, Trieu V, Yao R, Labao E, De T, Soon-Shiong P (2004) 601 increased transport of nanoparticle albumin-bound paclitaxel (ABI-007) by endothelial gp60-mediated caveolar transcytosis: a pathway inhibited by Taxol. Eur J Cancer Suppl 2:182. https://doi.org/10.1016/S1359-6349(04)80609-8

Desai N, Trieu V, Yao Z, Louie L, Ci S, Yang A, Tao C, De T, Beals B, Dykes D et al (2006) Increased antitumor activity, intratumor paclitaxel concentrations, and endothelial cell transport of cremophor-free, albumin-bound paclitaxel, ABI-007, compared with cremophor-based paclitaxel. Clin Cancer Res 12:1317–1324. https://doi.org/10.1158/1078-0432.CCR-05-1634

Farokhzad OC, Langer R (2009) Impact of nanotechnology on drug delivery. ACS Nano 3:16–20. https://doi.org/10.1021/nn900002m

Gordon AN, Granai C, Rose PG, Hainsworth J, Lopez A, Weissman C, Rosales R, Sharpington T (2000) Phase II study of liposomal doxorubicin in platinum- and paclitaxel-refractory epithelial ovarian cancer. J Clin Oncol 18:3093–3100. https://doi.org/10.1200/JCO.2000.18.17.3093

Halwani AA (2022) Development of pharmaceutical nanomedicines: from the bench to the market. Pharmaceutics 14(1):106. https://doi.org/10.3390/pharmaceutics14010106; PMID: 35057002; PMCID: PMC8777701
He H, Liu L, Morin EE, Liu M, Schwendeman A (2019) Survey of clinical translation of cancer nanomedicines—lessons learned from successes and failures. Acc Chem Res 52:2445–2461. https://doi.org/10.1021/acs.accounts.9b00228
Healthcare Nanotechnology (Nanomedicine) Market Size and Forecast (2023). https://www.verifiedmarketresearch.com/product/healthcare-nanotechnology-nanomedicine-market/
Hobson DW (2009) Commercialization of nanotechnology. Wiley Interdiscip Rev Nanomed Nanobiotechnol 1(2):189–202. https://doi.org/10.1002/wnan.28
Klein K, Stolk P, De Bruin ML, Leufkens H, Crommelin D, De Vlieger J (2019) The EU regulatory landscape of non-biological complex drugs (NBCDs) follow-on products: observations and recommendations. Eur J Pharm Sci 133:228–235. https://doi.org/10.1016/j.ejps.2019.03.029
Kumar R, Dalvi SV, Siril PF (2020) Nanoparticle-based drugs and formulations: current status and emerging applications. ACS Appl Nano Mater 3:4944–4961. https://doi.org/10.1021/acsanm.0c00606
Nohynek GJ, Lademann J, Ribaud C, Roberts MS (2007) Grey goo on the skin? Nanotechnology, cosmetic and sunscreen safety. Crit Rev Toxicol 37(3):251–277
Peer D, Karp JM, Hong S, Farokhzad OC, Margalit R, Langer R (2007) Nanocarriers as an emerging platform for cancer therapy. Nat Nanotechnol 2:751–760. https://doi.org/10.1038/nnano.2007.387
Pillai G (2014) Nanomedicines for cancer therapy: an update of FDA approved and those under various stages of development. SOJ Pharm Pharm Sci 1:13. https://doi.org/10.15226/2374-6866/1/2/00109
Pillai G, Ceballos-Coronel ML (2013) Science and technology of the emerging nanomedicines in cancer therapy: a primer for physicians and pharmacists. SAGE Open Med 1:13759. https://doi.org/10.1177/2050312113513759
Prabhakar U, Maeda H, Jain RK, Sevick-Muraca EM, Zamboni W, Farokhzad OC, Barry ST, Gabizon A, Grodzinski P, Blakey DC (2013) Challenges and key considerations of the enhanced permeability and retention effect for nanomedicine drug delivery in oncology. Cancer Res 73:2412–2417. https://doi.org/10.1158/0008-5472.CAN-12-4561
Prasad M, Lambe UP, Brar B, Shah I, Manimegalai J, Ranjan K, Rao R, Kumar S, Mahant S, Khurana SK et al (2018) Nanotherapeutics: an insight into healthcare and multi-dimensional applications in medical sector of the modern world. Biomed Pharmacother 97:1521–1537. https://doi.org/10.1016/j.biopha.2017.11.026
Rogers EM (1995) Diffusion of innovation. The Free Press, New York, NY
Schroeder A, Heller DA, Winslow MM, Dahlman JE, Pratt GW, Langer R, Jacks T, Anderson DG (2011) Treating metastatic cancer with nanotechnology. Nat Rev Cancer 12:39–50. https://doi.org/10.1038/nrc3180
Shahverdi AR, Fakhimi A, Shahverdi HR, Minaian S (2007) Synthesis and effect of silver nanoparticles on the antibacterial activity of different antibiotics against Staphylococcus aureus and Escherichia coli. Nanomedicine 3(2):168–171. https://doi.org/10.1016/j.nano.2007.02.001; Epub 2007 Apr 30. PMID: 17468052
Sreeramana A, Shubhrajyotsna A (2016) Nanotechnology innovations and commercialization—opportunities, challenges reasons for delay; https://mpra.ub.uni-muenchen.de/102891/
Vlachou E, Chipp E, Shale E, Wilson YT, Papini R, Moiemen NS (2007) The safety of nanocrystalline silver dressings on burns: a study of systemic silver absorption. Burns 33(8): 979–985. https://doi.org/10.1016/j.burns.2007.07.014; Epub 2007 Oct 23. PMID: 17959313
Von Hoff DD, Ramanathan RK, Borad MJ, Laheru DA, Smith LS, Wood TE, Korn RL, Desai N, Trieu V, Iglesias JL et al (2011) Gemcitabine Plusnab-paclitaxel is an active regimen in patients with advanced pancreatic cancer: a phase I/II trial. J Clin Oncol 29:4548–4554. https://doi.org/10.1200/JCO.2011.36.5742

Limitations of Nanocarriers Such as Cell and Tissue Toxicity, Genotoxicity, Scale-Up of Nanomaterials

8

Dilawar Hassan, Ayesha Sani, and Dora I. Medina

Abstract

Due to drug resistance by the bacteria and systematic toxicity of classical medicines, a new range of medicines is being researched, termed as "Nanomedicine". Nanomedicine is the use of nanomaterials for the diagnosis, prognosis and treatment of diseases. Nanomaterials are such materials, which have at least one dimension in Nanoscale range (one billionth of a metre). Nanomaterials are gaining worth due to large surface area and greater surface to volume ratio. Furthermore, the nanomaterials are being researched in the field of Nanobiotechnology, due to enhanced optical, physiochemical and biomedical applications. The nanomaterials are being used for their potential anti-cancer, anti-bacterial, anti-fungal and drug loading capability. While, to make target specific, nanomaterials are surface modified with anti-bodies, proteins and siRNA for enhanced drug delivery and reduced systematic toxicity, still, there are various aspects need to be answered, including toxicity and other limitations faced by nanomedicine. The current chapter is to investigate various limitations faced by nanomaterials when being used for drug delivery purposes.

Keywords

Nanomedicine · Nanotoxicity · Drug delivery · Nanocarriers · Genotoxicity · Scale-up · Tissue toxicity

D. Hassan (✉) · A. Sani
Tecnologico de Monterrey, School of Engineering and Sciences, Atizapan de Zaragoza, Estado de Mexico, Mexico

D. I. Medina
Tecnologico de Monterrey, Institute of Advanced Materials for Sustainable Manufacturing, Monterrey, Nuevo Leon, Mexico

F. A. Khan (ed.), *Nano Drug Delivery for Cancer Therapy*,
https://doi.org/10.1007/978-981-99-6940-1_8

8.1 Introduction

Diseases are leading cause of deaths, and the number is inkling every year. According to a survey, 13.7 million deaths, only in 2019, around the globe were related to pathogenic bacterial infections (Ikuta et al. 2022), while, ten million deaths, reported in year 2020, were caused by some type of cancer, globally (Sung et al. 2021). With evolution of science, in past, varieties of plants were employed to fight the diseases related to infections, due to the presence of a long range of phytochemicals in them, but to their mode of action was very slow. Since, traditionally used medications for treatment of various bacterial and fungal infections showed efficiency in past, but with time, bacteria (Panáček et al. 2018) and fungus evolved and created resistance against the used medicines. To overcome this resistance and scientific development lead to production of synthetic medicines, which were much more affective with a quick mode of action (Makabenta et al. 2021). Though these results were portraying desired results, but on the contrary, the fungus and bacteria had been getting stronger with time also (Morrison and Zembower 2020). Furthermore, classically used medicines for chemotherapeutic purposes have been known for their cytotoxicity and potential against the cancers (Alyassin et al. 2020). But due to this cytotoxic nature, the drugs have systematic toxicity also. The chemotherapeutic drugs are not target-specific and cause damage to normal and healthy cell lines also, besides killing the cancerous cells. To overcome these issues of bacterial and fungal resistance and systematic toxicity, new line of medicines is being research for the last decade (Khalil et al. 2018). With that, pharmaceutical research is also pursuing methods to enhance drug availability, and drug distribution to obtain better results without using higher concentrations or stronger medicine formulations. A new line of medicines is being researched due to their potential applications, called nanomedicines, to answer the issues of classical methods.

Nanomedicines make use of nanomaterials' anti-fungal, anti-bacterial and anticancer characteristics for medical applications. The ability of the nanomaterials to cure several other ailments, such as neurodegenerative disorders, have also been researched. Iron oxide (IO) (Hassan et al. 2018), nickel oxide (NiO) (Sani et al. 2021), chromium oxide (CrO) (Hassan et al. 2019), zinc oxide (ZnO) (Mendes et al. 2022), cobalt oxide (CoO) (Sani et al. 2023) and other metal oxides, both as bare nanoparticles and as drug carriers, are nanoparticles utilized for medical reasons. Figure 8.1 shows the various types of Nanoparticles/nanocarriers and their physical properties. Polymeric nanoparticles, such as lipid nano-emulsions and Poly (lactic acid-co-glycolic)-Polyethylene glycol (PLGA-PEG) (Simón-Vázquez et al. 2022), are another class of nanoparticles being researched for their potential to deliver drugs. When compared to previously used medications, nanomedicines have demonstrated greater toxicity against illnesses, revolutionizing the science of pharmacology (Vallorz et al. 2021). They have also demonstrated a method to prevent achieving stronger pharmaceutical and discourage the use of huge doses (Weng et al. 2021). Nanomaterials have been used for the treatment of various types of cancers by various treatments methods, one such method is induced hyperthermia for

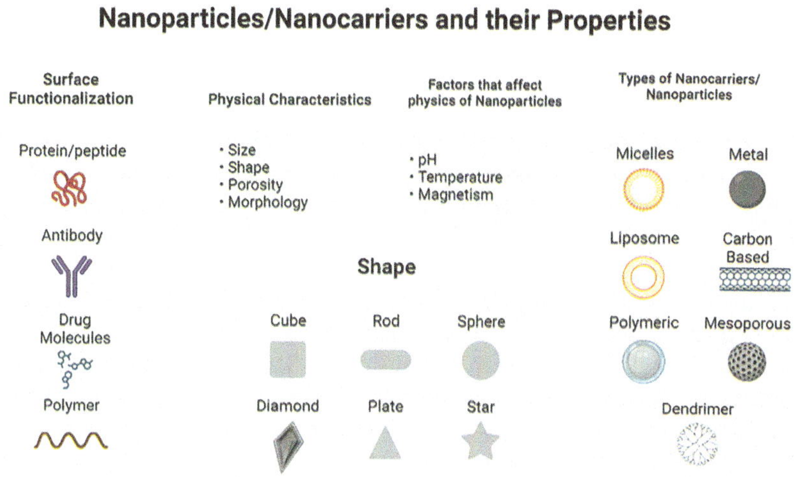

Fig. 8.1 Various types of Nanoparticles/nanocarriers, their physical properties and surface modifications

neutralizing cancerous cells (Kang et al. 2020). Another method makes use of LASER for release of drugs at specific site (Xue et al. 2021).

Though nanoscale provide materials with dramatic properties (Deng et al. 2021), but with that comes a new set of issues that limit the full potential use of nanomaterials as nanomedicines (Yang et al. 2019). Living body is a complex system, carrying various obstacles for any foreign body to reach any specific organ or part. Furthermore, due to small size and larger surface area, the nanomaterials pose enhanced toxicity to the body they are being used against. Genotoxicity (Ansari et al. 2019), muscle toxicity (Huang et al. 2019), cell toxicity (Mohammed et al. 2019) and scale-up (Soni et al. 2020) are issues that limit the use of nanomaterials as medicines in real life, though a lot of attention is being paid to overcome them, but these issues still stand. Furthermore, to declare a potential candidate for the next generation of medicines, focus must be paid to the evolution of nanomaterials, that should be target-specific, environmentally friendly, with greater bioavailability, enhanced retention time and lower toxicity.

Here in this chapter, we will discuss about nanomaterials that are being employed for medicinal purposes, their medicinal properties, but the chapter will place the greatest emphasis on the toxicity of nanomaterials and other problems that are restricting their usage. The chapter will concentrate on several toxicities (cell toxicity, genotoxicity, scale-up and tissue toxicity) and what steps have been done, particularly in recent years, to address these limitations.

8.2 Nanomaterials and Their Biomedical Applications

In nanoscale realm, the materials behave differently, since their bands become clearer, therefore, they have varied optical, physiochemical and biomedical properties, compared to their bulk counterparts. For a long time now, nanomaterials have been researched as a potential candidate for future of medicine. Nanomaterials have been studied for anti-bacterial, anti-fungal, anti-cancer, anti-leishmanial, anti-inflammatory and many more biomedical applications. These biomedical properties of nanomaterials heavily depend on the size and shape of nanoparticles (Dilawar et al. 2019). Most commonly surface morphology and shape of the nanomaterials is critical point in determining the potential anti-cancer effect of the nanomaterials (Steckiewicz et al. 2019). Research published by Sani, A., and group described the synthesis of Nickel Oxide nanoparticles. The group reported the diameter of nanoparticles to have an average of ~17 nm. The synthesized material was used for anti-bacterial potential against 12 pathogenic, waterborne, bacterial strains, antioxidant potential, anti-leishmania potential, haemolytic potential, and finally anti-cancer potential against HepG2 (hepatocellular carcinoma). The study elaborated that biosynthesized NiO nanomaterials had excellent cytotoxicity against bacterial, leishmanial and liver cancer cell lines, furthermore, the nanoparticles caused minimal lysis of human blood samples during haemolytic study (Sani et al. 2021). Though nanomaterials are being widely used for biomedical studies, yet a lot of research has been needed to discover actual mode of action of nanomaterials towards a living cell. Various routes have been reported to cause cellular death, these include (1) blockage and/or rapturing of cell wall/membrane, (2) enzyme inactivation, (3) causing damage and/or blocking nuclear membrane, (4) damaging DNA/RNA and (5) destabilizing ribosomes and damaging proteins (Rahman et al. 2019). Figure 8.2 contains pictorial description about possible interaction of nanoparticles with human blood components when introduced to human body via intravenous route. Since the bare nanoparticles are target less and can enter organ, therefore, a new approach is being investigated that loads a drug inside the nanoparticle, in such cases termed as "nanocarrier", and surface functionalized with an antibody or a compound that could take the nanocarrier to exact point to delivery, rather than wondering around in the entire body. The nanocarrier can be a metallic, core-shell structured, or polymeric nanoparticle. An article published by Zidan, A., et al., reported the synthesis of Poly(-D,L lactic-*co*-glycolic acid) PLGA nanocarriers loaded with doxorubicin (DOX) for their anti-cancer application against Ehrlich Ascites Carcinoma (EAC) in albino mice models. The group discovered that diameter of loaded nanocarrier was 106.77 ± 22.15 nm. The nanocarriers showed impressive results against the tested cancer cell lines, with enhanced cytotoxicity compared to DOX only (Zidan et al. 2023). Furthermore, Esfandyari-Manesh, M., and group synthesized Paclitaxel (PTX) encapsulated PLGA nanocarriers and surface modified them with Human Serum Albumin (HSA), that provided stealth properties to nanocarriers in sneaking out from reticuloendothelial system (RES). Group measured the size of nanoparticles using DLS and calculated the size of PTX encapsulated PLGA nanocarriers to be $\sim 190 \pm 10$ nm while after conjugation of

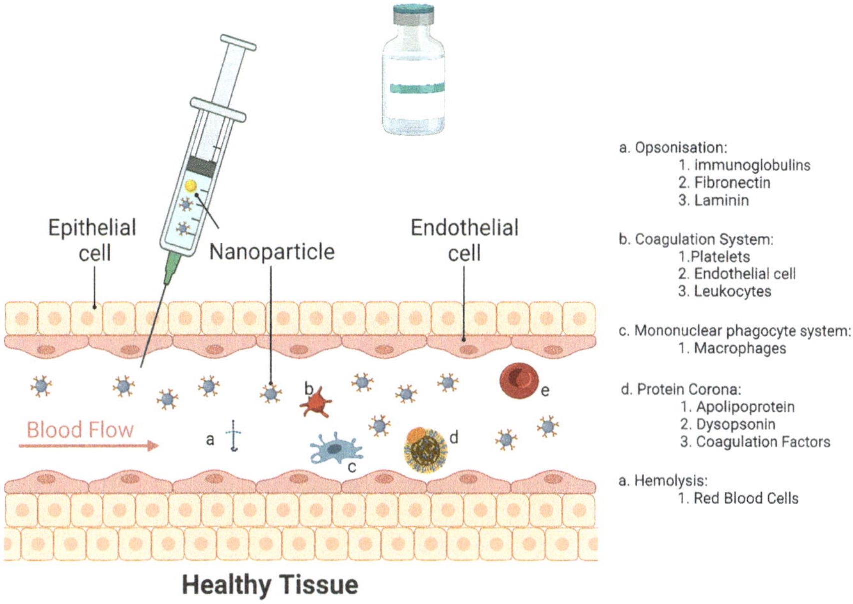

Fig. 8.2 Possible interaction between Nanocarriers/Nanoparticles and human blood, when introduced via intravenously

HSA, there was an increase of ~25 ± 5 nm was observed in their diameter, whereas zeta potential (ZE) value was found to be -6 mV. The drug release study showed that drug was released in sustained manner at more acidic medium (pH 7.4) compared to pH 5.5, up to 33 days (Esfandyari-Manesh et al. 2015). Although nanocarriers provide with various advantages over classical use of drugs, still there remains a large gap of issues that need to be answered. There are various issues associated with nanocarriers may include but not limited to high cost of production, scale-up, drug loading efficiencies, their toxicity and many more (Liu et al. 2020).

8.3 Limitations Faced by Drug Loaded Nanocarriers

Though drug loaded nanocarriers have revolutionized the field of nanobiotechnology and have provided with efficient drug delivery approaches for the treatment of various diseases, especially in the field of cancer research, still there are various obstacles that await the nanocarriers once they are introduced to a diseased body (Reinholz et al. 2018). Majorly the first hurdles, the nanocarriers are meet with, are immune system (IS) and blood–brain barrier (BBB) (Galstyan et al. 2019). Though these render the movement of nanocarriers inside the human or animal body, the nanocarriers are also known to cause toxicities against the IS and BBB. The structure and toxicities of nanocarriers against the IS and BBB and briefly described below.

8.3.1 Immune System (IS) and Nanocarriers

Immune system is defensive system of body (Islam et al. 2022). Its responsibility is to conserve the integrity of body by defending it against invading chemical and pathogenic alien bodies. Specially bacteria and viruses and other damages to human body (Kraus and Gruber 2021). IS constitutes of soluble components and cellular collections. There is a detailed and complex system that IS follows if it detects a pathogen. For instance, if a bacterium enters the human body, IS analyses its molecular structure, evaluates the tissue damage caused by it and then gets services of various innate immune system (IIS) cells to get rid of them. Simultaneously, dendritic cells (DCs) identify/sense pathogen caused damage and sample the invading pathogenic cell and transport it to lymph nodes. In the lymph node, DCs present antigen to T cells which help in the elimination of the pathogenic cell (Lung et al. 2020).

IS is further classified in two types: (1) innate immune system (IIS) (Costa da Silva et al. 2021) and (2) adaptive immune system (Fontana et al. 2018). Innate Immune System:

8.3.1.1 Innate Immune System

Innate immune system (IIS) is regarding as the first line of defence of body against an invading and/or attacking bacteria, virus and/or any pathogen (McKechnie and Blish 2020). IIS has various types of cells, including lymphoid and myeloid, which usually originate from bone marrow. Lymphoid includes innate lymphoid cells and natural killers (NK), while myeloid includes DCs, neutrophils, macrophages and monocytes. In addition to being first defence line, IIS also maintains homeostasis of tissues (Mattiola and Diefenbach 2023).

8.3.1.2 Adaptive Immune System

In contrast to IIS, which is very rapid and first defensive line, adaptive immune system (AIS) is very specialized and target specific immune system. AIS takes from days to weeks to develop for every specific type of pathogen (Blanchard et al. 2020). Unlike IIS, AIS is rooted on highly specialized class of white blood cells (WBCs) known as T cells and B cells. Furthermore, AIS consists of CD4 T cells, B cells, CD8 T cells, NK cells and $\gamma\delta$ T cells (Bomfim et al. 2017). T cells are further classified into Th1; Th2; Treh (regulatory T cells) and Th22 cells (Sompayrac 2022).

Since the primary function of IS is to eliminate the recognized foreign bodies, therefore, a weak IS can affect the quality of life gravely. Furthermore, due to large surface to volume ratio, nanocarriers may cause immunotoxicity also. Major contributors to these toxicity issues of nanocarriers are associated to their shape, size, larger surface area and specially their reactive oxygen species (ROS) production property. ROS are known to cause oxidative stress on cells causing them to start apoptosis (Ahmad et al. 2021). Furthermore, these nanocarriers also cause inflammation, since they are regarded as a foreign body by the immune system. It is also known that nanoparticles with size measurements greater than 100 nm are easily eliminated by the immune cells (Aljabali et al. 2023). Figure 8.3 shows the

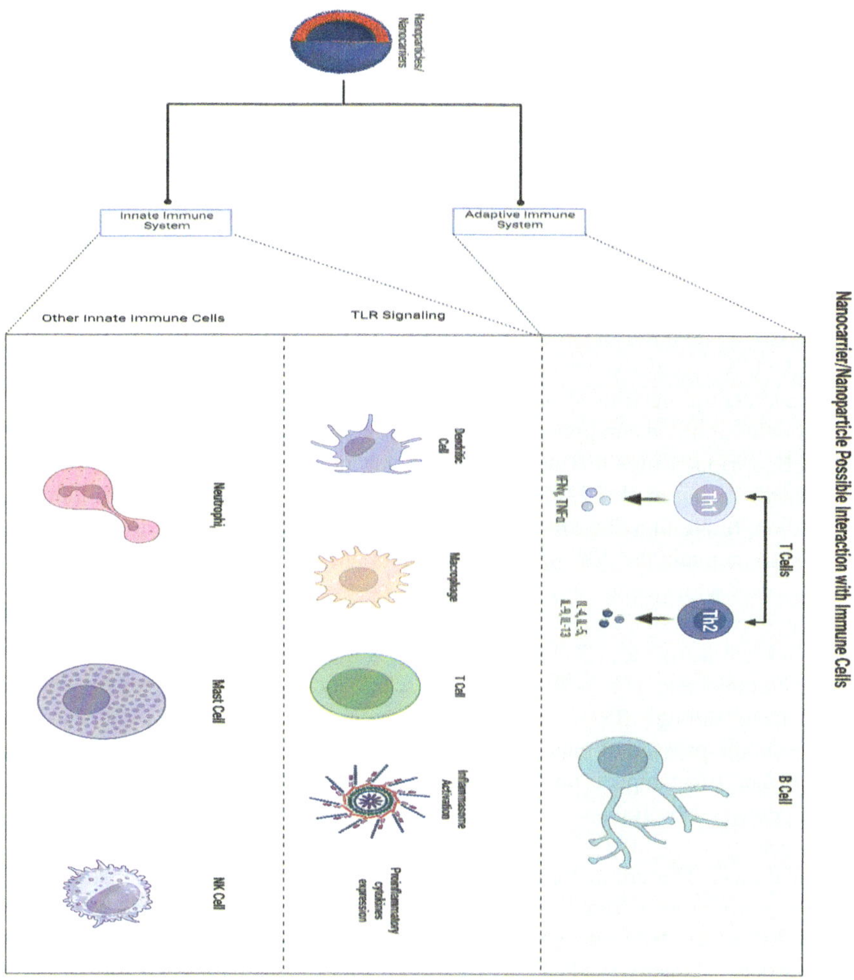

Fig. 8.3 Possible interaction between innate and adaptive immune system with metal-based nanoparticles/nanocarriers

schematics of nanocarriers/ nanoparticle's interaction with immune system. Various types of nanocarriers are used to deliver drugs to targeted cites including carbon, polymer and metal and metal oxide based nanocarriers to be more common. Furthermore, there is vast gap needed to be filled by research to know how nanoparticles do behave when they come in contact with human blood. Immunotoxic impacts of nanocarriers are broad spectrum based on the roue of administration and size; shape of nanocarriers, ranging from lungs, systematic damage, to liver and acute inflammation. If the nanocarriers are introduced intravenously, the nanocarriers can come in undesired contact with various blood components, including red blood cells (RBCs), white blood cells (WBCs), macrophages and blood proteins which can alter the blood components, resulting is posing a threat to biocompatibility, safety and biodegradability of the drug carrying nanoparticles (Hannon et al. 2019).

8.3.2 Blood–Brain Barrier (BBB)

Nanocarriers are not only suitable to deliver medicine to mutated tumour in lower parts of body (i.e., lungs, liver, kidneys, etc.) but due to small size and altered and enhanced photo and physiochemical properties, they have been studied for delivering drugs to upper part of body (Brain) also. Nanocarriers are being studied for drug delivery to treat various neurodegenerative diseases, since they have been noticed to cross the blood–brain barrier (BBB). BBB is a protective barrier that safeguards the brain and central nervous system (CNS) from being coming contact with larger or alien bodies or blood components. Due to highly selective and semipermeable nature of BBB, various drugs that are meant to reach brain to treat various diseases are precluded (Zhang et al. 2021). The main function of BBB is to maintain transport regulation, tissue function and protection of brain and CNS, further the duties also include to maintain the flow of essential metabolites that include water, glucose and amino acids with maintaining the brain homeostasis (Obermeier et al. 2013). The BBB is structured of capillary basement membrane (BM), endothelial cells of CNS, pericytes and astrocytes (Ballabh et al. 2004). Figure 8.4a shows the schematics of BBB structure and Fig. 8.4b shows the schematics of transportation of various components through BBB. Although there have been various developments and studies on the passage of nanocarriers moving across the BBB, on various animals, but still their toxic impacts on brain, CNS and BBB are still long from understood. A study published by Rabanel, J.-M., et al. reports the use of PEGylated PLA (Poly lactic acid) for to study their bioavailability across the BBB and their entry into neuronal cells. The group found that 90 and 100 nm sized polymeric nanocarriers were easily moving across the BBB of zebra fish, and found that after 24 h, ~80% of zebra fish eggs hatched when exposed to 133 μg/mL concentration of the nanocarriers (Rabanel et al. 2020). But still, a vast study is needed to assess the toxicity of the nanocarriers and nanoparticles across the BBB. To move freely across the BBB, nanoparticles/nanocarriers may disrupt the tight connections between endothelial cells, which may result in a temporary or permanent dysfunction of endothelial cell junctions, and there may be uncontrolled flow of other unwanted material also, that can disrupt the normal functionality of brain and CNS, resulting in causing toxicity (Ceña and Játiva 2018). Even though the above-mentioned study about the PEGylated PLA nanocarriers resulted in the hatching of ~80% of zebra fish eggs, yet, these were not loaded with any drug, the question arises that what if the drug crossing through the BBB, may be more toxic, and how much would be the safe concentration of drug and nanoparticles that can cross through the BBB without causing a permanent damage and with zero or minimal toxicity? Furthermore, a study conducted by Sharma, H. S., et al., evaluated the permeability of Silver (Ag) and Copper (Cu) nanoparticles, through BBB, in rat models. The group mentioned that all studies were carried out by following the guidelines provided

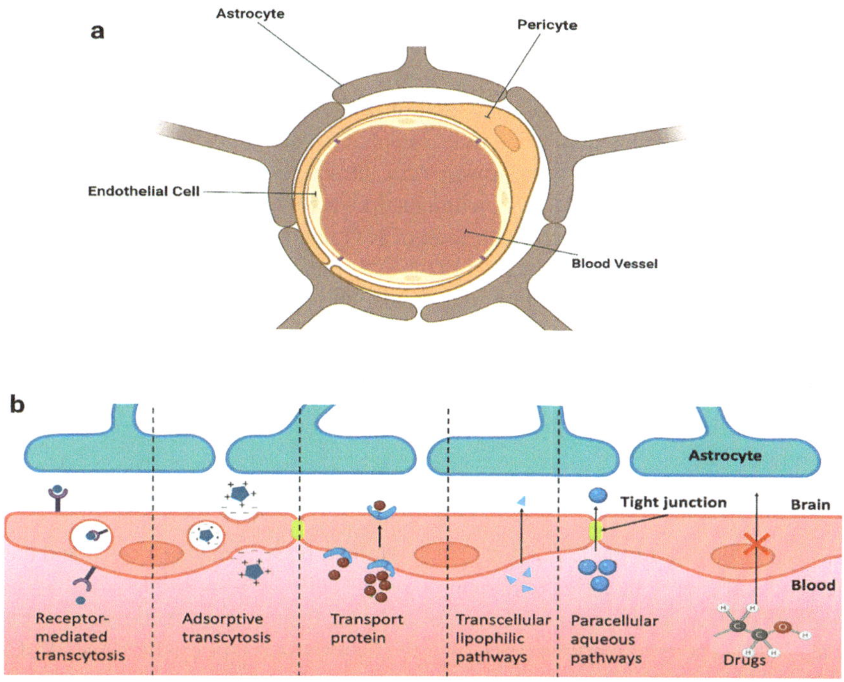

Fig. 8.4 (**a**) schematic structure of Blood–Brain Barrier (BBB) and (**b**) shows schematics of component transportation across healthy BBB

by National Institute of Health (NIH) of the United States of America (USA). The group revealed that the nanoparticles penetrated the BBB of rat models and caused alterations in myelin and glial cells other than causing nerve cell damages in various regions of brain (Ikuta et al. 2022). All these issues combined together limit the use of nanoparticles for drug carrying purposes, unless these issues are answered.

8.4 Toxicity of Nanocarriers

Although the major hindrances in studying or approving the nanocarriers for medical use are IS and BBB, but there are reports on various other toxicities of nanocarriers also, in lab study stages. The metal, metal oxide and polymeric nanocarriers have now been studied for their biomedical applications, for quite a time now. Various studies report the bare and drug loaded nanoparticles have potential in treating various fatal diseases, and reduce the systematic toxicity of drugs, but still, when cytotoxicity of nanocarriers remains a major question. Although various studies, when performed haemolytic studies, declared the use of nanoparticles to be safe for use in lower concentration, but the change and alteration in behaviour of nanoparticles/nanocarriers, when inside a living body, is yet to be investigated.

For instance, a study by Khalil, A. T., and group reported the genesis of Nickel Oxide (NiO) nanoparticles using distilled water-based extract of *Sageretia thea* leaves extract and studied them for various in vitro biomedical applications, i.e., anti-fungal, anti-cancer, anti-bacterial, anti-oxidant and anti-leishmanial. The results revealed very high potential of synthesized NiO nanoparticles against all tested strains and cell lines and even showed very attractive antioxidant properties, but when studied for haemolytic assay on human drawn blood, the nanoparticles showed ~22% of lysis at 200 μg/mL concentration of NiO nanoparticles (Khalil et al. 2018). There are various toxicities that can limit the use of nanoparticles and nanocarriers in their evolution from lab trials to human trials, and those toxicities can be against either cell, tissue and gene, other than previously explained toxicities against IS and BBB. Here we will discuss the toxicity of nanocarriers against cells and tissues and genotoxicity, one by one.

8.4.1 Cellular Toxicity of Nanocarriers

Cell is a basic unit of life (Zhu et al. 2019). Human body is made of cells, and they are food and energy factories of human body, the cells are responsible of delivering food and oxygen to all organs and parts of body for their proper functioning (Kumar et al. 2020). There are various types of cells, majorly two types, RBCs and WBCs. The function of RBCs is to carry oxygen and energy to organs (Scindia et al. 2019), while the function of WBCs is majorly defending body against infections and bacteria, fungal or viral attacks (Teoh 2023). If the WBCs somehow fail to contain the pathogenic bacterial or fungal cells or viral attack, the red blood cells get infected or mutated thus resulting in a sick or ill body (Bhattacharyya 2021). Therefore, to treat these diseases, various drugs are introduced that are toxic to attacking bacterial or fungal cells and/or virally mutated cells. Usually, anti-bacterial and anti-fungal drugs are not much toxic to human cells, but their misuse may cause various fatal side effects, like mutation in liver, etc. It is a well-known fact that now a days bacteria are becoming resistant to many available anti-bacterial drugs, since they have natural phenomenon to create resistance against any foreign entity, therefore, now either strong dose (i.e., up to 2 g per dose) or new generation of anti-microbial drugs are being prescribed. Various countries have put strong rules against the prescription of anti-bacterial drugs, but in many countries, especially third world countries, anti-bacterial drugs can be availed without production of prescription. On the other hand, anti-viral (i.e., anti-cancer) drugs are toxic to both mutated and normal healthy cell. Their major function is to kill a cell, regardless of diseased or healthy.

Therefore, a new range of medicines, Nanomedicine, are being researched to be evolution is the field of pharmacy and pharmacology. Nanomedicine are medicines that utilize nanomaterials for prognosis, diagnosis and treatment of various diseases. Nanomaterials can be used bare for introduction of toxicity to bacterial, fungal and cancerous cell or they can be used as a carrier for taking a drug to fungal, cancerous and/or bacterial cell. If the nanomaterial is opted to carry drug, it needs to be surface

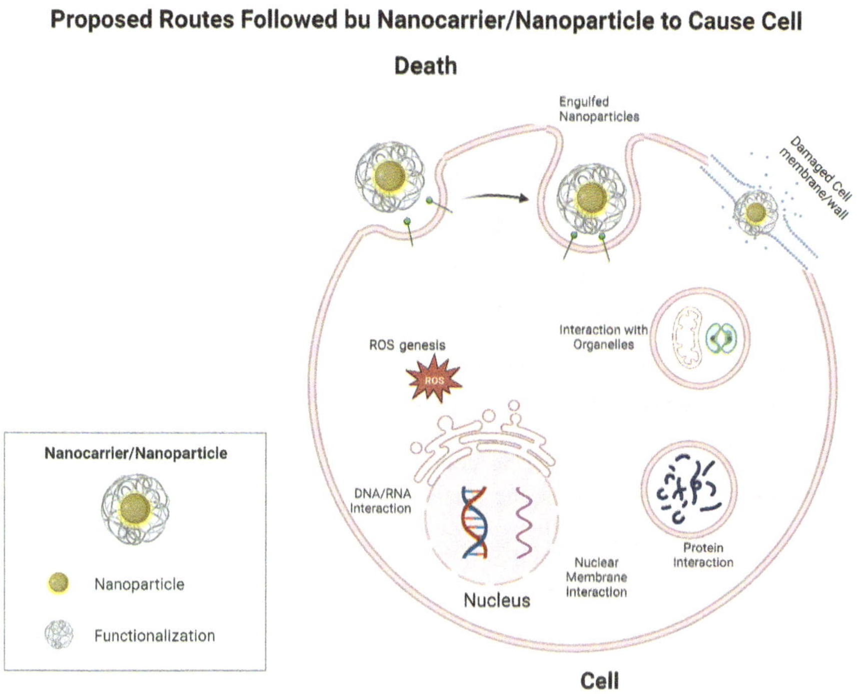

Fig. 8.5 Proposed routes followed by a nanocarrier or a nanoparticle to enter a cell and interact with the cell to cause cell death

functionalized with it or drug could be loaded inside the nanomaterial. In other terms such nanomaterials which carry a drug are termed as nanocarriers. Nanoparticles and nanocarriers used for their biomedical potential can be either, metal and metal oxide or polymeric. There are various possible ways by which a nanomaterial or nanocarriers can cause cell death. In a bacterial cell, a nanoparticle may follow one or more paths to kill it. The possible paths can be (1) generation of ROS, (2) Damage the electron transport chain, (3) Peptidoglycan disruption, (4) Cell membrane and/or cell wall disruption, (5) enzyme inactivation, (6) Protein denaturing, (7) DNA/RNA damaging and/or (8) Nuclear membrane damage, as shown in Fig. 8.5. A recent study published by Sani A., et al., reported green chemical synthesis of NiO nanoparticles by reducing the salt of Ni with the help of plant extract. The group then used the bio-modulated NiO nanoparticles against Hepatocellculr carcinoma (HepG2) cell lines and studied the toxic impact of biogenic nanoparticles. Group revealed that the possible route of cell death was due to ROS, while the group explains that there can be any possible route followed by nanoparticle ought to cause apoptosis (Sani et al. 2021). In another study, Khan, I., and group reported the synthesis of PLGA nanocarriers functionalized with Berberine and Doxirubicin (DOX) for co-delivery and study their toxicity against T47D and MDA-MB-231 breast cancer cell lines. Group defined the method of nanoparticles causing toxicity in cancer cell lines to be ROS generation and found out that nanoparticles were causing high rate of ROS generation at 25 μM concentration against MDA-MB-231 cell lines while against the T47D, the ROS generation was comparatively lower (Khan et al. 2019). Furthermore, Vaz-Ramos, J., and group generated Iron Oxide (IO) cored Silica (SiO_2) nanoparticles functionalized with Epirubicin (EPI) to study their toxicity against the HepG2 cell lines and rate of drug release over a period of 48 h. The group reported that EPI loaded IO.SiO_2 nanocarriers showed enhanced drug release even after 48 h of study while % release eached 38%, whereas, the nanocarriers apoptosis of almost ~90% cells when introduced to HepG2 cell lines with concentration of 100 μg/mL (Vaz-Ramos et al. 2020).

Although all results mentioned above show good toxicity against the cancer cell lines but in case of Sani A., et al., and Khan I., et al., reported the nanoparticles and Nanocarriers to be safe against the human blood drawn RBCs at very low concentration, while, as the nanoparticle concentration increased, the toxicity showed elevated results also. Whereas, the Vaz-Ramos, J., and group did not publish any report on toxicity of their synthesized nanoparticles against the normal and healthy blood cells. Furthermore, all studies published to date report the in vitro toxicity of nanoparticles and nanocarriers against healthy human blood, hence leaving a big question on feasibility of nanocarriers and nanoparticles to be used for the treatment on humans.

8.4.2 Gene Toxicity

Another possible rout nanocarriers and nanoparticles opt to cause death of a living cell (weather cancerous or healthy) can be by targeting the gene and causing genotoxicity. It is already a well-established fact that nanoparticles are known to generate Reactive Oxygen and Nitrogen Species (RONS) and during the treatment it is mandatory to keep a track on RONS concentrations being generated (Chen et al. 2023). If there is any imbalance in riddance and generation of RONS, and the RONS start accruing, they can cause damage to DNA, intracellular lipids and/or proteins, resulting in genotoxicity (Moldogazieva et al. 2018). And if this imbalance happens, it can trigger a chain reaction, resulting in immunogenicity and acute inflammation (Yang and Merlin 2023). Other than that it is also reported that various metal nanoparticles can catalyse the generation of reactive hydroxyl radicals and oxygen driven metabolites that engender lipid peroxidation (Paciorek et al. 2020), which can result in malfunction of various cellular organelles including plasma membrane, mitochondria and endoplasmic reticulum (ER) (Husain and Mahmood 2019).

Cells, in routine, produce superoxide dismutase (SOD), reduced glutathione (GSH) and α-lipoic acid to defuse RONS that can also cause oxidative stress and stop cell from countering RONS, in return causing genotoxicity (Yang and Merlin 2023). It has been found in studies also that ZnO nanoparticles with concentration of 50 mg/kg concentration largely upset the SOD genesis and causing acute intestinal damages in studied rats (Abbasalipourkabir et al. 2015). In addition, it has also been

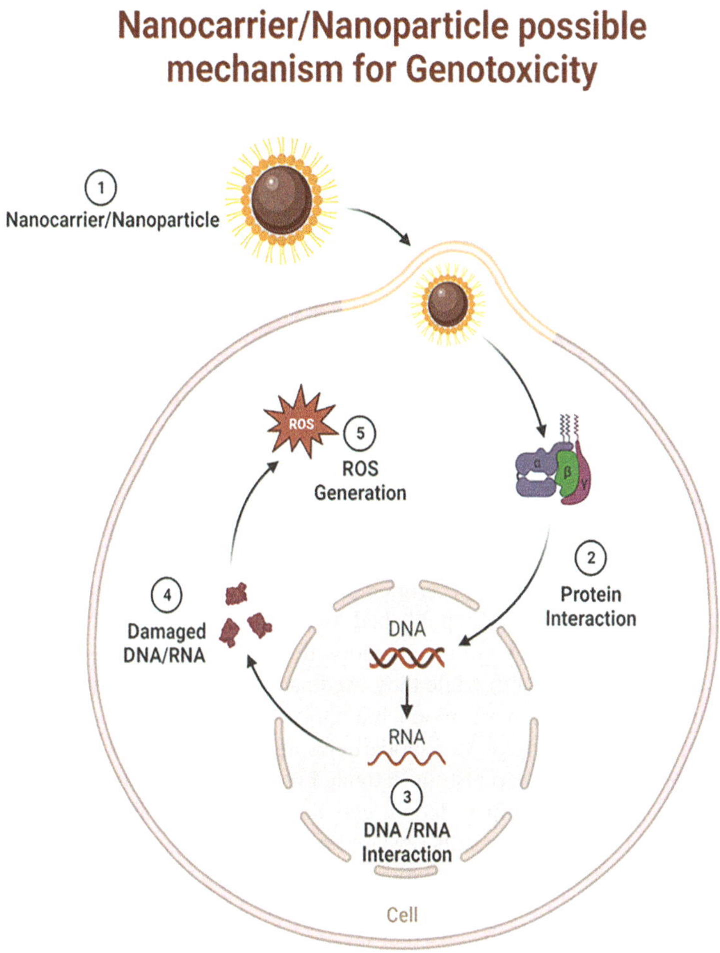

Fig. 8.6 Possible mechanisms of Nanocarriers/nanoparticle for their genotoxicity potential

found that Titania (TiO_2) nanoparticles gravely disrupt the ration between oxidized glutathione (GSSG) and GSH causing elevated concentration of ROS in hepatic cells resulting in lipid peroxidation, and rise in micronucleus frequency—that is one of the potential signs for genotoxicity (Shakeel et al. 2016). Whereas, it has also been found in case of copper (Cu) driven CuO nanoparticles that they tend to disturb the transmembrane potential of mitochondria and reduce the GSH levels in blood, resulting in altering the normal function of macrophages (Abudayyak et al. 2020). According to literature there can be three major types of mechanisms that a nanoparticle/nanocarrier can follow to cause genotoxicity in a cell (Landsiedel et al. 2022). Figure 8.6 shows possible mechanisms of Nanocarriers/nanoparticle for their genotoxicity potential.

- Direct interaction between nanocarriers/nanoparticles and DNA (primary mechanisms).
- The interaction between nanocarriers/nanoparticles with molecules that take part in cell division or DNA replication (primary indirect mechanisms).
- Nanocarriers/nanoparticles caused genesis of ROS that damage or disrupt the DNA (secondary mechanisms).

The toxicity of nanocarriers largely depend upon the rate of uptake of nanocarriers inside the cell, whereas the rate of uptake depends on the size of nanocarriers (Landsiedel et al. 2009). As reported by El-Shorbagy, H., and group, used Zinc Oxide (ZnO) nanoparticles bought from Sigma Aldrich and used them for studying their effect on *P53, Bcl2* and *Bax* genes for apoptotic induction in Ehrlich solid carcinoma (ESC) bearing mice models, by feeding them 50, 300 and 500 mg of nanoparticles per kg of their weight for a week continuously after every 24 h. The results revealed incline in the *P53* and *Bax* genes while decline in *Bcl2* genes with significant shrink in the tumour size of ZnO nanoparticles introduced mice. The tumour size was 46% smaller in size compared to control. Furthermore, histopathological analysis revealed massive infiltration of mononuclear inflammatory cells in the portal triad and vacuolar degeneration of hepatocytes (El-Shorbagy et al. 2019).

Another commentary study published by Wang, Y., and group reports the genotoxicity evaluation of Gold (Au) nanoparticles. The group reported the size of Au nanoparticles to be 14 nm while they used various concentration (0.1 to 100 μg/mL) against Human peripheral blood lymphocytes (PBLs), and 6.2 to 50 μg/mL concentration of 12 nm sized Au nanoparticles against BALB mice 3 T3 fibroblast cells. Furthermore, the group also investigated the genotoxicity of 10, 30 or 60 nm sized Au nanoparticles against HepG2 cell lines at concentration of 10 ppm via following comet assay. The results revealed noticeable damage in DNA and oxidative DNA damage. While in vivo study involved Rats and fish. The study conducted on rats revealed DNA damage, while the rats were fed with Au nanoparticles for 28 days. For the fish, the results raised DNA damage even when fish were exposed to 80 μg/L concentration of nanoparticles for 72 h (Wang et al. 2020).

Another study carried by Schulz et al., involved the introduction of Au nanoparticles to rats lugs via intratracheal installation for 72 h, though the study explained that there was no significant damage in DNA of the rats exposed to Au nanoparticles compared to control rats, but the question yet remains is the time of exposure. There should be various routes of study to explain the impact of different concentration of varied size and shaped nanoparticles over varied units of time (Schulz et al. 2012). Therefore, nanocarriers and nanoparticles, even after being very beneficial towards treatments of various fatal diseases, are still being seen as a potential hazard since their safety and toxicity is not fully known, which limit their progress.

8.4.3 Accumulation in Organs

Apart from toxicity, another issue that limits the progression of nanocarriers/nanoparticles being risen to the status of next generation of medicine is their accumulation inside organs. There have been reports of nanoparticles getting accumulated in liver, lungs, lymph nodes, brain, kidneys and others. In liver the nanoparticles tend to accumulate due to macrophage (Gustafson et al. 2015). Every drug or chemical that is entering a living body have to pass through the liver, in the same way, nanocarriers also find their way inside the liver (Boey and Ho 2020). More than 30% of all administered nanocarriers or nanomedicine into any animal body tend to get accumulated in the liver, resulting in enhanced toxicity to liver and reduce targeting application for treatment of any disease (Zhang et al. 2016). Jiang, X., and group published research about the synthesis of Indocyanine Green (ICG) and GSH loaded Au nanoparticles and studied their accumulation inside BALB/c mice models. The researchers found that 70% of all administered ICG-GSH-AU nanoparticles tend to accumulate in liver while their concentration decrease with time (Jiang et al. 2019).

Other than liver, nanocarriers have also been reported to be accumulated inside other organs as well. Though the longer retention time gives additional benefits to nanocarriers, i.e., increased bioavailability, but in return can also cause enhanced toxicity. Brain is also one of the most important organs of a human body. There have been various research studies to produce nanomedicines that can cross the highly selective BBB to treat various neurodegenerative diseases. Many of the studies have reported promising results, but their tendency to accumulate inside brain raises a major question about their toxicity. In previous topics we have already discussed the damage that nanocarriers may cause to BBB. It is impossible to quantitatively discern between endothelium accumulation and transit into the parenchyma from the published data on nanoparticle accumulation in the brain in vivo (Betzer et al. 2017). The factors of size and shape may impact the accumulation and retention rate of nanoparticles inside brain cells (Nowak et al. 2020). Kolhar and group reported the accumulation of anti-TfR anti-body functionalized poly-styrene (PS) nanoparticles with two different shapes, i.e., rods and spherical, and found the rod shaped nanoparticles tend to accumulate in higher concentration into brain compared to spherical shaped PS nanoparticles (Kolhar et al. 2013). while in another study by Arnida, M. et al., the accumulation of sphere shaped PEGylated Au nanoparticles was higher in brain compared to rod shaped counterparts (Arnida et al. 2011).

Other than these organs there is a study carried by Bartucci, R., et al., that reports the accumulation of PS nanoparticles functionalized carboxylate into liver lungs and kidneys. The size of nanoparticles was measured to be 40 nm while their accumulation was studies on C57BL/6 J mice models. The group found accumulation of nanoparticles inside all studied organs and reported that the rate of nanoparticles accumulation inclined with the increase in time. Furthermore, the group reported 40% of nanoparticles in first hour of exposure to lungs tissues while about 10% of the nanoparticles accumulated in liver in the same time. While after 2 days of

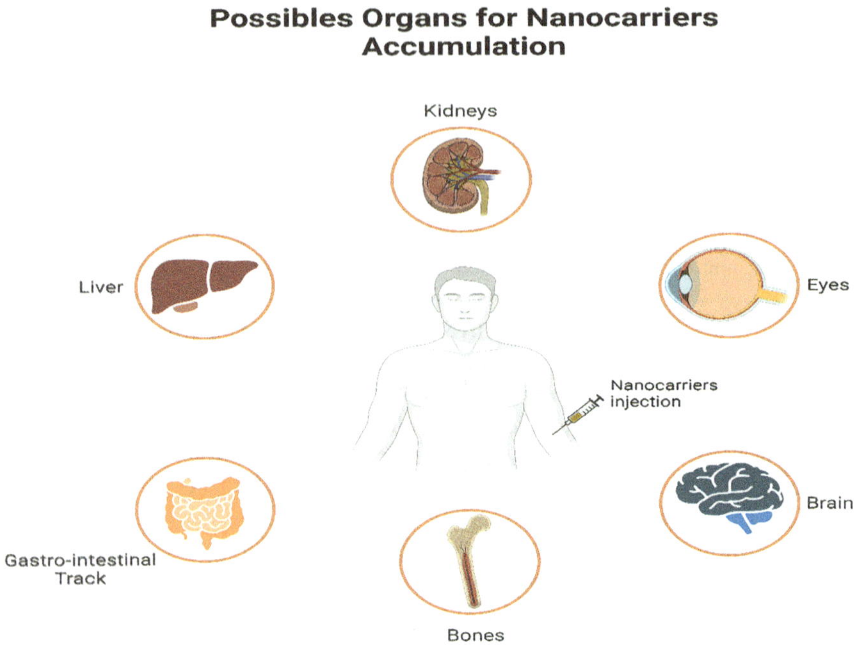

Fig. 8.7 Possible organs, where nanoparticles/nanocarriers can accumulate

exposure more than 80% of liver, lungs and kidneys macrophages contain PS nanoparticles in them. Whereas the concentration of PS nanoparticles accumulation inside kidney macrophages was less in comparison to liver and lungs (Bartucci et al. 2020).

Apart from nanocarriers accumulation in liver, lungs, kidneys and brain, the nanocarriers can get accumulated inside lymphatics, bines and eyes (Busatto et al. 2019). Figure 8.7 shows various organs, where nanoparticles/nanocarriers can possibly accumulate.

8.5 Scale-Up of Nanocarriers

Nanotechnology is interfacing technology with many other technologies, for instance, nano-chemistry (Li and Jin 2020), nano-electronics (Meteab et al. 2023), nanobiotechnology (Hassan et al. 2018), nanomedicine (Sani et al. 2023) and many more. Currently vast majority of researchers are turning to nanotechnology due to dramatic changes in the properties of materials when they reach nano realm (Cheng et al. 2021). A large number of those researchers is working the biomedical applications of nanomaterials (Abd Elkodous et al. 2019), focusing on the genesis of nanomaterials that can carry drug to reach various parts of body to treat diseases with decreased systematic toxicity (Sharifi-Rad et al. 2021). But before going to use them for acquired properties, materials must be brought to nanoscale. There are two major approaches that can be employed to reach nano realm: (1) Bottom-Up and (2) Top-Down (Camargos and Rezende 2021). In the Bottom-Up approach atoms or molecules are joined together to reach the nano size, whereas in case of Top-Down approach, bulk sized materials are broken down or chemically reduced to reach nano scale range. In the Top-Down approach there are various methods to get desired material of nanoscale range that may include chemical degradation of bulk materials, biological degradation (Ijaz et al. 2021), physical degradation (Salem and Fouda 2021), or biochemical degradation (Hassan et al. 2019) of bulk materials. Controlling materials at atomic or molecular scale is difficult (Khajetoorians et al. 2019) and needs high economical and energy resources (Jia et al. 2019) while to break down material from bulk is easier. Most widely used method to break a bulk material to obtain nanomaterial is co-precipitation method (Selvaraju et al. 2019) (chemical method), while it has its drawn backs also. Major drawbacks include the generation of bio- and environmentally toxic by-products (Sani et al. 2021). Now a days, a new approach is being research to obtain desired materials i.e., biochemical, or green chemical method (Khalil et al. 2018). The green chemical method employs the use of plant extracts for the reduction of metal salts of desired material.

The generation of materials at lab scale for small scale research is easy but scaling up of that same material at industrial scale is quite difficult (Wu et al. 2020). There are various issues that can be faced while trying to scale-up the synthesis of nanomaterials, that may include, but not limited to, economic issues, manpower, safety issues, toxicity issues to workers and most majorly environmental impacts. Other than that controlling the size and shape of nanocarriers/nanoparticles is quite difficult and challenging since minor pressure and temperature changes can cause variations. Furthermore, it is highly unlikely to reproduce the same sized and shaped nanomaterials under same conditions (Paliwal et al. 2014). Other than that, the % yield of production is also a major hindrance in scaling up the production of nanocarriers. Furthermore, the raw materials required for desired materials also varies in acquisition price.

Even though the production of nanocarriers could be started, the question about the toxicological impacts on the works remain a major question. Until today, the safety and bio- and environmental degradability of nanocarriers/nanoparticles remains unclear (Klaper 2020), therefore, getting approvals for hiking up the production of nanomaterials at large scale discouraged. Furthermore, the degradation of nanomaterials inside environment is also needing to be answered (Klaper 2020). It is still unknown that how nanomaterials behave when they escape into the environment and how deep they penetrate our environment and what is their shelf life before they get degraded completely. Furthermore, their behaviour towards the micro-organism living in the environment and other living things like animals and plants are also the aspects that need a lot of attention. Therefore, all these factors combined hinder the development and scaling up of nanomaterials/nanocarriers for their use in the field of medical industry.

8.6 Conclusion

Nanomaterials when loaded with a drug or molecule are called as nanocarriers. Even though the nanocarriers have many advantages over the classically used medicines, but there are various factors that breaking the speed of their development and growth. The factors like the toxicity of nanocarriers when a living system is exposed to them need high priority attentions, since their impacts and behaviour when they enter a living system are unknown. Their can be various factors that can impact the action and targeting properties of nanocarriers when they enter a living body. The first hindrance nanocarriers will face is immune system. Immune system has the job of capturing and eliminating the alien entities entering living body, since nanocarriers are not part of living body, the immune system treats them as a foreign body and tries to eliminate them, or else, captures and encapsulates them, which limit the mode of action and bioavailability if the nanocarriers. Second barrier nanoparticles can face inside an animal or human body is BBB, the BBB is major protector of brain and nervous system, if breached there can be various grave issues to the exposed human or animal body. Furthermore, nanocarriers tend to produce various toxicities to cells, genes, tissues, muscles and organs followed by the issue of nanocarriers accumulation inside organs and muscles largely question the development of nanocarriers and act as an obstacle towards their growths. Even if all these issues are ignored, the scale-up of nanocarriers remains a major issue also, since societal impacts of nanocarriers/nanomaterials are unknown. Therefore, before going to test nanocarriers on human or enhancing their production, all these issues must be answered.

References

Abbasalipourkabir R, Moradi H, Zarei S, Asadi S, Salehzadeh A, Ghafourikhosroshahi A, Mortazavi M, Ziamajidi N (2015) Toxicity of zinc oxide nanoparticles on adult male Wistar rats. Food Chem Toxicol 84:154–160

Abd Elkodous M, El-Sayyad GS, Abdelrahman IY, El-Bastawisy HS, Mohamed AE, Mosallam FM, Nasser HA, Gobara M, Baraka A, Elsayed MA, El-Batal AI (2019) Therapeutic and diagnostic potential of nanomaterials for enhanced biomedical applications. Colloids Surf B: Biointerfaces 180:411–428

Abudayyak M, Guzel E, Özhan G (2020) Cupric oxide nanoparticles induce cellular toxicity in liver and intestine cell lines. Adv Pharm Bull 10(2):213

Ahmad S, Idris RAM, Wan Hanaffi WN, Perumal K, Boer JC, Plebanski M, Jaafar J, Lim JK, Mohamud R (2021) Cancer nanomedicine and immune system—interactions and challenges. Front Nanotechnol 3:681305

Aljabali AA, Obeid MA, Bashatwah RM, Serrano-Aroca Á, Mishra V, Mishra Y, El-Tanani M, Hromić-Jahjefendić A, Kapoor DN, Goyal R, Naikoo GA, Tambuwala MM (2023) Nanomaterials and their impact on the immune system. Int J Mol Sci 24(3):2008

Alyassin Y, Sayed EG, Mehta P, Ruparelia K, Arshad MS, Rasekh M, Shepherd J, Kucuk I, Wilson PB, Singh N, Chang M-W, Fatouros DG, Ahmad Z (2020) Application of mesoporous silica nanoparticles as drug delivery carriers for chemotherapeutic agents. Drug Discov Today 25(8): 1513–1520

Ansari MO, Parveen N, Ahmad MF, Wani AL, Afrin S, Rahman Y, Jameel S, Khan YA, Siddique HR, Tabish M (2019) Evaluation of DNA interaction, genotoxicity and oxidative stress induced by iron oxide nanoparticles both in vitro and in vivo: attenuation by thymoquinone. Sci Rep 9(1):1–14

Arnida MM, Ray A, Peterson CM, Ghandehari H (2011) Geometry and surface characteristics of gold nanoparticles influence their biodistribution and uptake by macrophages. Eur J Pharm Biopharm 77(3):417–423

Ballabh P, Braun A, Nedergaard M (2004) The blood–brain barrier: an overview: structure, regulation, and clinical implications. Neurobiol Dis 16(1):1–13

Bartucci R, Paramanandana A, Boersma YL, Olinga P, Salvati A (2020) Comparative study of nanoparticle uptake and impact in murine lung, liver and kidney tissue slices. Nanotoxicology 14(6):847–865

Betzer O, Shilo M, Opochinsky R, Barnoy E, Motiei M, Okun E, Yadid G, Popovtzer R (2017) The effect of nanoparticle size on the ability to cross the blood–brain barrier: an in vivo study. Nanomedicine 12(13):1533–1546

Bhattacharyya SE (2021) Intelligent modeling, prediction, and diagnosis from epidemiological data: COVID-19 and beyond. CRC, Boca Raton, FL

Blanchard N, Salvioni A, Robey EA (2020) In: Weiss LM, Kim K (eds) Toxoplasma gondii, 3rd edn. Academic Press, pp 1107–1146

Boey A, Ho HK (2020) All roads lead to the liver: metal nanoparticles and their implications for liver health. Small 16(21):2000153

Bomfim GF, Rodrigues FL, Carneiro FS (2017) Are the innate and adaptive immune systems setting hypertension on fire? Pharmacol Res 117:377–393

Busatto S, Pham A, Suh A, Shapiro S, Wolfram J (2019) Organotropic drug delivery: synthetic nanoparticles and extracellular vesicles. Biomed Microdevices 21(2):46

Camargos CHM, Rezende CA (2021) Antisolvent versus ultrasonication: bottom-up and top-down approaches to produce lignin nanoparticles (LNPs) with tailored properties. Int J Biol Macromol 193:647–660

Ceña V, Játiva P (2018) Nanoparticle crossing of blood–brain barrier: a road to new therapeutic approaches to central nervous system diseases. Nanomedicine 13(13):1513–1516

Chen Q, Nan Y, Yang Y, Xiao Z, Liu M, Huang J, Xiang Y, Long X, Zhao T, Wang X, Huang Q, Ai K (2023) Nanodrugs alleviate acute kidney injury: manipulate RONS at kidney. Bioact Mater 22:141–167

Cheng Z, Li M, Dey R, Chen Y (2021) Nanomaterials for cancer therapy: current progress and perspectives. J Hematol Oncol 14(1):85

Costa da Silva M, Vieira Rocha C, Bañobre-López M, Gallo J (2021) Stimulation and suppression of the innate immune system through nanotechnology. ACS Appl Nano Mater 4(3):2303–2316

Deng Y, Wang M, Zhuang Y, Liu S, Huang W, Zhao Q (2021) Circularly polarized luminescence from organic micro–/nano-structures. Light Sci Appl 10(1):76

Dilawar H, Sidra A, Amber Rehana S, Saima QM (2019) Chapter 5. In: Ishaq A, Tingkai Z (eds) Ion beam techniques and applications. IntechOpen, Rijeka

El-Shorbagy H, Eissa S, Sabet S, El-Ghor A (2019) Apoptosis and oxidative stress as relevant mechanisms of antitumor activity and genotoxicity of ZnO-NPs alone and in combination with N-acetyl cysteine in tumor-bearing mice. Int J Nanomedicine 14:3911–3928

Esfandyari-Manesh M, Mostafavi SH, Majidi RF, Koopaei MN, Ravari NS, Amini M, Darvishi B, Ostad SN, Atyabi F, Dinarvand R (2015) Improved anticancer delivery of paclitaxel by albumin surface modification of PLGA nanoparticles. DARU J Pharm Sci 23(1):28

Fontana F, Figueiredo P, Bauleth-Ramos T, Correia A, Santos HA (2018) Immunostimulation and immunosuppression: nanotechnology on the brink. Small Methods 2(5):1700347

Galstyan A, Markman JL, Shatalova ES, Chiechi A, Korman AJ, Patil R, Klymyshyn D, Tourtellotte WG, Israel LL, Braubach O, Ljubimov VA, Mashouf LA, Ramesh A, Grodzinski ZB, Penichet ML, Black KL, Holler E, Sun T, Ding H, Ljubimov AV, Ljubimova JY (2019) Blood–brain barrier permeable nano immunoconjugates induce local immune responses for glioma therapy. Nat Commun 10(1):3850

Gustafson HH, Holt-Casper D, Grainger DW, Ghandehari H (2015) Nanoparticle uptake: the phagocyte problem. Nano Today 10(4):487–510

Hannon G, Lysaght J, Liptrott NJ, Prina-Mello A (2019) Immunotoxicity considerations for next generation cancer nanomedicines. Adv Sci 6(19):1900133

Hassan D, Khalil AT, Saleem J, Diallo A, Khamlich S, Shinwari ZK, Maaza M (2018) Biosynthesis of pure hematite phase magnetic iron oxide nanoparticles using floral extracts of Callistemon viminalis (bottlebrush): their physical properties and novel biological applications. Artif Cells Nanomed Biotechnol 46(sup1):693–707

Hassan D, Khalil AT, Solangi AR, El-Mallul A, Shinwari ZK, Maaza M (2019) Physiochemical properties and novel biological applications of Callistemon viminalis-mediated α-Cr2O3 nanoparticles. Appl Organomet Chem 33(8):e5041

Huang L-H, Sun X-Y, Ouyang J-M (2019) Shape-dependent toxicity and mineralization of hydroxyapatite nanoparticles in A7R5 aortic smooth muscle cells. Sci Rep 9(1):1–18

Husain N, Mahmood R (2019) Copper(II) generates ROS and RNS, impairs antioxidant system and damages membrane and DNA in human blood cells. Environ Sci Pollut Res 26(20): 20654–20668

Ijaz M, Zafar M, Iqbal T (2021) Green synthesis of silver nanoparticles by using various extracts: a review. Inorg Nano-Met Chem 51(5):744–755

Ikuta KS, Swetschinski LR, Robles Aguilar G, Sharara F, Mestrovic T, Gray AP, Davis Weaver N, Wool EE, Han C, Gershberg Hayoon A, Aali A, Abate SM, Abbasi-Kangevari M, Abbasi-Kangevari Z, Abd-Elsalam S, Abebe G, Abedi A, Abhari AP, Abidi H, Aboagye RG, Absalan A, Abubaker Ali H, Acuna JM, Adane TD, Addo IY, Adegboye OA, Adnan M, Adnani QES, Afzal MS, Afzal S, Aghdam ZB, Ahinkorah BO, Ahmad A, Ahmad AR, Ahmad R, Ahmad S, Ahmad S, Ahmadi S, Ahmed A, Ahmed H, Ahmed JQ, Ahmed Rashid T, Ajami M, Aji B, Akbarzadeh-Khiavi M, Akunna CJ, Al Hamad H, Alahdab F, Al-Aly Z, Aldeyab MA, Aleman AV, Alhalaiqa FAN, Alhassan RK, Ali BA, Ali L, Ali SS, Alimohamadi Y, Alipour V, Alizadeh A, Aljunid SM, Allel K, Almustanyir S, Ameyaw EK, Amit AML, Anandavelane N, Ancuceanu R, Andrei CL, Andrei T, Anggraini D, Ansar A, Anyasodor AE, Arabloo J, Aravkin AY, Areda D, Aripov T, Artamonov AA, Arulappan J, Aruleba RT, Asaduzzaman M, Ashraf T, Athari SS, Atlaw D, Attia S, Ausloos M, Awoke T, Ayala Quintanilla BP, Ayana TM, Azadnajafabad S, Azari Jafari A et al (2022) Global mortality associated with 33 bacterial pathogens in 2019: a systematic analysis for the global burden of disease study 2019. Lancet 400(10369):2221–2248

Islam MA, Haque MA, Rahman MA, Hossen F, Reza M, Barua A, Marzan AA, Das T, Kumar Baral S, He C, Ahmed F, Bhattacharya P, Jakariya M (2022) A review on measures to rejuvenate immune system: natural mode of protection against coronavirus infection. Front Immunol 13:837290

Jia X, Khan W, Wu Z, Choi J, Yip ACK (2019) Modern synthesis strategies for hierarchical zeolites: bottom-up versus top-down strategies. Adv Powder Technol 30(3):467–484

Jiang X, Du B, Zheng J (2019) Glutathione-mediated biotransformation in the liver modulates nanoparticle transport. Nat Nanotechnol 14(9):874–882

Kang JK, Kim JC, Shin Y, Han SM, Won WR, Her J, Park JY, Oh KT (2020) Principles and applications of nanomaterial-based hyperthermia in cancer therapy. Arch Pharm Res 43(1): 46–57

Khajetoorians AA, Wegner D, Otte AF, Swart I (2019) Creating designer quantum states of matter atom-by-atom. Nat Rev Phys 1(12):703–715

Khalil AT, Ovais M, Ullah I, Ali M, Shinwari ZK, Hassan D, Maaza M (2018) Sageretia thea (Osbeck.) modulated biosynthesis of NiO nanoparticles and their in vitro pharmacognostic, antioxidant and cytotoxic potential. Artif Cells Nanomed Biotechnol 46(4):838–852

Khan I, Joshi G, Nakhate KT, Ajazuddin, Kumar R, Gupta U (2019) Nano-co-delivery of berberine and anticancer drug using PLGA nanoparticles: exploration of better anticancer activity and in vivo kinetics. Pharm Res 36(10):149

Klaper RD (2020) The known and unknown about the environmental safety of nanomaterials in commerce. Small 16(36):2000690

Kolhar P, Anselmo AC, Gupta V, Pant K, Prabhakarpandian B, Ruoslahti E, Mitragotri S (2013) Using shape effects to target antibody-coated nanoparticles to lung and brain endothelium. Proc Natl Acad Sci 110(26):10753–10758

Kraus RF, Gruber MA (2021) Neutrophils—from bone marrow to first-line defense of the innate immune system. Front Immunol 12:767175

Kumar P, Mahajan P, Kaur R, Gautam S (2020) Nanotechnology and its challenges in the food sector: a review. Mater Today Chem 17:100332

Landsiedel R, Kapp MD, Schulz M, Wiench K, Oesch F (2009) Genotoxicity investigations on nanomaterials: methods, preparation and characterization of test material, potential artifacts and limitations—many questions, some answers. Mutat Res 681(2):241–258

Landsiedel R, Honarvar N, Seiffert SB, Oesch B, Oesch F (2022) Genotoxicity testing of nanomaterials. Wiley Interdiscip Rev Nanomed Nanobiotechnol 14(6):e1833

Li Y, Jin R (2020) Seeing ligands on nanoclusters and in their assemblies by X-ray crystallography: atomically precise nanochemistry and beyond. J Am Chem Soc 142(32):13627–13644

Liu Y, Yang G, Jin S, Xu L, Zhao C-X (2020) Development of high-drug-loading nanoparticles. ChemPlusChem 85(9):2143–2157

Lung P, Yang J, Li Q (2020) Nanoparticle formulated vaccines: opportunities and challenges. Nanoscale 12(10):5746–5763

Makabenta JMV, Nabawy A, Li C-H, Schmidt-Malan S, Patel R, Rotello VM (2021) Nanomaterial-based therapeutics for antibiotic-resistant bacterial infections. Nat Rev Microbiol 19(1):23–36

Mattiola I, Diefenbach A (2023) Regulation of innate immune system function by the microbiome: consequences for tumor immunity and cancer immunotherapy. Semin Immunol 66:101724

McKechnie JL, Blish CA (2020) The innate immune system: fighting on the front lines or fanning the flames of COVID-19? Cell Host Microbe 27(6):863–869

Mendes CR, Dilarri G, Forsan CF, Sapata VDMR, Lopes PRM, de Moraes PB, Montagnolli RN, Ferreira H, Bidoia ED (2022) Antibacterial action and target mechanisms of zinc oxide nanoparticles against bacterial pathogens. Sci Rep 12(1):2658

Meteab MH, Hashim A, Rabee BH (2023) Controlling the structural and dielectric characteristics of PS-PC/Co2O3-SiC hybrid nanocomposites for nanoelectronics applications. SILICON 15(1): 251–261

Mohammed YH, Holmes A, Haridass IN, Sanchez WY, Studier H, Grice JE, Benson HAE, Roberts MS (2019) Support for the safe use of zinc oxide nanoparticle sunscreens: lack of skin penetration or cellular toxicity after repeated application in volunteers. J Investig Dermatol 139(2):308–315

Moldogazieva NT, Mokhosoev IM, Feldman NB, Lutsenko SV (2018) ROS and RNS signalling: adaptive redox switches through oxidative/nitrosative protein modifications. Free Radic Res 52(5):507–543

Morrison L, Zembower TR (2020) Antimicrobial resistance. Gastrointest Endosc Clin N Am 30(4): 619–635

Nowak M, Brown TD, Graham A, Helgeson ME, Mitragotri S (2020) Size, shape, and flexibility influence nanoparticle transport across brain endothelium under flow. Bioeng Transl Med 5(2): e10153

Obermeier B, Daneman R, Ransohoff RM (2013) Development, maintenance and disruption of the blood-brain barrier. Nat Med 19(12):1584–1596

Paciorek P, Żuberek M, Grzelak A (2020) Products of lipid peroxidation as a factor in the toxic effect of silver nanoparticles. Materials 13(11):2460

Paliwal R, Babu RJ, Palakurthi S (2014) Nanomedicine scale-up technologies: feasibilities and challenges. AAPS PharmSciTech 15(6):1527–1534

Panáček A, Kvítek L, Smékalová M, Večeřová R, Kolář M, Röderová M, Dyčka F, Šebela M, Prucek R, Tomanec O, Zbořil R (2018) Bacterial resistance to silver nanoparticles and how to overcome it. Nat Nanotechnol 13(1):65–71

Rabanel J-M, Piec P-A, Landri S, Patten SA, Ramassamy C (2020) Transport of PEGylated-PLA nanoparticles across a blood brain barrier model, entry into neuronal cells and in vivo brain bioavailability. J Control Release 328:679–695

Rahman S, Rahman L, Khalil AT, Ali N, Zia D, Ali M, Shinwari ZK (2019) Endophyte-mediated synthesis of silver nanoparticles and their biological applications. Appl Microbiol Biotechnol 103(6):2551–2569

Reinholz J, Landfester K, Mailänder V (2018) The challenges of oral drug delivery via nanocarriers. Drug Deliv 25(1):1694–1705

Salem SS, Fouda A (2021) Green synthesis of metallic nanoparticles and their prospective biotechnological applications: an overview. Biol Trace Elem Res 199(1):344–370

Sani A, Hassan D, Khalil AT, Mughal A, El-Mallul A, Ayaz M, Yessimbekov Z, Shinwari ZK, Maaza M (2021) Floral extracts-mediated green synthesis of NiO nanoparticles and their diverse pharmacological evaluations. J Biomol Struct Dyn 39(11):4133–4147

Sani A, Murad A, Hassan D, Channa GM, El-Mallul A, Medina DI (2023) Photo-catalytic and biomedical applications of one-step, plant extract-mediated green-synthesized cobalt oxide nanoparticles. Environ Sci Pollut Res 30(8):20736–20745

Schulz M, Ma-Hock L, Brill S, Strauss V, Treumann S, Gröters S, van Ravenzwaay B, Landsiedel R (2012) Investigation on the genotoxicity of different sizes of gold nanoparticles administered to the lungs of rats. Mutat Res 745(1):51–57

Scindia PY, Leeds MDJ, Swaminathan MDS (2019) Iron homeostasis in healthy kidney and its role in acute kidney injury. Semin Nephrol 39(1):76–84

Selvaraju C, Karthick R, Veerasubam R (2019) The modification of structural, optical and antibacterial activity properties of rare earth gadolinium-doped ZnO nanoparticles prepared by co-precipitation method. J Inorg Organomet Polym Mater 29(3):776–782

Shakeel M, Jabeen F, Shabbir S, Asghar MS, Khan MS, Chaudhry AS (2016) Toxicity of nano-titanium dioxide (TiO2-NP) through various routes of exposure: a review. Biol Trace Elem Res 172(1):1–36

Sharifi-Rad J, Quispe C, Butnariu M, Rotariu LS, Sytar O, Sestito S, Rapposelli S, Akram M, Iqbal M, Krishna A, Kumar NVA, Braga SS, Cardoso SM, Jafernik K, Ekiert H, Cruz-Martins-N, Szopa A, Villagran M, Mardones L, Martorell M, Docea AO, Calina D (2021) Chitosan nanoparticles as a promising tool in nanomedicine with particular emphasis on oncological treatment. Cancer Cell Int 21(1):318

Simón-Vázquez R, Tsapis N, Lorscheider M, Rodríguez A, Calleja P, Mousnier L, de Miguel Villegas E, González-Fernández Á, Fattal E (2022) Improving dexamethasone drug loading and efficacy in treating arthritis through a lipophilic prodrug entrapped into PLGA-PEG nanoparticles. Drug Deliv Transl Res 12(5):1270–1284

Sompayrac LM (2022) How the immune system works. John Wiley & Sons, Chichester

Soni G, Kale K, Shetty S, Gupta MK, Yadav KS (2020) Quality by design (QbD) approach in processing polymeric nanoparticles loading anticancer drugs by high pressure homogenizer. Heliyon 6(4):e03846

Steckiewicz KP, Barcinska E, Malankowska A, Zauszkiewicz Pawlak A, Nowaczyk G, Zaleska-Medynska A, Inkielewicz-Stepniak I (2019) Impact of gold nanoparticles shape on their cytotoxicity against human osteoblast and osteosarcoma in in vitro model. Evaluation of the safety of use and anti-cancer potential. J Mater Sci Mater Med 30(2):22

Sung H, Ferlay J, Siegel RL, Laversanne M, Soerjomataram I, Jemal A, Bray F (2021) Global cancer statistics 2020: GLOBOCAN estimates of incidence and mortality worldwide for 36 cancers in 185 countries. CA Cancer J Clin 71(3):209–249

Teoh TT (2023) Convolutional neural networks for medical applications. Springer Nature Singapore, Singapore, pp 53–68

Vallorz EL, Blohm-Mangone K, Schnellmann RG, Mansour HM (2021) Formoterol PLGA-PEG nanoparticles induce mitochondrial biogenesis in renal proximal tubules. AAPS J 23(4):88

Vaz-Ramos J, Cordeiro R, Castro MMCA, Geraldes CFGC, Costa BFO, Faneca H, Durães L (2020) Supercritically dried superparamagnetic mesoporous silica nanoparticles for cancer theranostics. Mater Sci Eng C 115:111124

Wang Y, Zhang H, Shi L, Xu J, Duan G, Yang H (2020) A focus on the genotoxicity of gold nanoparticles. Nanomedicine 15(4):319–323

Weng Q, Sun H, Fang C, Xia F, Liao H, Lee J, Wang J, Xie A, Ren J, Guo X, Li F, Yang B, Ling D (2021) Catalytic activity tunable ceria nanoparticles prevent chemotherapy-induced acute kidney injury without interference with chemotherapeutics. Nat Commun 12(1):1436

Wu L-P, Wang D, Li Z (2020) Grand challenges in nanomedicine. Mater Sci Eng C 106:110302

Xue S, Zhou X, Sang W, Wang C, Lu H, Xu Y, Zhong Y, Zhu L, He C, Ma J (2021) Cartilage-targeting peptide-modified dual-drug delivery nanoplatform with NIR laser response for osteoarthritis therapy. Bioact Mater 6(8):2372–2389

Yang C, Merlin D (2023) Challenges to safe nanomedicine treatment. Nanomaterials (Basel) 13(7): 1171

Yang W, Liang H, Ma S, Wang D, Huang J (2019) Gold nanoparticle based photothermal therapy: development and application for effective cancer treatment. Sustain Mater Technol 22:e00109

Zhang Y-N, Poon W, Tavares AJ, McGilvray ID, Chan WCW (2016) Nanoparticle–liver interactions: cellular uptake and hepatobiliary elimination. J Control Release 240:332–348

Zhang W, Sigdel G, Mintz KJ, Seven ES, Zhou Y, Wang C, Leblanc RM (2021) Carbon dots: a future blood–brain barrier penetrating nanomedicine and drug nanocarrier. Int J Nanomedicine 16:5003–5016

Zhu Y, Liu X, Ding X, Wang F, Geng X (2019) Telomere and its role in the aging pathways: telomere shortening, cell senescence and mitochondria dysfunction. Biogerontology 20(1):1–16

Zidan A, El Saadany AA, El Maghraby GM, Abdin AA, Hedya SE (2023) Potential cardioprotective and anticancer effects of carvedilol either free or as loaded nanoparticles with or without doxorubicin in solid Ehrlich carcinoma-bearing mice. Toxicol Appl Pharmacol 465: 116448

Health and Environmental Hazards Associated with the Synthesis of Nanomaterials-Respiratory Diseases, Government Regulations, Ethical Issues

9

Khalid Umar Fakhri and Firdos Alam Khan

Abstract

Recently, there has been growing interest in nanomaterials, especially for their applications in the biomedical field. These nanomaterials possess a small size together with astonishing physicochemical properties, which makes them most suitable for therapeutic applications. Different types of nanomaterials, such as metallic, carbon-based, fluorescent, and polymer-based, have been used in tissue engineering, drug delivery, and other therapeutic applications. Despite many useful applications, nanomaterials possess some harmful effects on the human body, vegetation, and also on the environment. Nanomaterials can enter cells or tissues and organs because of their nano-size, large surface area, and better penetrability, which may cause serious health issues (inhibition of cell proliferation, cell wall damage, modification of function and morphology of organelles, oxidative stress, genotoxicity, and cell death) in living organisms including humans. In this chapter, we have aimed to provide the harmful effects of nanomaterials as per published research papers. The topics related to nanomaterial's toxicity, mode of action and penetration of nanomaterials in the human body are discussed. We have also discussed the government regulations, ethical issues related to nanomaterials toxicity and possible health hazards.

K. U. Fakhri
Department of Biosciences, Jamia Millia Islamia, New Delhi, India

F. A. Khan (✉)
Department of Stem Cell Research, Institute for Research and Medical Consultations (IRMC), Imam Abdulrahman Bin Faisal University, Dammam, Saudi Arabia
e-mail: fakhan@iau.edu.sa

F. A. Khan (ed.), *Nano Drug Delivery for Cancer Therapy*,
https://doi.org/10.1007/978-981-99-6940-1_9

Keywords

Health and environmental hazards · Synthesis of nanocarriers · Nano drug delivery · Respiratory diseases · Government regulations · Ethical issues

9.1 Introduction

Nanoparticles are tiny structures composed of atoms, with dimensions falling between the sizes of ions (10^-10 m) and larger materials. Their uniqueness stems from their small atom count, leading to a substantial proportion of atoms being present on or near the particle's surfaces. This prominent surface presence makes notable alterations to the particle's atomic, electronic, and magnetic properties and physical and chemical characteristics, setting it apart from bulk materials. Additionally, the surfaces of nanoparticles themselves can exhibit distinct attributes (Banfield and Zhang 2001). Nanoparticles have been used in many areas, such as agriculture, cosmetics, medicine, drug delivery, water purification, the oil industry, etc. Presently, nanotechnology serves as a valuable instrument for delving into the most intricate aspects of medical science through various avenues such as imaging, sensing, precise administration of drugs, genetic material delivery systems, and the creation of synthetic implants. Modern pharmaceutical advancements include nanoparticle forms comprising polymers, metals, or ceramics, capable of addressing ailments like cancer and combating human pathogens like bacteria.

Building upon nanotechnology, the development of remarkably effective indicators and accurate, quantifiable detection tools holds the potential to revolutionize patient care. These innovations can significantly impact medical management by enabling early diagnosis and meticulous monitoring of treatments. As a result, patients' quality of life can be enhanced, and mortality rates in conditions such as cancer and Alzheimer's can be reduced (Nasimi and Haidari 2013). While nanoparticles bring numerous benefits to the medical domain, their appealing characteristics can also influence their potential for toxicity within biological systems, encompassing aspects like absorption, distribution, metabolism, and clearance. These toxicological considerations underscore the need to consider size, shape, surface chemistry, and chemical composition when producing nanoparticles. These concerns also extend to their impact on cellular interactions, the endocytic pathway, and the absorption process. Consequently, nanoparticles can trigger cytotoxic mechanisms that disrupt cellular equilibrium (Mauricio et al. 2018). In this chapter, we have aimed to provide the harmful effects of nanomaterials as per published research papers. The topics related to nanomaterial's toxicity, mode of action, and penetration of nanomaterials in the human body are discussed. We have also discussed the government regulations, ethical issues related to nanomaterials toxicity and possible health hazards.

9.2 Types of Nanoparticles

Nanoparticles are categorized according to shape, size, and chemical attributes. These classifications are determined by their physical and chemical properties. Based on their dimensions, nanomaterials can be classified into zero-dimensional (0-D), one-dimensional (1-D), two-dimensional (2-D), and three-dimensional (3-D). In the 0-D category, nanomaterials possess all three of their dimensions within the nanoscale range, such as quantum dots, fullerenes, and nanoparticles. The 1-D class comprises nanomaterials with one dimension extending beyond the nanoscale. Examples encompass nanotubes, nanofibers, nanorods, nanowires, and nanohorns. Nanomaterials in the 2-D grouping have two dimensions exceeding the nanoscale. Instances include nanosheets, nanofilms, and nanolayers. Three-dimensional nanomaterials (3-D) or bulk nanomaterials do not possess confinement to the nanoscale in any dimension. This category encompasses bulk powders, dispersions of nanoparticles, arrays of nanowires and nanotubes, and similar structures (Joudeh and Linke 2022).

Fullerenes and carbon nanotubes (CNTs) are prominent categories within carbon-based nanoparticles. Fullerenes consist of nanomaterials characterized by spherical hollow structures, representing various allotropes of carbon. They have garnered considerable attention in the commercial sphere due to their exceptional attributes, such as electrical conductivity, robustness, unique structure, electron affinity, and adaptability. On the other hand, CNTs are elongated, tube-like formations measuring around 1–2 nm in diameter. Their diameter dictates their metallic or semiconducting nature. Structurally, they resemble the rolling of a graphite sheet upon itself (Khan et al. 2019). Due to its outstanding antioxidant potential, Fullerene holds significant promise as a fundamental component in numerous dermatological and skincare formulations (Mousavi et al. 2017). There is a strong focus on investigating the mechanisms behind the selective targeting of tumors using biofunctionalized CNTs. This heightened interest is driven by their remarkable capacity to convert near-infrared (NIR) laser radiation into heat. This inherent characteristic presents fresh opportunities for creating innovative immunoconjugates tailored for cancer phototherapy. These novel constructs exhibit notable potential and effectiveness in precisely eradicating cancerous cells through selective thermal ablation (Iancu and Mocan 2011).

Metal nanoparticles are composed solely of metal precursors and exhibit distinct optoelectrical characteristics owing to the well-established phenomenon of localized surface plasmon resonance (LSPR). Notably, alkali and noble metals such as Copper (Cu), Silver (Ag), and Gold (Au) nanoparticles exhibit a broad absorption band within the visible region of the electromagnetic solar spectrum. In contemporary materials research, precise control over the synthesis of metal nanoparticles in terms of their facets, sizes, and shapes holds significant importance. Leveraging their advanced optical attributes, metal nanoparticles find utility across diverse research domains. Gold nanoparticles, for instance, are commonly employed to coat surfaces for scanning electron microscopy (SEM) sampling. This coating effectively enhances the flow of electrons, consequently contributing to acquiring high-quality SEM images (Khan et al. 2019).

Polymers are large molecules composed of numerous repeating units, forming a chain-like molecular structure with a wide range of compositions, structures, and characteristics. This diversity in compositions, structures, and attributes makes polymers a valuable component in nanoparticle systems, enabling the creation of nanoparticles tailored to specific biomedical purposes. The primary application of polymeric nanoparticles lies in drug delivery, while they also find utility in tasks like bioimaging and biosensing assays (Morena-Vega et al. 2012). Semiconductor nanomaterials exhibit captivating physical and chemical attributes along with practical functionalities that distinguish them from conventional bulk substances and molecular materials. Noteworthy features of these materials include concentrated emission spectra with limited width, uninterrupted absorption ranges, robust chemical and photobleaching stability, processability, and surface adaptability (Sahu 2019).

Lipid-based nanoparticles, such as liposomes, solid lipid nanoparticles (SLN), and nanostructured lipid carriers, have garnered significant interest in the realms of drug development and cancer therapy. These nanoparticles can convey hydrophobic and hydrophilic compounds while exhibiting minimal to negligible toxicity. Furthermore, they enhance drug efficacy by prolonging drug activity through extended half-lives and controlled release mechanisms. The primary constituents of these nanoparticles consist of phospholipids, which arrange themselves in a bilayer configuration owing to their amphipathic nature. When exposed to water, they assemble into vesicles, enhancing the solubility and stability of anticancer medications when incorporated within their framework (García-Pinel et al. 2019). Despite many applications of nanoparticles, it is important to explore the toxicities induced by them because of continuous applications.

9.3 Harmful Effects of Nanomaterials

The speedy growth of nanomaterials production and synthesis has raised safety concerns for human health and the surrounding environment. The unique properties of nanomaterials make them the most promising candidates for various biomedical applications but at the same time, they pose a potential threat to the biological functions of the human body and produce hazardous effects (Soares and Soares 2021). Humans are exposed to nanomaterials intentionally (Table 9.1) through the usage of nanomedicine and also get exposed to them unintentionally through industrial production and synthesis of nanomaterials (Lombi et al. 2019; Klaper 2020). The biological application of nanomaterials may cause damage to human cells and tissues as they can easily penetrate inside human cells due to their small sizes. The unintentional impact of nanomaterials, such as their application in environmental remediation processes (Guerra et al. 2018; Qian et al. 2020) and in the agricultural (Servin et al. 2015; Usman et al. 2020), is noteworthy. Examples of unintentional release of nanomaterials are when products such as paints, cosmetics,

Table 9.1 Toxic effects of nanomaterials

Name of nanomaterials	Toxic effect	Reference
Carbon nanotubes and fullerenes	May lead to long disease such as granulomas, fibrosis and lung cancer upon inhalation	http://www.hse.gov.uk/
Titanium oxide	May distribute to different part of the body and may develop acute systemic toxicity	http://www.nicnas.gov.au/
Zinc oxide	Induce oxidative stress	http://www.nicnas.gov.au/
Cerium oxide	Toxic and genotoxicity in human dermal fibroblasts	http://www.nicnas.gov.au/
Nano-silver, cooper, aluminum	Induce neurotoxic effect on the brain	http://www.nicnas.gov.au/

and sunscreens use different types of nanomaterials (Sun et al. 2016; Wu et al. 2020), textile industries during the washing of nano-textiles (Yetisen et al. 2016). It was estimated 63–91% that of the global by-product of nanomaterials reach landfills, whereas 8–28% are released into soils and the remaining released into 0.4–0.7% in natural water bodies, and 0.1–1.5% are released into the atmosphere (Keller et al. 2013).

9.4 Impact of Nanomaterials-Induced Toxicity

Humans are constantly exposed to various sources of nanoparticles, and the current method of evaluation of toxicity induced by nanoparticles is not always fully representative of real-life scenarios (Forest 2021). The interaction of nanoparticles with human cells and tissues is difficult to assess in living individuals, which is responsible for combined toxicity (Forest 2021). Certainly, the biological effects generated by nanoparticles are usually evaluated focusing on individual nanoparticles (Carlsten 2018; Rosário et al. 2020), whereas their interaction with co-contaminants can intensely impact the cells and tissues with stimulatory and inhibitory ways (Forest 2021; Abbas et al. 2020). The toxicity of individual nanoparticles can be easily measured, whereas it is difficult to measure toxicity induced by multiple or complex nanoparticles (Silins and Hogberg 2011; Zheng et al. 2012).

When dissolved in the solvent, certain nanomaterials release ions in the medium that can reach cells to produce a harmful effect after intracellular (Soares and Soares 2021). It has been reported that metal ions do not diffuse freely across the plasma membrane. Consequently, different membrane transport proteins (pumps and channels) are involved in their influx inside the cells (Argueello et al. 2012). There are certain metallic nanomaterial, where ions can play a major role in their inducing ecotoxicity especially in the case of Zinc Oxide nanomaterials induced cytotoxicity in (1) bacteria (Heinlaan et al. 2008; Li et al. 2011; Wang et al. 2016), (2) yeasts (Kasemets et al. 2009; Bayat et al. 2014), (3) crustaceans (Heinlaan et al. 2008;

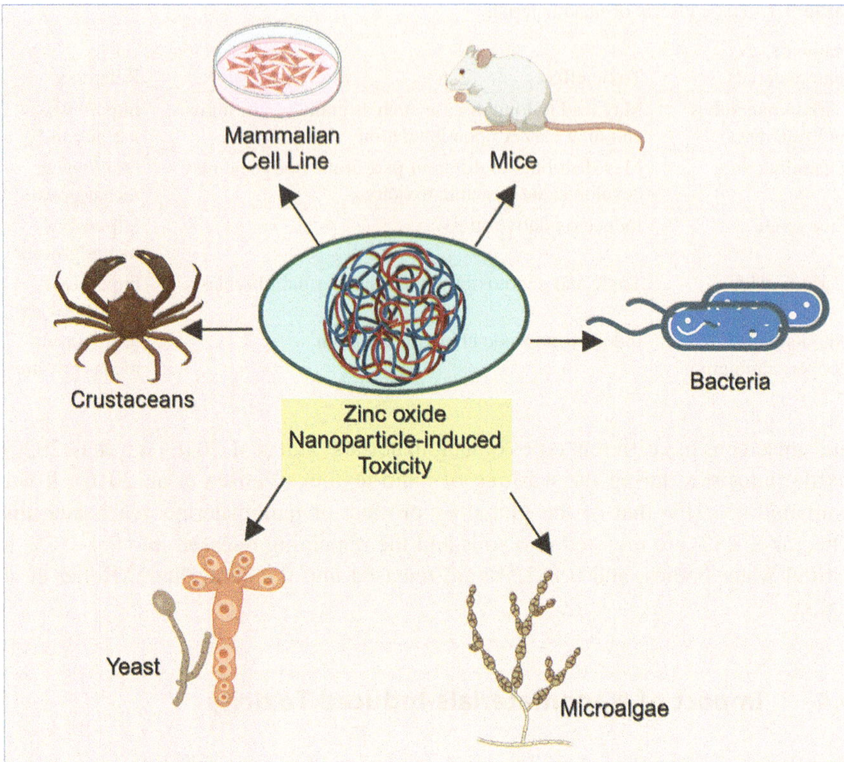

Fig. 9.1 Illustration of induction of toxicity of zinc oxide nanomaterials on different species

Wiench et al. 2009; Vimercati et al. 2020), (4) microalgae (Franklin et al. 2007; Miller et al. 2010; Lee and An 2013; Aravantinou et al. 2015; Schiavo et al. 2016) or (5) mammalian cell lines (Brunner et al. 2006; Zhang et al. 2012) or in vivo model (Wang et al. 2017). The transcriptomics, metabolomics, and lipidomics have confirmed that metal ions mediated the main toxicological pathways of Zinc oxide nanomaterials in lung epithelial cells (Dekkers et al. 2018). In addition, Cupper oxide nanomaterials toxicity were also reported in bacteria (Bondarenko et al. 2012), yeasts (Kasemets et al. 2009, 2013; Bayat et al. 2014; Bao et al. 2015), green algae (Aruoja et al. 2009; von Moos et al. 2015), crustaceans (Heinlaan et al. 2008; Jo et al. 2012) or human cell lines (Cohen et al. 2013; He et al. 2020). Metallic ions leaching effects was also detected in *P. subcapitata* and S. cerevisiae (Moriyama et al. 2019) (Fig. 9.1).

The harmful effects of quantum dots (QDs) are linked to releasing metal ions like cadmium, lead, and arsenic from their cores due to oxidation by environmental factors. These QDs can enter mitochondria, leading to structural alterations and organelle malfunctions. The introduction of Cadmium based QDs into cells prompts the creation of free Cd2+ ions, resulting in oxidative stress (Singh et al. 2012;

Ambrosone et al. 2012). Furthermore, the interaction of nanoparticles with cells can damage the cytoskeleton. For instance, titanium dioxide (TiO_2) nanoparticles can cause tubulin, a protein, to undergo structural alterations and hinder its assembly into microtubules. This interference disrupts processes like cellular transport, division, and movement. In human umbilical vein endothelial cells (HUVECs), cytoskeletal damage impedes the development of adhesive complexes that connect the cytoskeleton with the extracellular matrix. As a result, the formation of the vascular network is disrupted. Moreover, the toxicity of nanoparticles can disrupt processes such as cell specialization and the creation of proteins while also triggering the activation of inflammation-related genes and the production of inflammatory substances. It's crucial to highlight that typical defense mechanisms don't impact nanoparticles. Notably, larger PEGylated nanoparticles are taken up more effectively by macrophages than smaller ones, causing nanoparticles to accumulate within the body (Walkey et al. 2012). Research has revealed that superparamagnetic iron oxide nanoparticles can hinder or even completely suppress the differentiation of stem cells into bone-forming cells (osteogenic differentiation). These nanoparticles also prompt the generation of signaling molecules and tumor antigens. Additionally, the interaction between nanoparticles and cells heightens the expression of genes responsible for lysosome formation, disrupting their function and impeding protein synthesis. An investigation into the harmful effects of nanoparticles with varying compositions on lung epithelial cells and human tumor cell lines demonstrated that nanoparticles prompt the production of inflammatory mediators, such as IL8 (Sukhanova et al. 2018). According to Park and Park (2009), who conducted studies in controlled cell environments and living organisms, exposure to silicon nanoparticles leads to an increased expression of proinflammatory cytokines like interleukin 1 beta (IL-1β) and tumor necrosis factor-alpha (TNFα).

The cytotoxic effects triggered by nanomaterials arise from their interaction with cellular components at the nanomaterial surface. As the particle diameter diminishes, the surface area undergoes exponential expansion. Consequently, even with identical compositions, nanomaterials can exhibit significantly varying levels of cytotoxicity due to particle size and surface reactivity. Moreover, particle size significantly influences how nanomaterials are delivered within cells and distributed in living organisms. In evaluating nanomaterial cytotoxicity, it is crucial to consider both chemical attributes and size-dependent effects, along with the extent of such size-dependent impacts (Lefebvre et al. 2015). To elicit cytotoxicity and provoke an inflammatory response in animal models, nanoparticles must traverse the epithelial barrier. Here, nanoparticle size emerges as a pivotal determinant. In scenarios involving the inhalation of nanoparticles, these particles penetrate deeply into the lung parenchyma. Diverse nanoparticle sizes exhibit distinct distribution patterns in the respiratory tract, a phenomenon influenced by factors like Stoke's and Reynold's numbers. Initial distribution occurs predominantly in the gas phase, but upon inhalation, nanoparticles transition into the liquid phase present in respiratory fluids (Gliga et al. 2014).

In a recent study by Braakhuis et al. (2016), the cytotoxicity induced by inhaled silver nanoparticles of varying sizes was evaluated. Silver nanoparticles measuring

18, 34, 60, and 160 nm were generated and administered to rats at different concentrations. Following exposure, the rats were sacrificed to measure the silver nanoparticle content in their lungs. The results showed that the 18 and 34 nm silver nanoparticles triggered dose-dependent expression of lactate dehydrogenase (LDH), an indicator of cell damage, after 24 h. Conversely, the 60 and 160 nm nanoparticles showed no dose-dependent cell damage. Despite an overall higher presence of the larger 60 and 160 nm nanoparticles in the lungs, the alveoli contained more of the 18 and 34 nm nanoparticles. The authors suggested that the increased surface area of the nano-scaled particles was the primary factor contributing to the toxicological effects of the silver nanoparticles.

Oxidative stress generally occurs when an imbalance between the generation of reactive species (RS) and the level of antioxidant defenses, either enzymatic or non-enzymatic (Halliwell and Gutteridge 2015) in the cells. There are different types of nanomaterials, such as monoxide nanomaterials that possess a pro-oxidant capability that is responsible for the production of reactive oxygen species (ROS) (Nel et al. 2006). Moreover, reactive oxygen species ROS production can occur when metallic ions induce cell death in cells. Oxidative stress is related to the impairment of biological molecules such as lipids, carbohydrates, proteins and DNA (Halliwell and Gutteridge 2015) due to cytotoxicity induced by nanomaterials (Nel et al. 2006; Xia et al. 2006). Metal ions released from nanoparticles (NPs) have been demonstrated to participate in redox cycling and chemical catalysis through reactions like the Fenton reaction (H_2O_2 + Fe_2+ → Fe_3+ + HO− + ·OH) or Fenton-like reactions (Ag + H_2O_2 + H+ = Ag+ + ·OH + H_2O). The separated metal ions, such as Ag+, also lead to the deactivation of cellular enzymes, disruption of membrane structures, interference with electron transport processes, lowered redox potential levels, impaired mitochondrial membrane potentials (MMP), and heightened accumulation of ROS within cells (Yu et al. 2020).

Lipids, particularly unsaturated fatty acids, hold significance as crucial cellular components, playing vital roles in both cell membrane structure and function. Nanoparticles exhibit a strong attraction to the cell membrane, where they can trigger the generation of ROS and induce lipid peroxidation in the outer membrane. This alteration in fatty acid composition within the cell membrane can potentially lead to heightened cell permeability, causing unregulated movement of NPs from the external environment into the cytoplasm. This infiltration may pave the way for ongoing cellular harm. Upon entering the cell, internalized NPs initiate subsequent rounds of ROS bursts. The excessive ROS accumulation subsequently results in the rupture of organelle membranes, leading to the release of organelle contents (Foster et al. 2011), deactivation of cell receptors (Hirano et al. 2008), the liberation of lactate dehydrogenase (LDH), and further irreversible cellular damage (Gu et al. 2014).

Consequently, ROS production by various monoxide nanomaterials may cause cell oxidative disturbances, cell membrane damage, reduced mitochondrial function, chromatin condensation, DNA damage, and cell death via apoptotic pathway were reported in bacteria (Kumar et al. 2011; Rodea et al. 2012), yeasts (Zhang et al. 2016; Babele et al. 2018; Sousa et al. 2018a), freshwater and marine microalgae (Rodea

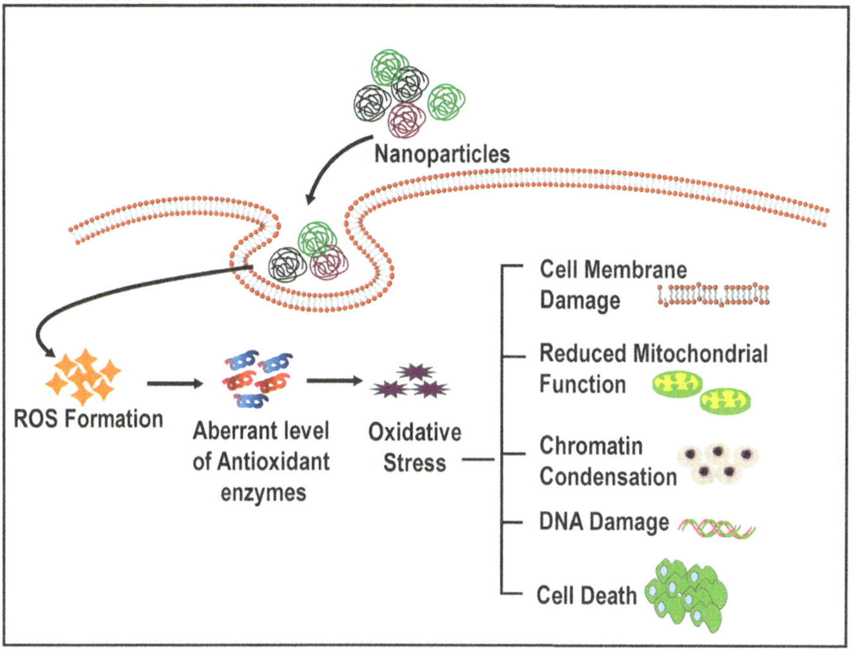

Fig. 9.2 Illustration of exposure of nanomaterials to human cells leading to apoptotic-induced cell death

et al. 2012; von Moos et al. 2015; Xia et al. 2015; Suman et al. 2015; Oukarroum et al. 2017; Sousa et al. 2018b), carp (Cyprinus carpio) larva (Naeemi et al. 2020), and human cell lines (Karlsson et al. 2008; Park et al. 2008; Lu et al. 2015; Rajiv et al. 2016; Subramaniam et al. 2020) (Fig. 9.2).

ROS target the hydrophobic segments of amino acids, disrupting peptide bonds and interfering with the proper functioning of these proteins. Another outcome of oxidative damage is protein carbonylation. Proteins affected by carbonylation form chemically irreversible aggregates that cannot be broken down by proteasomes, ultimately causing a permanent loss of function in these proteins. Gurunathan et al. (2019) demonstrated that platinum nanoparticles (PtNPs) could intensify ROS production and elevate levels of carbonylated proteins. This, in turn, hindered the proliferation of osteosarcoma cells and contributed to the process of apoptosis. In a separate instance, nanoparticles derived from combustion and friction (CFDNPs) were found to accumulate in the brains of young adults with Alzheimer's disease. This accumulation likely spurred the generation of ROS, resulting in protein misfolding, aggregation, and the formation of fibrils. Additionally, research by Pelgrift and Friedman (2013) indicated that magnesium nanoparticles (Mg NPs) might impede gene transcription or directly harm proteins.

9.5 Methods of Evaluating Nanomaterial-Induced Toxicities

Two different methods are used to evaluate the toxicities induced by the nanomaterials: in vitro and in vivo. In both cases, the impact of nanomaterials is evaluated by exposing the cells or tissues or individuals (animals or humans) to various concentrations of the nanomaterials and examining their effects using cellular, molecular and genetic techniques. Control samples (without nanomaterials) are included in the studies to compare the effects of both methods. The main parameters used to evaluate toxicity induced by the nanomaterials are bio-distribution and cellular uptake of nanomaterials and the assessment of the induction of toxicity on the cell damage, DNA condensation, mitochondrial damage, cell membrane disruption, and cell death (Hillegass et al. 2010; Kumar et al. 2017; Domb et al. 2021).

9.5.1 In Vitro Method of Testing

The application of in vitro model is based on the cell lines or primary cell lines developed from animal and human tissue biopsies. These cell lines are preferred due to their homogeneity and stability during the in vitro testing and help get reproducible results. These cell lines are generated from different organs/tissues of the animals or humans, either from normal or diseased individuals. These cell lines could be cancerous cells or normal cells, which can be artificially immortalized, with a high rate of growth and proliferation, making them easy to cultivate. These cells are available in large quantities at an economical cost (Drasler et al. 2017; Joris et al. 2013). Furthermore, different cell types can be used for different applications, for example, cells from the immune system such as macrophages, epithelial cells, fibroblasts, cells and cells from lung, cells are different from cells of the gastrointestinal tract, and neurons. Moreover, lung and gastrointestinal cells are frequently used for testing the inhalation and ingestion of nanomaterials (Drasler et al. 2017). These cell lines are commercially available from the American Cell Culture Type (ATCC), USA with reasonable price. There are limitations for the 2D culture cell lines as they don't represent the actual tissues or organs, and to overcome this probleb, researchers have developed 3D cell lines which represent the actual tissues and organs and however these 3D cell lines are expensive and donor variations is another issues associated with this (Rach et al. 2014; Lacroix et al. 2018; Kooter et al. 2019; Bessa et al. 2021).

To examine in vitro toxicity of the nanomaterials, different methods are used where cell viability, cell death or rate of cell proliferation are measured after the treatments of the nanomaterials. The most widely used in vitro methods of nanomaterials testing are quantitative colorimetric or fluorimetric assays such as MTT (3-(4,5-dimethylthiazol-2-yl)-2,5-diphenyltetrazolium bromide) assay, XTT (tetrazolium dye) assay, WST-1 assay, Alamar blue staining, neutral red uptake staining. The in vitro tests also include LDH, trypan blue assays whereas cell division or proliferation can be assessed through clonogenic assays or cell counting.

Furthermore, programmed cell death or apoptosis assays such as TUNEL assay, caspase activation or apoptosis/necrosis assays can be used to toxicity induced by nanomaterials (Hillegass et al. 2010; Kumar et al. 2017; Park et al. 2009; Takhar and Mahant 2011; Gupta and Xie 2018; Savage et al. 2019; Tirumala et al. 2021).

The oxidative stress is recognized as the chief mechanism of nanomaterial-induced toxicity (Kumar et al. 2017). It has been reported that treatment of nanomaterials caused the production of ROS-reactive oxygen species and RNS-reactive nitrogen species in the treated cells (Kumar et al. 2017). It has been reported that cells normally generate ROS to maintain normal metabolism or homeostasis, then overproduction of ROS can interfere cell signal transduction pathways. The ROS overproduction also induced programmed cell death or apoptosis and DNA damage (Naz et al. 2020; Kumar et al. 2017; Tirumala et al. 2021).

Genotoxicity is the possible injury made by nanomaterials to the DNA of the cells (Drasler et al. 2017). There are two types of genotoxicity primary genotoxicity and secondary genotoxicity. In primary genotoxicity, the genetic daismage caused by nanomaterials themselves, whereas in secondary genotoxicity, nanomaterials do not interact directly with the target cell then produce an inflammatory response in neighboring cells, which resulted the over-production of the ROS which finally caused the DNA damage. It must be remembered that primary genotoxicity may also mediated by ROS overproduction to damage the DNA (Drasler et al. 2017; Achawi et al. 2021). To evaluate DNA damage, gene mutations, and chromosomal damage different DNA markers are used (Kohl et al. 2020) and these include in vitro mammalian cell gene mutation assay and the in vitro mammalian cell micronucleus assay respectively (Drasler et al. 2017; El Yamani et al. 2017). The micronucleus assay is also used to assess the chromosomal loss in the nanomaterials treated cells. In addition, the comet assay is widely used to detect the genotoxic potential of nanomaterials, which help to detect allow single- and double-stranded DNA breaks in individual cells (Bajpayee et al. 2019).

There are many advantages of using in vitro models in testing nanomaterial-induced toxicities. The application of cell lines from variety of tissues and organs help to study the impact of the nanomaterials on individual type of cells. The in vitro method also help to screen the effects of a very small quantity of nanomaterials and studies can be conducted with different concentration and with time intervals such as 6 h, 12 h, 24 h, 48 h, 72 h or 96-h post-treatment of nanomaterials (Takhar and Mahant 2011; Fröhlich and Salar-Behzadi 2014). Another advantage of using the in vitro method is that the experiment can be performed under controlled testing conditions; they help reduce the experiment-to-experiment variability (Russell and Burch 1959). The in vitro method is based on the "3R" rule, which was introduced in the year 1959 aiming to Reduce, Refine, and Replace animal experiments (Fröhlich 2018), which is now commonly fortified by international legislation (Fröhlich 2018). In 2010, the European Commission requested the partial and even full replacement of animal studies in drug testing (European Commission 2010) as the in vitro method is easy, faster, and economical (Kumar et al. 2011).

9.5.2 In Vivo Method of Testing

The leading benefit of in vivo models is that they can mimic human physiology and allow the systemic evaluation of the effects activated by nanomaterials. The impact of nanomaterials can also well examine the body's antioxidant defense and body tissue repair mechanism. Certainly, in vivo testing is the most technically relevant model as it can capture the effects of nanomaterials on the body tissues or organisms. Another key advantage of in vivo models is that they allow the measurement of nanomaterials' different human physiological impacts through absorption, distribution, metabolism, and excretion in the clinical. In vivo, testing has been initiated for a range of nanomaterials mainly exposed to humans by the oral and dermal routes. For the respiratory inhalation of nanomaterials, the application of intra-tracheal instillation is employed in rodents (rats and mice), as these rodents are commonly used species for toxicity testing. In addition, as the in vivo testing is a very expensive affairs, it's technically not possible to evaluate all nanomaterials through in vivo method. It has been estimated that $1.18 billion is needed for comprehensive in vivo long-term testing in the animals' studies (Choi et al. 2009). Another negative aspect of animal testing is species differences as different animals produce varied physiological responses. For example, in the animals, the autophagy inhibitor 3-methyladenine decreased acute lung injury triggered by poly-amidoamine dendrimers, but it lacks clinical usefulness due to the low stability of the nanomaterials (Jud et al. 2013). Moreover, there is a difference in the immune response to the nanomaterials in animals and human which is another issue with in vivo testing (Wheeler et al. 1990).

But there are few specific experiments which are not possible to conduct in vitro testing such as retention of inhaled particles and long-term studies cannot be performed in vitro. The use of in vivo testing where nanomaterials with size $\leq$100 nm can be inhaled and can be retained in the body for longer period of time, for example, 75% of 100 nm carbon nanomaterials were retained for more than 48 h in hamster lungs (Geiser and Kreyling 2010). Explanations for the prolonged retention include deep penetration of nanomaterials into the mucus lung lining layer. In both cases, interaction with airway lung cells and probability of transmigration is increased. Astonishingly, not all nanomaterials reserved in the lung cells but they migrate to lymph nodes or other body tissues (Kreyling et al. 2009). One study in dogs showed that inhalation of nanomaterials appeared in the lung epithelium, where macrophage-mediated clearance occurred and about 90% of the inhaled nanomaterials were recovered in the lung (Semmler-Behnke et al. 2007). The capability of macrophages to swallow nanomaterials is lower for nanomaterials than for micro materials (Geiser et al. 1990). It has been reported that after 24 h of nanomaterial inhalation, only 1.7% of macrophages contained nanomaterials compared to 12–15% with micro materials (Geiser et al. 1994).

9.6 Reduction of Nanoparticles Toxicity

Comprehending the mechanism behind nanoparticle toxicity is essential to guide the development of nanoparticles that have a reduced impact on the environment. Redesign strategies should be chosen considering both the primary mode of toxicity and the preservation of the nanomaterial's intended functionality. To diminish interactions with cell surfaces, nanoparticles can be designed with a negative surface charge, employ ligands like polyethylene glycol to mitigate protein binding, or adopt shapes that discourage cell surface attachment. To curb the dissolution of nanoparticles into harmful ions, toxic components can be substituted with less harmful elements possessing similar traits. The nanoparticle's surface can be capped with a protective shell; its morphology can be tailored to minimize surface area and dissolution, or chelating agents can be introduced to the nanomaterial's surface. To lower the generation of ROS, the material's band gap can be adjusted by utilizing different elements or doping. An outer shell layer can be added to prevent direct contact with the nanoparticle's core, and antioxidant molecules can be affixed to the nanoparticle surface. During the nanoparticle redesign, validating that the chosen strategy effectively reduces toxicity to organisms in relevant environmental settings is imperative. Additionally, it's crucial to ensure that the modified nanomaterial still retains the crucial physicochemical properties that initially warranted its incorporation into a product or device (Buchman et al. 2019).

Nanotechnology-enabled formulations for pesticides have emerged as a promising avenue to enhance pesticide efficiency while mitigating their environmental impact. In a recent study, researchers focused on optimizing the utilization of prochloraz, an active pesticide ingredient, by encapsulating it within mesoporous silica nanoparticles (MSNs). The MSNs were functionalized with chitosan to enhance controlled release capabilities, covalently linked to the nanoparticle surface via a silane coupling agent. This resulted in the synthesis of MSNs-chitosan@prochloraz nanoparticles. The investigation revealed that these nanoparticles exhibited a notable loading efficiency of prochloraz (25.4% w/w) and effectively improved the stability of the pesticide under light exposure. The nanoparticles demonstrated remarkable dual-responsive attributes, reacting sensitively to esterase activity and pH changes. This responsiveness was associated with controlled release behavior. The biological assessment revealed that the interplay of fruit-produced acid and enzymes efficiently triggered the opening of the "gate" enforced by chitosan, facilitating an on-demand release of the pesticide in response to esterase activity and pH fluctuations. Comparative analysis with prochloraz emulsifiable concentrate demonstrated that applyingMSNs-chitosan@prochloraz nanoparticles before harvesting led to extended efficacy and enhanced antifungal activity against citrus diseases. Importantly, these nanoparticle-based formulations exhibited significantly reduced toxicity to zebrafish, with toxicity levels over six times lower compared to prochloraz technical, suggesting MSNs-chitosan@prochloraz nanoparticles as a sustainable and eco-friendly preharvest treatment in agricultural applications (Liang et al. 2018).

Prior research has demonstrated the adverse effects of copper oxide nanoparticles (CuO) on the growth of duckweed (*Landoltia punctata*), a prominent aquatic plant consumed by waterfowl and commonly found in agricultural runoff channels in temperate climates. Nonetheless, previous investigations of CuO nanoparticle toxicity toward duckweed have predominantly been carried out in systems devoid of dissolved organic matter. In this present study, the inhibition of duckweed growth was found to be closely linked to the concentration of aqueous Cu_2+. The most pronounced growth inhibition stemmed from aqueous $CuCl_2$, and for nanoparticle systems, the inhibitory effect increased as the size of CuO particles decreased.

Evaluations of CuO NP dissolution in ½ Hoagland's solution revealed that dissolution rates escalated with diminishing particle size. Furthermore, Suwannee River Humic and Fulvic acids (HA; FA) amplified the dissolution process. However, the study's findings indicate that both Humic acid and, to a lesser degree, Fulvic acid actually mitigate the toxicity of CuO NPs and free ionized Cu toward duckweed. This amelioration is attributed to the inhibition of copper availability facilitated by forming Cu-dissolved organic matter complexes. These outcomes align with predictions from speciation modeling software, suggesting alterations in copper speciation driven by dissolved organic matter, thereby influencing toxicity within natural environments (Rippner et al. 2018). The dissolution of ZnO nanoparticles and the subsequent release of Zn_2+ ions trigger a cascade of sublethal and lethal cellular responses, which can be mitigated through iron doping. Introducing iron alters the particle matrix and decelerates the particle dissolution rate. To assess whether iron doping of ZnO particles also leads to reduced toxic effects in vivo, toxicological investigations were conducted using rodent and zebrafish models. Initially, a fresh batch of ZnO nanoparticles doped with 1–10 wt % of Fe was synthesized. Iron doping, akin to the effects observed with the metal chelator DTPA, counteracted the inhibitory effects of Zn_2+ on zebrafish hatching. In a murine model utilizing oropharyngeal aspiration, iron-doped particles were associated with decreased polymorphonuclear cell counts and lowered IL-6 mRNA production. Furthermore, the doped particles elicited reduced expression of heme oxygenase 1 in the murine lung. In rat studies involving intratracheal instillation, iron doping correlated with diminished polymorphonuclear cell counts, LDH levels, and albumin levels. Collectively, the aforementioned data underline that iron doping presents a plausible and safe design approach to attenuate ZnO toxicity in both animals and the environment (Xia et al. 2011).

TiO_2@Ag nanoparticles (TiAgNps) exhibited notable antileishmanial effects. However, while a lethal dose of TiAgNps was found to be toxic for *L. tropica* promastigotes, it also posed harm to macrophage cells. Additionally, concentrations of TiAgNps that were non-toxic had limited effectiveness in inhibiting *L. tropica* promastigotes and amastigotes. Interestingly, when combined with other antileishmanial compounds, such as meglumine antimoniate (MA) at non-toxic concentrations, the efficacy of both agents appeared to increase while their toxicities decreased. The combinations of TiAgNps and MA at concentrations that did not induce toxicity exhibited remarkable effectiveness against both promastigotes and amastigotes. In contrast, the application of MA alone did not display any inhibitory effects. The combined treatments reduced *L. tropica* promastigote proliferation, contrasting with the outcomes seen with sole MA application. The efficacy of the TiAgNps-MA combinations significantly surpassed that of using MA alone at non-toxic concentrations. These results suggest a strong potential for the development of novel antileishmanial drugs targeting leishmaniasis (Abamor et al. 2017).

9.7 Nanomaterials and Regulations for Preventing Health Hazards

Similar to many emerging technologies, including nascent medicine and medical devices, there is a growing discussion surrounding potential negative outcomes from using nanoscale particles. As nanotechnology becomes more prevalent, it becomes crucial to thoroughly comprehend the risks associated with exposure to nanoparticles, the various pathways through which they enter the body, and the molecular processes underlying any potential toxicity. These minuscule particles can infiltrate the body through the skin, respiratory system, or digestive tract, accumulating in multiple organs and potentially triggering unfavorable biological responses by altering the physicochemical characteristics of living matter at the nanoscale. Moreover, the toxicity of nanoparticles is contingent on factors like their persistence within the body and whether they are eliminated from different entry points, as well as the host's ability to effectively counteract and eliminate these particles. Recent investigations have pinpointed nanoparticles as culprits for toxicity in various organs. Consequently, it is prudent to assess the balance between risks and benefits when considering the incorporation of nanoparticles into technological or medical advancements. For this purpose, many global organizations have outlined the guidelines (Medina et al. 2009).

The goal of these global organizations such as the Royal Society and Royal Academy of Engineering (2004), NIOSH (2005), Jacobstein (2006), CEST (2008), Tomellini and Giordani (2008), Luigi (2009), Nanocyl (2009), NNI (2011), Forloni (2012), VCI (2012), BASF (2013), and BIAC (2013). (Schulte et al. 2014) to provide the rules and guidelines for the successful production of the nanomaterials for various applications. Preferably, the idea of responsible development of nanomaterials is to prevent the harmful effects of the nanomaterials to humans and the environment. The front-line people who are directly exposed to the nanomaterials are researchers, and industry workers who are working in both government laboratories and private companies. These people are the first ones who are exposed to the potential hazards of nanomaterials during the research and development, manufacturing, production, recycling, and clearance of nanomaterials. It has been found that workers often have the highest exposure in the development of nanomaterials when nanomaterials are produced. If exposure to nanomaterials has harmful effects on the workers, then in that case, the production process is not being conscientiously developed. Because of these issues, occupational safety and health is the basis of responsible nanomaterials production (Maynard and Kuempel 2005; Schulte and Salamanca-Buentello 2007; Seaton et al. 2010).

There is an ethical requirement for the workers who are working in the production of nanomaterials, and these workers have the right to have a safe and clean work environment (Gewirth 1986). Consequently, the safety of workers has been recognized as a basic human right as per the 2008 Seoul Declaration on Safety and Health at Work (ILO 2008). Importantly, these rights bring proportionate responsibilities for employers and government authorities to protect workers from harm effects of nanomaterials. These guidelines were codified in legal methods such as the Occupational Safety and Health Act, the Mine Safety and Health Act, and the Toxic Substances Control Act in the United States. In addition, similar legislation was also prepared globally, such as WHO (1994), BAuA (2007), Nanosafe (2008), ISO (2009), Japan NIOSH (2009), Saner and Pelley (2009), (Murashov et al. 2011; Nakanishi 2011a, b). Principal criteria for the responsible development of nanotechnology must be improved to protect workers and consumers from exposure to nanomaterials (Kreider and Halperin 2011).

References

Abamor ES, Allahverdiyev AM, Bagirova M, Rafailovich M (2017) Meglumine antımoniate-TiO2@ag nanoparticle combinations reduce toxicity of the drug while enhancing its antileishmanial effect. Acta Trop 169:30–42. https://doi.org/10.1016/j.actatropica.2017.01.005

Abbas Q, Yousaf B, Amina AMU, Munir MAM, El-Naggar A, Rinklebe J, Naushad M (2020) Transformation pathways and fate of engineered nanoparticles (ENPs) in distinct interactive environmental compartments: a review. Environ Int 138:105646. https://doi.org/10.1016/j.envint.2020.105646

Achawi S, Huot L, Nesslany F, Pourchez J, Simar S, Forest V, Feneon B (2021) Exploring graphene-based materials' genotoxicity: inputs of a screening method. Nanotoxicology 15:1279–1294. https://doi.org/10.1080/17435390.2021.2018734

Ambrosone A, Mattera L, Marchesano V, Quarta A, Susha AS, Tino A et al (2012) Mechanisms underlying toxicity induced by CdTe quantum dots determined in an invertebrate model organism. Biomaterials 33(7):1991–2000

Aravantinou AF, Tsarpali V, Dailianis S, Manariotis ID (2015) Effect of cultivation media on the toxicity of ZnO nanoparticles to freshwater and marine microalgae. Ecotoxicol Environ Saf 114:109–116. https://doi.org/10.1016/j.ecoenv.2015.01.016

Argueello JM, Raimunda D, Gonzalez-Guerrero M (2012) Metal transport across biomembranes: emerging models for a distinct chemistry. J Biol Chem 287:13510–13517. https://doi.org/10.1074/jbc.R111.319343

Aruoja V, Dubourguier H-C, Kasemets K, Kahru A (2009) Toxicity of nanoparticles of CuO, ZnO and TiO2 to microalgae Pseudokirchneriella subcapitata. Sci Total Environ 407:1461–1468. https://doi.org/10.1016/j.scitotenv.2008.10.053

Babele PK, Thakre PK, Kumawat R, Tomar RS (2018) Zinc oxide nanoparticles induce toxicity by affecting cell wall integrity it pathway, mitochondrial function and lipid homeostasis in Saccharomyces cerevisiae. Chemosphere 213:65–75. https://doi.org/10.1016/j.chemosphere.2018.09.028

Bajpayee M, Kumar A, Dhawan A (2019) The comet assay: assessment of in vitro and in vivo DNA damage. Methods Mol Biol 2031:237–257. https://doi.org/10.1007/978-1-4939-9646-9_12

Banfield JF, Zhang Z (2001) Nanoparticles in the environment. Rev Mineral Geochem 44(1):1–58

Bao S, Lu Q, Fang T, Dai H, Zhang C (2015) Assessment of the toxicity of CuO nanoparticles by using Saccharomyces cerevisiae mutants with multiple genes deleted. Appl Environ Microbiol 81:8098–8107. https://doi.org/10.1128/AEM.02035-15

BASF (2013) Nanotechnology code of conduct; http://www.basf.com/group/corporate/en/about-basf/strategy/purpose-principles-values/code-of-conduct/index. Accessed 16 Aug 2013

BAuA (Bundesanstalt für Arbeitsschutz und Arbeitsmedizin) (2007) Guidance for handling and use of nanomaterials at the workplace. http://www.baua.de/en/Topics-from-A-to-Z/Hazardous-Substances/Nanotechnology/pdf/guidance.pdf?__blob=publicationFile&v=2. Accessed 16 Aug 2013

Bayat N, Rajapakse K, Marinsek-Logar R, Drobne D, Cristobal S (2014) The effects of engineered nanoparticles on the cellular structure and growth of Saccharomyces cerevisiae. Nanotoxicology 8:363–373. https://doi.org/10.3109/17435390.2013.788748

Bessa MJ, Brandão F, Fokkens P, Cassee FR, Salmatonidis A, Viana M, Vulpoi A, Simon S, Monfort E, Teixeira JP et al (2021) Toxicity assessment of industrial engineered and airborne process-generated nanoparticles in a 3D human airway epithelial in vitro model. Nanotoxicology 15:542–557. https://doi.org/10.1080/17435390.2021.1897698

Bondarenko O, Ivask A, Kaekinen A, Kahru A (2012) Sub-toxic effects of CuO nanoparticles on bacteria: kinetics, role of cu ions and possible mechanisms of action. Environ Pollut 169:81–89. https://doi.org/10.1016/j.envpol.2012.05.009

Braakhuis HM, Cassee FR, Fokkens PH, de la Fonteyne LJ, Oomen AG, Krystek P, de Jong WH, van Loveren H, Park MV (2016) Identification of the appropriate dose metric for pulmonary inflammation of silver nanoparticles in an inhalation toxicity study. Nanotoxicology 10(1): 63–73. https://doi.org/10.3109/17435390.2015.1012184

Brunner TJ, Wick P, Manser P, Spohn P, Grass RN, Limbach LK, Bruinink A, Stark WJ (2006) In vitro cytotoxicity of oxide nanoparticles: comparison to asbestos, silica, and the effect of particle solubility. Environ Sci Technol 40:4374–4381. https://doi.org/10.1021/es052069i

Buchman JT et al (2019) Understanding nanoparticle toxicity mechanisms to informredesign strategies to reduce environmental impact. Acc Chem Res 52(6):1632–1642

Business Industrial Advisory Committee to the OECD (BIAC) (2013) Responsible development of nanotechnology turning vision into reality; http://www.biac.org/news/2013/1303_BIAC_Vision_Paper_on_Nanotechnology.htm. Accessed 26 Jul 2013

Carlsten C (2018) Synergistic environmental exposures and the airways capturing complexity in humans. Chest 154:918–924. https://doi.org/10.1016/j.chest.2018.06.004

CEST (commission de l' Ethique de la science et de la Technologie) (2008) Ethics, risk, and nanotechnology: responsible approaches dealing with risk. In: Allhoff F, Lin P (eds) Nanotechnology and society. Springer, DV, pp 75–89

Choi JY, Ramachandran G, Kandlikar M (2009) The impact of toxicity testing costs on nanomaterial regulation. Environ Sci Technol 43:3030–3034

Cohen D, Soroka Y, Ma'or Z, Oron M, Portugal-Cohen M, Bregegere FM, Berhanu D, Valsami-Jones E, Hai N, Milner Y (2013) Evaluation of topically applied copper(II) oxide nanoparticle cytotoxicity in human skin organ culture. Toxicol In Vitro 27:292–298. https://doi.org/10.1016/j.tiv.2012.08.026

Dekkers S, Williams TD, Zhang J, Zhou J, Vandebriel RJ, la Fonteyne LJJ, Gremmer ER, He S, Guggenheim EJ, Lynch I, Cassee FR, De Jong WH, Viant MR (2018) Multi-omics approaches confirm metal ions mediate the main toxicological pathways of metal-bearing nanoparticles in lung epithelial A549 cells. Environ Sci 5:1506–1517

Domb AJ, Sharifzadeh G, Nahum V, Hosseinkhani H (2021) Safety evaluation of nanotechnology products. Pharmaceutics 13:1615. https://doi.org/10.3390/pharmaceutics13101615

Drasler B, Sayre P, Steinhäuser KG, Fink A, Rothen-Rutishauser B (2017) In vitro approaches to assess the hazard of nanomaterials. NanoImpact 8:99–116. https://doi.org/10.1016/j.impact.2017.08.002

El Yamani N, Collins A, Rundén-Pran E, Fjellsbø LM, Shaposhnikov S, Zielonddiny S, Dusinska M (2017) In vitrogenotoxicity testing of four reference metal nanomaterials, titanium dioxide, zinc oxide, cerium oxide and silver: towards reliable hazard assessment. Mutagenesis 32:117–126. https://doi.org/10.1093/mutage/gew060

European Commission (2010) European Union directive 2010/63/EU of the European Parliament and of the council of 22 September 2010 on the protection of animals used for scientific purposes. European Commission, Brussels

Forest V (2021) Combined effects of nanoparticles and other environmental contaminants on human health—an issue often overlooked. NanoImpact 23:100344. https://doi.org/10.1016/j.impact.2021.100344

Forloni G (2012) Responsible nanotechnology development. J Nanopart Res 14:1007

Foster HA, Ditta IB, Varghese S, Steele A (2011) Photocatalytic disinfection using titanium dioxide: spectrum and mechanism of antimicrobial activity. Appl Microbiol Biotechnol 90:1847–1868

Franklin NM, Rogers NJ, Apte SC, Batley GE, Gadd GE, Casey PS (2007) Comparative toxicity of nanoparticulate ZnO, bulk ZnO, and ZnCl2 to a freshwater microalga (Pseudokirchneriella subcapitata): the importance of particle solubility. Environ Sci Technol 41:8484–8490. https://doi.org/10.1021/es071445r

Fröhlich E (2018) Comparison of conventional and advanced in vitro models in the toxicity testing of nanoparticles. Artif Cells Nanomed Biotechnol 46:1091–1107. https://doi.org/10.1080/21691401.2018.1479709

Fröhlich E, Salar-Behzadi S (2014) Toxicological assessment of inhaled nanoparticles: role of in vivo, ex vivo, in vitro, and in silico studies. Int J Mol Sci 15:4795–4822. https://doi.org/10.3390/ijms15034795

García-Pinel B et al (2019) Lipid-based nanoparticles: application and recent advances in cancer treatment. Nanomaterials (Basel) 9(4):638

Geiser M, Kreyling WG (2010) Deposition and biokinetics of inhaled nanoparticles. Part Fibre Toxicol 7:2

Geiser M, Cruz-Orive LM, Im HV, Gehr P (1990) Assessment of particle retention and clearance in the intrapulmonary conducting airways of hamster lungs with the fractionator. J Microsc 160:75–88

Geiser M, Baumann M, Cruz-Orive LM, Im HV, Waber U, Gehr P (1994) The effect of particle inhalation on macrophage number and phagocytic activity in the intrapulmonary conducting airways of hamsters. Am J Respir Cell Mol Biol 10:594–603

Gewirth A (1986) Human-rights and the workplace. Am J Ind Med 9(1):31–40

Gliga AR, Skoglund S, Wallinder IO, Fadeel B, Karlsson HL (2014) Size-dependent cytotoxicity of silver nanoparticles in human lung cells: the role of cellular uptake, agglomeration and ag release. Part Fibre Toxicol 11:11. https://doi.org/10.1186/1743-8977-11-11

Gu L, Li Q, Quan X, Cen Y, Jiang X (2014) Comparison of nanosilver removal by flocculent and granular sludge and short- and long-term inhibition impacts. Water Res 58:62–70

Guerra FD, Attia MF, Whitehead DC, Alexis F (2018) Nanotechnology for environmental remediation: materials and applications. Molecules 23:1760. https://doi.org/10.3390/molecules23071760

Gupta R, Xie H (2018) Nanoparticles in daily life: applications, toxicity and regulations. J Environ Pathol Toxicol Oncol 37:209–230. https://doi.org/10.1615/JEnvironPatholToxicolOncol.2018026009

Gurunathan S, Jeyaraj M, Kang MH, Kim JH (2019) Tangeretin-assisted platinum nanoparticles enhance the apoptotic properties of doxorubicin: combination therapy for osteosarcoma treatment. Nanomaterials (Basel) 9:1089

Halliwell B, Gutteridge JMC (2015) Oxidative stress and redox regulation: adaptation, damage, repair, senescence, and death. In: Free radicals in biology and medicine, 5th edn. Oxford University Press, Oxford, pp 199–283

He H, Zou Z, Wang B, Xu G, Chen C, Qin X, Yu C, Zhang J (2020) Copper oxide nanoparticles induce oxidative DNA damage and cell death via copper ion-mediated p38 MAPK activation in vascular endothelial cells. Int J Nanomedicine 15:3291–3302. https://doi.org/10.2147/IJN.S241157

Heinlaan M, Ivask A, Blinova I, Dubourguier H-C, Kahru A (2008) Toxicity of nanosized and bulk ZnO, CuO and TiO2 to bacteria Vibrio fischeri and crustaceans Daphnia magna and Thamnocephalus platyurus. Chemosphere 71:1308–1316. https://doi.org/10.1016/j.chemosphere.2007.11.047

Hillegass JM, Shukla A, Lathrop SA, MacPherson MB, Fukagawa NK, Mossman BT (2010) Assessing nanotoxicity in cells in vitro. Wiley Interdiscip Rev Nanomed Nanobiotechnol 2: 219–231. https://doi.org/10.1002/wnan.54

Hirano S, Kanno S, Furuyama A (2008) Multi-walled carbon nanotubes injure the plasma membrane of macrophages. Toxicol Appl Pharmacol 232:244–251

Iancu C, Mocan L (2011) Advances in cancer therapy through the use of carbon nanotube-mediated targeted hyperthermia. Int J Nanomedicine 6:1675–1684. https://doi.org/10.2147/IJN.S23588

ILO (2008) Seoul Declaration on Safety and Health at Work (2008). www.seouldeclaration.org. Accessed 16 Aug 2013

ISO (2009) Guide to safe handling and disposal of manufactured nanomaterials, Draft report. ISO/TC 229 Nanotechnologies working group 3: health, safety and the environment, project group 6. International Organization for Standardization, Seattle

Jacobstein N (2006) Foresight guidelines for responsible nanotechnology development; https://www.foresight.org/guidelines/ForesightGuidelinesV6.pdf. Accessed 16 Aug 2013

Jo HJ, Choi JW, Lee SH, Hong SW (2012) Acute toxicity of ag and CuO nanoparticle suspensions against Daphnia magna: the importance of their dissolved fraction varying with preparation methods. J Hazard Mater 227:301–308. https://doi.org/10.1016/j.jhazmat.2012.05.066

Joris F, Manshian BB, Peynshaert K, de Smedt SC, Braeckmans K, Soenen SJ (2013) Assessing nanoparticle toxicity in cell-based assays: influence of cell culture parameters and optimized models for bridging the in vitro–in vivo gap. Chem Soc Rev 42:8339–8359. https://doi.org/10.1039/c3cs60145e

Joudeh N, Linke D (2022) Nanoparticle classification, physicochemical properties, characterization, and applications: a comprehensive review for biologists. J Nanobiotechnol 20:262

Jud C, Clift MJ, Petri-Fink A, Rothen-Rutishauser B (2013) Nanomaterials and the human lung: what is known and what must be deciphered to realise their potential advantages? Swiss Med Wkly 143:w13758

Karlsson HL, Cronholm P, Gustafsson J, Moeller L (2008) Copper oxide nanoparticles are highly toxic: a comparison between metal oxide nanoparticles and carbon nanotubes. Chem Res Toxicol 21:1726–1732

Kasemets K, Ivask A, Dubourguier H-C, Kahru A (2009) Toxicity of nanoparticles of ZnO, CuO and TiO2 to yeast Saccharomyces cerevisiae. Toxicol Vitr 23:1116–1122. https://doi.org/10.1016/j.tiv.2009.05.015

Kasemets K, Suppi S, Kuennis-Beres K, Kahru A (2013) Toxicity of CuO nanoparticles to yeast Saccharomyces cerevisiae BY4741 wild-type and its nine isogenic single-gene deletion mutants. Chem Res Toxicol 26:356–367. https://doi.org/10.1021/tx300467d

Keller AA, McFerran S, Lazareva A, Suh S (2013) Global life cycle releases of engineered nanomaterials. J Nanoparticle Res 15:1. https://doi.org/10.1007/s11051-013-1692-4

Khan I et al (2019) Nanoparticles: properties, applications and toxicities. Arab J Chem 12(7): 908–931

Klaper RD (2020) The known and unknown about the environmental safety of nanomaterials in commerce. Small 2000690:2000690. https://doi.org/10.1002/smll.202000690

Kohl Y, Rundén-Pran E, Mariussen E, Hesler M, El Yamani N, Longhin EM, Dusinska M (2020) Genotoxicity of nanomaterials: advanced in vitro models and high throughput methods for human Hazard assessment—a review. Nano 10:1911. https://doi.org/10.3390/nano10101911

Kooter I, Ilves M, Gröllers-Mulderij M, Duistermaat E, Tromp PC, Kuper F, Kinaret P, Savolainen K, Greco D, Karisola P et al (2019) Molecular signature of asthma-enhanced sensitivity to CuO nanoparticle aerosols from 3D cell model. ACS Nano 13:6932–6946. https://doi.org/10.1021/acsnano.9b01823

Kreider T, Halperin W (2011) Engineered nanomaterials: learning from the past, planning for the future. J Occup Environ Med 53(6):S108–S112

Kreyling WG, Semmler-Behnke M, Seitz J, Scymczak W, Wenk A, Mayer P, Takenaka S, Oberdorster G (2009) Size dependence of the translocation of inhaled iridium and carbon nanoparticle aggregates from the lung of rats to the blood and secondary target organs. Inhal Toxicol 21:55–60

Kumar A, Pandey AK, Singh SS, Shanker R, Dhawan A (2011) Engineered ZnO and TiO2 nanoparticles induce oxidative stress and DNA damage leading to reduced viability of Escherichia coli. Free Radic Biol Med 51:1872–1881. https://doi.org/10.1016/j.freeradbiomed.2011.08.025

Kumar V, Sharma N, Maitra SS (2017) In vitro and in vivo toxicity assessment of nanoparticles. Int Nano Lett 7:243–256. https://doi.org/10.1007/s40089-017-0221-3

Lacroix G, Koch W, Ritter D, Gutleb A, Larsen ST, Loret T, Zanetti F, Constant S, Chortarea S, Rothen-Rutishauser B et al (2018) Air–liquid interfacein vitromodels for respiratory toxicology research: consensus workshop and recommendations. Appl Vitr Toxicol 4:91–106. https://doi.org/10.1089/aivt.2017.0034

Lee W-M, An Y-J (2013) Effects of zinc oxide and titanium dioxide nanoparticles on green algae under visible, UVA, and UVB irradiations: no evidence of enhanced algal toxicity under UV pre-irradiation. Chemosphere 91:536–544. https://doi.org/10.1016/j.chemosphere.2012.12.033

Lefebvre DE, Venema K, Gombau L, Valerio LG Jr, Raju J, Bondy GS, Bouwmeester H, Singh RP, Clippinger AJ, Collnot EM, Mehta R, Stone V (2015) Utility of models of the gastrointestinal tract for assessment of the digestion and absorption of engineered nanomaterials released from food matrices. Nanotoxicology 9(4):523–542. https://doi.org/10.3109/17435390.2014.948091

Li M, Zhu L, Lin D (2011) Toxicity of ZnO nanoparticles to Escherichia coli: mechanism and the influence of medium components. Environ Sci Technol 45:1977–1983. https://doi.org/10.1021/es102624t

Liang Y et al (2018) Preparation of MSNs-chitosan@Prochloraznanoparticles for reducing toxicity and improving release properties of Prochloraz. ACS Sustain Chem Eng 6:8

Lombi E, Donner E, Dusinska M, Wickson F (2019) A one health approach to managing the applications and implications of nanotechnologies in agriculture. Nat Nanotechnol 14:523–531. https://doi.org/10.1038/s41565-019-0460-8

Lu S, Zhang W, Zhang R, Liu P, Wang Q, Shang Y, Wu M, Donaldson K, Wang Q (2015) Comparison of cellular toxicity caused by ambient ultrafine particles and engineered metal oxide nanoparticles. Part Fibre Toxicol 12:5. https://doi.org/10.1186/s12989-015-0082-8

Luigi F (2009) Responsible care and nanomaterials case study: nanocyl. In: Proceedings of the European responsible care conference; http://www.cefic.org/Documents/ResponsibleCare/04_Nanocyl.pdf. Accessed 3 Dec 2013

Mauricio MD et al (2018) Nanoparticles in medicine: a focus on vascular oxidative stress. Oxidative Med Cell Longev 2018:1

Maynard AD, Kuempel ED (2005) Airborne nanostructured particles and occupational health. J Nanopart Res 7(6):587–614

Medina C, Santos-Martinez MJ, Radomski A et al (2009) Nanoparticles: pharmacological and toxicological significance. Br J Pharmacol 150(5):552–558

Miller RJ, Lenihan HS, Muller EB, Tseng N, Hanna SK, Keller AA (2010) Impacts of metal oxide nanoparticles on marine phytoplankton. Environ Sci Technol 44:7329–7334. https://doi.org/10.1021/es100247x

Morena-Vega AI et al (2012) Polymeric and ceramic nanoparticles in biomedical applications. J Nanotechnol 2012:1

Moriyama A, Takahashi U, Mizuno Y, Takahashi J, Horie M, Iwahashi H (2019) The truth of toxicity caused by yttrium oxide nanoparticles to yeast cells. J Nanosci Nanotechnol 19:5418–5425. https://doi.org/10.1166/jnn.2019.16544

Mousavi SZ et al (2017) Fullerene nanoparticle in dermatological and cosmetic applications. Nanomedicine 13(3):1071–1087. https://doi.org/10.1016/j.nano.2016.10.002

Murashov V, Schulte PA, Geraci C, Howard J (2011) Regulatory approaches to worker protection in nanotechnology industry in the USA and European Union. Ind Health 49(3):280–296

Naeemi AS, Elmi F, Vaezi G, Ghorbankhah M (2020) Copper oxide nanoparticles induce oxidative stress mediated apoptosis in carp (Cyprinus carpio) larva. Gene Rep 19:100676. https://doi.org/10.1016/j.genrep.2020.100676

Nakanishi J (ed) (2011a) Risk assessment of manufactured nanomaterials: approaches. Overview of approaches and results. NEDO, Kawasaki City

Nakanishi J (ed) (2011b) Risk assessment of manufactured nanomaterials: carbon nanotubes (CNT). NEDO, Kawasaki City

Nanocyl (2009) Responsible care and nanomateials case study nanocyl. Presentation at European Responsible Care Conference, Prague; http://www.cefic.be/Files/Downloads/04_Nanocyl.pdf. Accessed 23 Apr 2010

Nanosafe (2008) The first-edition of the Nanosafe Conference. https://www.cea.fr/ceatech/pns/nanosafe/en/Pages/Nanosafe-Conference/Last%20Editions/Nanosafe-2008.aspx

Nasimi P, Haidari M (2013) Medical use of nanoparticles: drug delivery and diagnosis diseases. Int J Green Nanotechnol 1:194308921350697

Naz S, Gul A, Zia M (2020) Toxicity of copper oxide nanoparticles: a review study. IET Nanobiotechnol 14:1–13. https://doi.org/10.1049/iet-nbt.2019.0176

Nel A, Xia T, Madler L, Li N (2006) Toxic potential of materials at the nanolevel. Science 311:622–627. https://doi.org/10.1126/science.1114397

NIOSH (2005) Approaches to safe nanotechnology: An information exchange with NIOSH. US Department of Health and Human Services, Centers for Disease Control and Prevention, National Institute for Occupational Safety and Health, Draft Report for Public Review, Cincinnati, OH

NIOSH (2009). https://www.cdc.gov/niosh/docs/2010-135/pdfs/2010-135.pdf, https://www.cdc.gov/niosh/docs/2010-135/pdfs/2010-135.pdf

NNI (2011) National nanotechnology initiative strategic plan. National Nanotechnology Initiative; http://www.nano.gov/node/581. Accessed 3 Dec 2013

Oukarroum A, Zaidi W, Samadani M, Dewez D (2017) Toxicity of nickel oxide nanoparticles on a freshwater green algal strain of *Chlorella vulgaris*. Biomed Res Int 2017:1–8. https://doi.org/10.1155/2017/9528180

Park E-J, Park K (2009) Oxidative stress and pro-inflammatory responses induced by silica nanoparticles in vivo and in vitro. Toxicol Lett 184(1):18–25

Park E-J, Choi J, Park Y-K, Park K (2008) Oxidative stress induced by cerium oxide nanoparticles in cultured BEAS-2B cells. Toxicology 245:90–100. https://doi.org/10.1016/j.tox.2007.12.022

Park MV, Lankveld DP, van Loveren H, de Jong WH (2009) The status of in vitro toxicity studies in the risk assessment of nanomaterials. Nanomedicine 4:669–685. https://doi.org/10.2217/nnm.09.40

Pelgrift RY, Friedman AJ (2013) Nanotechnology as a therapeutic tool to combat microbial resistance. Adv Drug Deliv Rev 65:1803–1815

Qian Y, Qin C, Chen M, Lin S (2020) Nanotechnology in soil remediation-applications vs implications. Ecotoxicol Environ Saf 201:110815. https://doi.org/10.1016/j.ecoenv.2020.110815

Rach J, Budde J, Möhle N, Aufderheide M (2014) Direct exposure at the air-liquid interface: evaluation of anin vitroapproach for simulating inhalation of airborne substances. J Appl Toxicol 34:506–515. https://doi.org/10.1002/jat.2899

Rajiv S, Jerobin J, Saranya V, Nainawat M, Sharma A, Makwana P, Gayathri C, Bharath L, Singh M, Kumar M, Mukherjee A, Chandrasekaran N (2016) Comparative cytotoxicity and genotoxicity of cobalt (II, III) oxide, iron (III) oxide, silicon dioxide, and aluminum oxide nanoparticles on human lymphocytes in vitro. Hum Exp Toxicol 35:170–183. https://doi.org/10.1177/0960327115579208

Rippner DA, Green PG, Young TM, Parikh SJ (2018) Dissolved organic matter reduces CuO nanoparticle toxicity to duckweed in simulated natural systems. Environ Pollut 234:692–698. https://doi.org/10.1016/j.envpol.2017.12.014

Rodea-Palomares I, Gonzalo S, Santiago-Morales J, Leganes F, Garcia-Calvo E, Rosal R, Fernandez-Pinas F (2012) An insight into the mechanisms of nanoceria toxicity in aquatic photosynthetic organisms. Aquat Toxicol 122:133–143. https://doi.org/10.1016/j.aquatox.2012.06.005

Rosário F, Bessa MJ, Brandão F, Costa C, Lopes CB, Estrada AC, Tavares DS, Teixeira JP, Reis AT (2020) Unravelling the potential cytotoxic effects of metal oxide nanoparticles and metal (loid) mixtures on A549 human cell line. Nano 10:447. https://doi.org/10.3390/nano10030447

Royal Society and Royal Academy of Engineering (2004) Nanoscience and nanotechnologies: opportunities and uncertainties. Royal Society and Royal Academy of Engineering, London

Russell WMS, Burch RL (1959) The principles of humane experimental technique. Methuen, London

Sahu MK (2019) Semiconductor nanoparticles theory and applications. Int J Appl Eng Res 14(2): 491–494

Saner M, Pelley J (2009) International Approaches to the Regulatory Governance of Nanotechnology (April 1, 2009). Regulation Papers, April 2009, Available at SSRN: https://ssrn.com/abstract=1532390

Savage DT, Hilt JZ, Dziubla TD (2019) In vitro methods for assessing nanoparticle toxicity. In: Zhang Q (ed) Nanotoxicity. Methods in molecular biology, vol 1894. Humana Press, New York, NY, pp 1–29

Schiavo S, Oliviero M, Miglietta M, Rametta G, Manzo S (2016) Genotoxic and cytotoxic effects of ZnO nanoparticles for Dunaliella tertiolecta and comparison with SiO2 and TiO2 effects at population growth inhibition levels. Sci Total Environ 550:619–627. https://doi.org/10.1016/j.scitotenv.2016.01.135

Schulte PA, Salamanca-Buentello F (2007) Ethical and scientific issues of nanotechnology in the workplace. Environ Health Perspect 115(1):5–12

Schulte PA, Geraci CL, Murashov V, Kuempel ED, Zumwalde RD, Castranova V, Hoover MD, Hodson L, Martinez KF (2014) Occupational safety and health criteria for responsible development of nanotechnology. J Nanopart Res 16(1):2153. https://doi.org/10.1007/s11051-013-2153-9; Epub 2013 Dec 7. PMID: 24482607; PMCID: PMC3890581

Seaton A, Tran L, Aitken R, Donaldson K (2010) Nanoparticles, human health hazard and regulation. J R Soc Interface 7(Suppl 1):S119–S129

Semmler-Behnke M, Takenaka S, Fertsch S, Wenk A, Seitz J, Mayer P, Oberdorster G, Kreyling WG (2007) Efficient elimination of inhaled nanoparticles from the alveolar region: evidence for interstitial uptake and subsequent reentrainment onto airways epithelium. Environ Health Perspect 115:728–733

Servin A, Elmer W, Mukherjee A, la Torre-Roche R, Hamdi H, White JC, Bindraban P, Dimkpa C (2015) A review of the use of engineered nanomaterials to suppress plant disease and enhance crop yield. J Nanopart Res 17:1. https://doi.org/10.1007/s11051-015-2907-7

Silins I, Hogberg J (2011) Combined toxic exposures and human health: biomarkers of exposure and effect. Int J Environ Res Public Health 8:629–647. https://doi.org/10.3390/ijerph8030629

Singh BR, Singh BN, Khan W, Singh HB, Naqvi AH (2012) ROS-mediated apoptotic cell death in prostate cancer LNCaP cells induced by biosurfactant stabilized CdS quantum dots. Biomaterials 33(23):5753–5767

Soares EV, Soares HMVM (2021) Harmful effects of metal(loid) oxide nanoparticles. Appl Microbiol Biotechnol 105(4):1379–1394. https://doi.org/10.1007/s00253-021-11124-1; Epub 2021 Feb 1. PMID: 33521847; PMCID: PMC7847763

Sousa CA, Soares HMVM, Soares EV (2018a) Nickel oxide (NiO) nanoparticles induce loss of cell viability in yeast mediated by oxidative stress. Chem Res Toxicol 31:658–665. https://doi.org/10.1021/acs.chemrestox.8b00022

Sousa CA, Soares HMVM, Soares EV (2018b) Toxic effects of nickel oxide (NiO) nanoparticles on the freshwater alga Pseudokirchneriella subcapitata. Aquat Toxicol 204:80–90. https://doi.org/10.1016/j.aquatox.2018.08.022

Subramaniam VD, Murugesan R, Pathak S (2020) Assessment of the cytotoxicity of cerium, tin, aluminum, and zinc oxide nanoparticles on human cells. J Nanopart Res 22:1. https://doi.org/10.1007/s11051-020-05102-3

Sukhanova A et al (2018) Dependence of nanoparticle toxicity on their physical and chemical properties. Nanoscale Res Lett 13:44

Suman TY, Rajasree SRR, Kirubagaran R (2015) Evaluation of zinc oxide nanoparticles toxicity on marine algae *Chlorella vulgaris* through flow cytometric, cytotoxicity and oxidative stress analysis. Ecotoxicol Environ Saf 113:23–30. https://doi.org/10.1016/j.ecoenv.2014.11.015

Sun TY, Bornhoft NA, Hungerbuhler K, Nowack B (2016) Dynamic probabilistic modeling of environmental emissions of engineered nanomaterials. Environ Sci Technol 50:4701–4711. https://doi.org/10.1021/acs.est.5b05828

Takhar P, Mahant S (2011) In vitro methods for Nanotoxicity assessment: advantages and applications. Arch Appl Sci Res 3:389–403

Tirumala MG, Anchi P, Raja S, Rachamalla M, Godugu C (2021) Novel methods and approaches for safety evaluation of nanoparticle formulations: a focus towards in vitro models and adverse outcome pathways. Front Pharmacol 12:612659. https://doi.org/10.3389/fphar.2021.612659

Tomellini R, Giordani J (2008) Third international dialogue on responsible research and development of nanotechnology. European Commission, Brussels

Usman M, Farooq M, Wakeel A, Nawaz A, Cheema SA, Rehman HU, Ashraf I, Sanaullah M (2020) Nanotechnology in agriculture: current status, challenges and future opportunities. Sci Total Environ 721:137778. https://doi.org/10.1016/j.scitotenv.2020.137778

VCI (2012) Responsible production and use of nanomaterials; https://www.vci.de/Services/Leitfaeden/Seiten/Responsible-Production-and-Use-of-Nanomaterials.aspx. Accessed 14 Aug 2013

Vimercati L, Cavone D, Caputi A, De Maria L, Tria M, Prato E, Ferri GM (2020) Nanoparticles: an experimental study of zinc nanoparticles toxicity on marine crustaceans. General overview on the health implications in humans. Front Public Health 8:192. https://doi.org/10.3389/fpubh.2020.00192

von Moos N, Maillard L, Slaveykova VI (2015) Dynamics of sub-lethal effects of nano-CuO on the microalga *Chlamydomonas reinhardtii* during short-term exposure. Aquat Toxicol 161:267–275. https://doi.org/10.1016/j.aquatox.2015.02.010

Walkey CD, Olsen JB, Guo H, Emili A, Chan WCW (2012) Nanoparticle size and surface chemistry determine serum protein adsorption and macrophage uptake. J Am Chem Soc 134(4):2139–2147

Wang D, Lin Z, Wang T, Yao Z, Qin M, Zheng S, Lu W (2016) Where does the toxicity of metal oxide nanoparticles come from: the nanoparticles, the ions, or a combination of both? J Hazard Mater 308:328–334. https://doi.org/10.1016/j.jhazmat.2016.01.066

Wang C, Cheng K, Zhou L, He J, Zheng X, Zhang L, Zhong X, Wang T (2017) Evaluation of long-term toxicity of oral zinc oxide nanoparticles and zinc sulfate in mice. Biol Trace Elem Res 178(2):276–282. https://doi.org/10.1007/s12011-017-0934-1

Wheeler CW, Park SS, Guenthner TM (1990) Immunochemical analysis of a cytochrome P-450IA1 homologue in human lung microsomes. Mol Pharmacol 38:634–643

WHO (1994) Declaration on occupational health for all. https://www.who.int/publications/i/item/WHO_OCH_94.1

Wiench K, Wohlleben W, Hisgen V, Radke K, Salinas E, Zok S, Landsiedel R (2009) Acute and chronic effects of nano- and non-nano-scale TiO2 and ZnO particles on mobility and reproduction of the freshwater invertebrate Daphnia magna. Chemosphere 76:1356–1365. https://doi.org/10.1016/j.chemosphere.2009.06.025

Wu F, Seib M, Mauel S, Klinzing S, Hicks AL (2020) A citizen science approach estimating titanium dioxide released from personal care products. PLoS One 15:e0235988. https://doi.org/10.1371/journal.pone.0235988

Xia T, Kovochich M, Brant J, Hotze M, Sempf J, Oberley T, Sioutas C, Yeh JI, Wiesner MR, Nel AE (2006) Comparison of the abilities of ambient and manufactured nanoparticles to induce cellular toxicity according to an oxidative stress paradigm. Nano Lett 6:1794–1807. https://doi.org/10.1021/nl061025k

Xia T, Zhao Y, Sager T, George S, Pokhrel S, Li N, Schoenfeld D, Meng H, Lin S, Wang X, Wang M, Ji Z, Zink JI, Mädler L, Castranova V, Lin S, Nel AE (2011) Decreased dissolution of ZnO by iron doping yields nanoparticles with reduced toxicity in the rodent lung and zebrafish embryos. ACS Nano 5(2):1223–1235. https://doi.org/10.1021/nn1028482

Xia B, Chen B, Sun X, Qu K, Ma F, Du M (2015) Interaction of TiO_2 nanoparticles with the marine microalga *Nitzschia closterium*: growth inhibition, oxidative stress and internalization. Sci Total Environ 508:525–533. https://doi.org/10.1016/j.scitotenv.2014.11.066

Yetisen AK, Qu H, Manbachi A, Butt H, Dokmeci MR, Hinestroza JP, Skorobogatiy M, Khademhosseini A, Yun SH (2016) Nanotechnology in textiles. ACS Nano 10:3042–3068. https://doi.org/10.1021/acsnano.5b08176

Yu Z et al (2020) Reactive oxygen species-related nanoparticle toxicity in the biomedical field. Nanoscale Res Lett 15:115

Zhang H, Ji Z, Xia T, Meng H, Low-Kam C, Liu R, Pokhrel S, Lin S, Wang X, Liao Y-P, Wang M, Li L, Rallo R, Damoiseaux R, Telesca D, Maedler L, Cohen Y, Zink JI, Nel AE (2012) Use of metal oxide nanoparticle band gap to develop a predictive paradigm for oxidative stress and acute pulmonary inflammation. ACS Nano 6:4349–4368. https://doi.org/10.1021/nn3010087

Zhang W, Bao S, Fang T (2016) The neglected nano-specific toxicity of ZnO nanoparticles in the yeast Saccharomyces cerevisiae. Sci Rep 6:24839. https://doi.org/10.1038/srep24839

Zheng D, Wang N, Wang X, Tang Y, Zhu L, Huang Z, Tang H, Shi Y, Wu Y, Zhang M et al (2012) Effects of the interaction of TiO2 nanoparticles with bisphenol a on their physicochemical properties and in vitro toxicity. J Hazard Mater 199–200:426–432. https://doi.org/10.1016/j.jhazmat.2011.11.040

10 Future Trends and Innovation in Nano Drug Delivery for Cancer Therapy, Application of siRNA (Nanoparticle-Based RNA) Therapy, Ultrasound Linked Nano-Cancer Therapeutics, and Application of Exosomes-Based Cancer Therapy

Kanwal Abbasi, Kauser Siddiqui, Saeeda Bano, Samina Iqbal, and Shagufta A. Shaikh

Abstract

The chapter discusses the roles of nanoparticles and hybrid nanoparticles for drug delivery in chemotherapy, targeted therapy, and immunotherapy and describes the targeting mechanism of nanoparticle-based drug delivery as well as its function on reversing drug resistance, compared to conventional drugs, nanoparticle-based drug delivery has specific advantages, such as improved stability and biocompatibility, enhanced permeability, retention effect, and precise targeting. The application and development of nano-system, which incorporates the combined properties of different nanoparticles, has led this type of drug-carrier system to the next level. The chapter also focuses on combinations of siRNA with chemotherapeutic drug delivery systems for the treatment of cancer and gives an overview of several nanocarrier formulations in both research and clinical applications. The use of various combinations of ultrasounds and nano-systems for cancer treatment has been reviewed, the current chapter introduces the biogenesis and uptake of exosomes and compares different approaches for isolation and drug loading, focusing on the application and current challenges of exosomes as drug delivery vehicles in cancer therapy.

Keywords

Nano drug delivery · Cancer therapy · Application of siRNA · Nanoparticle-based RNA therapy · Ultrasound linked nano-cancer therapeutics · Application of exosomes-based cancer therapy

K. Abbasi (✉) · K. Siddiqui · S. Bano · S. Iqbal · S. A. Shaikh
Pakistan Council of Scientific and Industrial Research Karachi (PCSIR), Karachi, Pakistan

F. A. Khan (ed.), *Nano Drug Delivery for Cancer Therapy*,
https://doi.org/10.1007/978-981-99-6940-1_10

Abbreviation

AMF	Alternating magnetic field
BBB	Blood–brain barrier
BSA	Bovine serum albumin
CNTs	Carbon nanotubes
DHA	Docosahexaenoic acid
DOX	Doxorubicin
DTX	Docetaxel
EGFR	Epidermal growth factor receptor
EPR	Enhanced permeability and retention
FDA	Food and Drug Administration
FITC	Fluorescein isothiocyanate
FR	Folate receptor
HEP-mPEG	Heparin-functionalized monomethoxy PEG
IGF1	Insulin-like growth factor 1
IONPs	Iron oxide nanoparticles
Lf	Lactoferrin
MRI	Magnetic resonance imaging
MSNs	Mesoporous silica nanoparticles
MWCNTs	Multi-walled carbon nanotubes
NIR	Near-IR
PAMAM	Polyamide amine
PDDAC	Poly(diallyldimethylammonium chloride)
PDT	Photodynamic therapy
PDX	Patient-derived xenograft
PEG	Polyethylene glycol
PLA	Polylactic acid
PSMA	Prostate-specific membrane antigen
PVA	Polyvinyl alcohol
RITC	Rhodamine Bisothiocyanate
SCID	Severe combined
siRNA	Small interfering RNA
TMZ	Temozolomide

10.1 Introduction

Nanotechnology has been widely studied and used in cancer therapy because nanoparticles (NPs) can play an important role as a drug delivery system. Compared with conventional drugs, nanoparticle-based drug delivery has special advantages such as better stability and biocompatibility, better permeability and retention effect, and precise targeting. The application and development of hybrid nanoparticles, which combine the combined properties of different nanoparticles, have taken this type of drug delivery systems to the next level (Bahrami et al. 2017). Nanoparticle-based drug delivery systems have been shown to play a role in overcoming cancer-related drug resistance. Nanoparticles offer several advantages in cancer therapy and have been shown to be effective in solving the problem of delivering hydrophobic agents, protecting bio-sensitive cargo, and promoting controlled drug release.6-8 Nanoparticles are also reliable platforms for tumor surveillance and microenvironment-responsive technologies.

RNAi is an important therapeutic strategy to prevent gene overexpression in many diseases, including cancer. RNAi processes allow siRNA to recognize and degrade a homologous mRNA sequence in the cell. However, the limited potential of siRNA to target cells, poor cellular uptake, off-target effects, and their degradation by serum nucleases in the systemic circulation are factors limiting their therapeutic use. Nanocarriers for the delivery of siRNA cargo and their combined use with anticancer drugs, as well as their applications in cancer therapy, are undergoing a period of clinical translation (Dolly et al. 2023). The current status of siRNA therapies in clinical trials is explored, providing a systematic and comprehensive overview of siRNA-based therapies for effective anticancer therapy. In the current chapter we discussed the future trends and innovation in nano drug delivery for cancer therapy, application of siRNA (nanoparticle-based RNA) therapy.

10.2 NPs in Cancer Therapy

The NPs used in medical treatment usually have specific sizes, shapes, and surface characteristics as these three aspects have a major influence on the efficiency of the nano-drug delivery and thus control therapeutic efficacy.

The composition of nanoparticles can be varied, because the precursors can be biological lipids, dextran, lactic acid, phospholipids, chitosan or chemicals such as silica, carbon, metals, and various polymers. Interactions between biological components of nanoparticles are different from interactions with non-biological components. Biodegradable nanoparticles are necessary in drug delivery to transport the drug and release it at the destination. NPs with a diameter of 10–100 nm are generally considered suitable for cancer therapy due to their ability to efficiently deliver drugs and improve permeability and retention (EPR) effect (Venturoli and Rippe 2005). Smaller particles can easily leak from normal blood vessels (less than 1–2 nm) to damage normal cells and can be easily filtered by the kidneys (less than 10 nm in diameter), while particles less than 100 nm in diameter can be eliminated of the blood by phagocytes (Decuzzi et al. 2009).

Nanoparticles are essentially non-toxic, chemically stable and biocompatible materials, and this nature allows them to be chosen as an effective drug delivery vehicle. Nanotechnology and nanoparticles have attracted great interest in cancer therapy because they can provide better and targeted drug delivery systems to overcome the shortcomings of conventional chemotherapy. Limitations of chemotherapy include inefficient delivery of drugs to the target due to physical and

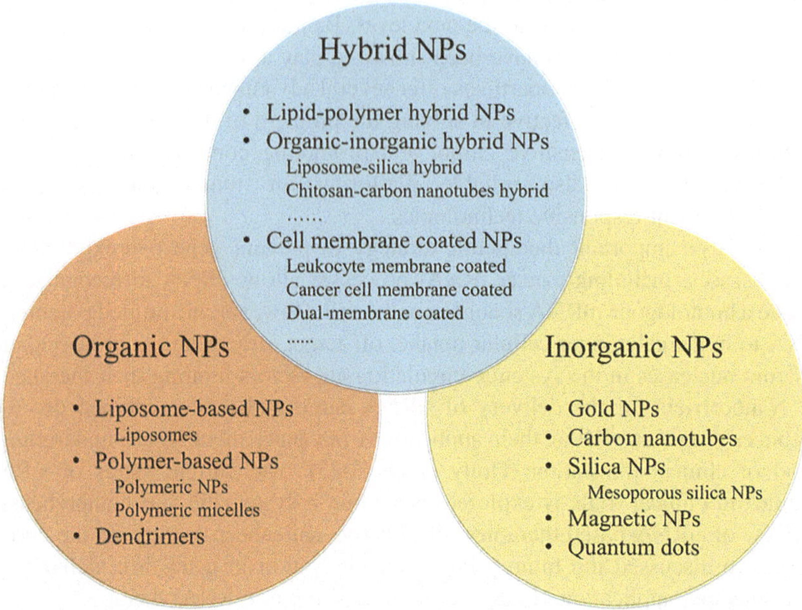

Fig. 10.1 Different types of nanoparticles (NPs) for cancer therapy. NPs applied to drug delivery systems include organic NPs, inorganic NPs, and hybrid NPs. The organic NPs contain liposome-based NPs, polymer-based NPs, and dendrimers. Among polymer-based NPs, polymeric NPs and polymeric micelles are common. The inorganic NPs consist of gold NPs (Au NPs), carbon nanotubes, silica NPs, magnetic NPs, and quantum dots. Hybrid NPs combine the advantages of different NPs, including lipid-polymer hybrid NPs, organic-inorganic hybrid NPs, and cell membrane-coated NPs (Yao et al. 2020)

biochemical barriers, and drug resistance at the tumor level due to cellular and non-cellular mechanisms that limit drug action, leading to recurrence and high mortality. The different properties of NP determine their therapeutic effect in cancer treatment. Different types of NP for cancer treatment are shown in Fig. 10.1 (Yao et al. 2020).

10.3 NPs Used for Drug Delivery in Cancer Therapy

Several innovative drug delivery strategies are currently being used to treat cancer, and new cancer therapies have been developed using a variety of nanomaterials, including organic and inorganic particles, as well as synthetically produced lipids, proteins, and polymers. Drug delivery encapsulated in NPs offers several advantages over direct delivery of refined chemotherapy drugs. These include improved drug delivery, protection of encapsulated drugs from degradation in the bloodstream, targeted drug delivery, effective therapy with reduced systemic toxicity, improved drug solubility, and drug pharmacodynamic and pharmacokinetic properties (Perrault et al. 2009; Wong et al. 2011; Yang et al. 2014).

10.3.1 Inorganic NPs for Drug Delivery in Cancer Therapy

Inorganic NCs have been shown to be very useful in cancer therapy due to their large surface area and volume and easy conjugation with various anticancer drugs (Table 10.1). In addition, biocompatibility, low toxicity, and the ability to control drug release facilitated the use of inorganic NPs, such as mesoporous silica NCs (MSNPs), gold NPs (AuNPs), magnetic NPs (MNPs), carbon nanotubes NPs (CNT-NP), graphene oxide, and quantum dots (QDs) as drug carriers (Kaushik et al. 2022).

The investigated inorganic NPs include gold NPs (AuNPs), carbon nanotubes (CNTs), quantum dots, magnetic NPs (MNPs), and silica NPs (SNPs). AuNPs are the most widely studied inorganic NPs, and gold core-based monolayer-protected mixed clusters are considered promising candidates for drug delivery system. The gold core is inert and nontoxic, and surface-functionalized AuNPs have been shown to enhance drug accumulation in tumors and overcome drug resistance. AuNPs are involved in multimodal cancer therapy, including gene therapy, photothermal therapy, and immunotherapy (Ali et al. 2021). Carbon nanotubes are tubular materials with unique biological, physical, and chemical properties that have been used to deliver anticancer agents such as doxorubicin, paclitaxel, and methotrexate RNA to a variety of cancers. CNTs generate heat when exposed to near-infrared radiation, which can be used for thermal ablation in cancer therapy (Debnath and Srivastava 2021). Mesoporous silica nanoparticle carriers are a type of SNPs that have a large internal pore volume, which allows them to encapsulate the maximum amount of anticancer drugs, and the supramolecular components act as a cap that allows the drugs to be trapped and released (Cheng et al. 2019; Lei et al. 2019). Porous silica NPs have shown great potential in immunotherapy due to their immunoadjuvant properties, including promoting antigen cross-presentation, lymphocyte polarization, and interferon-g (IFN-g) secretion (Fontana et al. 2017). Magnetic NPs (MNPs) used for drug delivery usually contain metal or metal oxide NPs. To improve stability and biocompatibility, MNPs are usually coated with organic materials, including polymers and fatty acids. They have been shown to have high efficacy in chemotherapy, gene therapy, and magnetic hyperthermia using MNPs to achieve thermal ablation of tumors, providing an alternative anticancer treatment (Basoglu et al. 2018; Mandriota et al. 2019).

10.3.1.1 Liposomes

A liposome, the first nanosized drug approved for clinical use, consists of an outer lipid layer and a core surrounding either a hydrophobic or hydrophilic drug. Liposomes can perform many functions by altering the structure of the lipid bilayer, including mimicking the biophysical properties of living cells (e.g., motility and deformation). which can help achieve more effective therapeutic drug delivery (Zylberberg and Matosevic 2016). Liposomes are a good platform for the in vivo

Table 10.1 Inorganic drug delivery NCs in cancer therapy

Material	Description of carrier	Material advantage	Specificity
CNT	Anti-P-glycoprotein antibody functionalized CNT-doxorubicin	Defeats multidrug resistance	Leukemia cells
CNT	Multi-walled CNT decorated with guanidinylated dendritic molecular transporters	Efficient DOX delivery	Prostate cancer cells
CNT	PEG-CNT complex	Mitochondrial targeting	Lung cancer cells
Layered double hydroxide NPs	Co-delivery of 5-fluorouracil and siRNAs	Prevents drug resistance and enhances cancer treatment	Various cancer cells
Layered double hydroxide NPs	Raloxifene intercalated into the interlayer gallery of LDH host	Improves therapeutic efficacy, reduction of adverse side effects	Solid tumors
Iron oxide NPs	Phospholipid-PEG-coated super- paramagnetic iron oxide NPs	Chemotherapy and hyperthermia treatment	Solid cancers
Magnetic NPs	Pluronic F127-anchored iron oxide NPs	Active and passive delivery of hydrophobic drugs	Folate-positive cancer cells
Magnetic NPs	Chitosan-coated superparamagnetic iron oxide NPs	Doxorubicin delivery	Ovarian cancer cells
Mesoporous silica NPs	Azobenzene-modified mesoporous silica for NIR-triggered anticancer drug delivery	Drug release rate can be controlled by varying the intensity and/or time	Solid tumor
Mesoporous silica NPs	Hyaluronic acid-capped mesoporous Silica NPs	Site-selective, controlled-release delivery	MDA-MB-231 and NIH3T3 cells
QDs	Riboflavin-targeting graphene quantum dots-PEG-benzofuran	High potency, improved dispersibility	Laryngeal, lung and colorectal cancer cells
QDs	Hyaluronic acid/ferrocene	Selective binding to CD44 receptors, redox-based drug delivery	Diverse range of cancer cells
QDs	(Fc)-anchored nitrogen-doped graphene QDs (Fc-GQD-HA)	Mono-dispersive capacity, elevated drug-loading	In vivo application

CNT carbon nanotubes, *NPs* nanoparticles, *QDs* quantum dots, *LDH* layered double hydroxides, *PEG* polyethylglycol

delivery of many antitumor drugs such as doxorubicin and paclitaxel, other chemotherapeutic agents, and nucleic acids. paclitaxel liposomes have better antitumor efficacy and better bioavailability compared to free paclitaxel. Liposomal doxorubicin has been shown to reduce cardiotoxicity in breast cancer with comparable efficacy. Liposome-based nanosystems have also provided an alternative to drug combination that can improve therapeutic efficacy (Gao et al. 2023) (Table 10.2).

10.3.1.2 Polymeric Nanoparticles

Polymer-based NPs are another type of NPs that have specific structural arrangements for drug delivery formed by different monomers. Polylactic-co-glycolic acid (PLGA), a common polymeric NP, involves the copolymerization of glycolic acid and lactic acid. Due to its better biocompatibility and biodegradability and EPR effect, PLGA is widely used as a drug delivery carrier, and the multiple functional groups on its surface improve the ability to load and transport therapeutic agents (Feltrin et al. 2022). Polyethylene glycol (PEG), also known as polyethylene oxide and polyoxyethylene, has a linear polymer structure. PEG nanoparticles can increase the systemic circulation of the drug in the body and reduce the immunogenicity, which leads to their widespread use in cancer treatment. Due to their non-immunogenic and biodegradable properties, PEG nanoparticles are considered "stealth" polymers and several, including DOXil, Asclera, and Movantik, have been approved by the US FDA (Nordstrom et al. 2021). Polyethyleneimine (PEI) is a cationic polymer composed of repeating linear and branched ethylamine units. Due to the strong positive charge on the surface of PEI, it is possible to condense nucleic acid through electrostatic interactions, which enables high transport efficiency of drugs in cancer treatment (Chou et al. 2018; Hailing et al. 2020). The multifunctional construction methods based on PEI are mainly mediated by covalent coupling amine groups. 70 For example, the amine groups of PEI can be covalently coupled with carboxyl, such as 4-(bromomethyl) phenylboronic acid (PBA) (Zhu et al. 2018). Cyclodextrins (CDs) are degradation products of amylase and consist of glycopyranoses that form cyclic oligosaccharides. CDs are widely used as delivery vectors in cancer therapy due to their hydrophilic properties and ability to form host-guest structures with drugs (Liu et al. 2021a, b). Chitosan (CS) is a natural cationic polymer composed of β-(1,4)-linked N-acetylglucosamine units produced by deacetylation of chitin in shellfish shells. Because CS nanoparticles have excellent biocompatibility and biodegradability, they can be used as a drug coating and drug delivery vector in cancer therapy (Wang et al. 2022). Dextran is a hydrophilic natural polymer formed by the condensation of glucose. The dextran chain consists of α-1,6-glucosidic bonds between glucose monomers, which have hydrophilic, bio-compatible and biodegradable properties, dextran can be widely used as a scaffold surface, for coating and delivery of vector for cancer treatment (Chenglong et al. 2020). Hyaluronic acid (HA) is a natural polymer composed of d-glucuronic acid and N-acetyl-d-glucosamine. HA is a biodegradable hydrophilic polymer material used as a biological scaffold in cancer therapy. HA has tumor-targeting, non-immunogenic, and biosafety properties, HA has been widely used in drug delivery systems for drug regulation and sustained release (Huang and Huang

Table 10.2 Organic drug delivery NCs in cancer therapy

Material	Description of carrier	Material advantage	Specificity
Liposomes	Liposomal doxorubicin	Improved delivery to site of disease; decrease in systemic toxicity of free drug	Ovarian cancer; multiple myeloma
Liposomes	Liposomal daunorubicin	Improved delivery to tumor site; lower systemic toxicity arising from side effects	Karposi's sarcoma
Liposomes	Genistein and plumbagin encapsulated nanoliposomes	Inhibition of cell metabolism	In vitro and in vivo prostate cancer
Liposomes	Folate-conjugated bovine serum albumin bound paclitaxel NPs	Increased solubility, cellular uptake; targeted specificity	Prostate cancer cells
Protein-based	Alpha mangostin loaded crosslinked silk fibroin-based NPs	Physicochemically stable, increased the drug's solubility	Colorectal and breast cancer
Protein-based	Noscapine-loaded human serum albumin NPs	High drug-loading efficiency (85–96%) and delivery of maximum quantity of drug to the tumor site	Breast cancer cells
Protein-based	Plasmid cDNA (pGL3) polyethyleneimine (PEI)-coated HSA NPs	Enhance endosomal escape	In vitro gene delivery application
Micelles	Polymeric methoxy-PEG-poly(D,L-lactide) micelle formulation of paclitaxel	Improved delivery to site of disease; decrease in systemic toxicity of free drug	Breast cancer; ovarian cancer
Micelles	Folate-PEG/Hyd-curcumin/C18-g-poly-succinimide	pH sensitive drug release	Colon cancer
Micelles	PEGylated prodrug nano-micelles	Glucose-sensitive	In vitro and in vivo anticancer activity
Polymeric Micelles	CD44v6-targeted polymeric micelles (PM) loaded with niclosmide	Increase drug safety	Efficacy against colorectal stem cells
Self-assembly	Aptamer-tethered DNA assembly	Stronger targeting ability, higher cellular uptake	Cancer cell imaging
Self-assembly	DNA-aptamer conjugated RNA-triple helix hydrogel	Efficient cellular uptake and enhanced nuclease resistance with superior biocompatibility	Triple negative breast cancer detection and treatment
Self-assembly	Folate-modified MPEG-PCL	Improved bioavailability, low toxicity, sustained drug release	Colorectal cancer mice model
Self-assembly	Folate receptor-targeted β-cyclodextrin (β-CD)	Biosafety, bioavailability, and improve curcumin drug loading capacity	Cervical cancer, fibroblast cells

2018). Thanks to the glucuronic acid and glucosamine contained in HA, multifunctional design strategies are also used in cancer treatment.

10.3.1.3 Peptides and Protein Nanoparticles

Peptide nanoparticles consist of nanoparticle-sized natural or synthetic amino acids. Peptide nanoparticles are widely used in cancer therapy, gene delivery, and targeted drug delivery due to peptide biodegradability and active groups, multifunctional peptide nanoparticles include physical adsorption and chemical covalent bonding with various drugs. Protein NPs can be incorporated into biodegradable polymers in a monolithic particle structure, such as microspheres, to enable controlled and sustained drug release, and can be produced using proteinaceous materials such as silk protein fibroin, bovine, human serum albumin, and gliadin. Peptides contain benzene or macrocyclic structures that allow peptide nanoparticles to interact with various metals (Rong et al. 2018). The fluorescent peptide nanoparticles with cyclic peptides modified them with Arginylglycylaspartic acid (RGD) to encapsulate epirubicin (EPI) to form a multifunctional RGD-f-PNPs/EPI complex through π–π stacking and electrostatic interactions. Due to the RGD surface modification, the prepared nanoparticles selectively targeted EC cells and promoted EPI internalization. Albumin is a commonly used protein nanoparticle in cancer therapy due to its high drug loading efficiency, with bovine serum albumin (BSA) and human serum albumin (HSA) commonly used in drug delivery systems (Fan et al. 2018).

10.3.2 Micelles NPs

Micelles are spherical nano-sized structures formed by the self-assembly of amphiphilic block copolymers with both hydrophilic and hydrophobic moieties in aqueous solution to form a hydrophobic core and a hydrophilic shell known as polymeric micelles. Micelles can form naturally at specific concentrations (critical micelle concentration (CMC)) and temperatures. In polymeric micelles, the hydrophobic core acts as a reservoir for hydrophobic drug incorporation, while the hydrophilic shell stabilizes the core and keeps the polymer and drug intact, making (Ghezzi et al. 2021). The micelles suitable for intravenous administration Genexol-PM [PEG-poly (D,Llactiide)-paclitaxel] was the first cremophor-free polymeric gel containing paclitaxel with no adverse effects (Kim et al. 2004).

10.3.3 Self-Assembled Drug NPs

Both organic and inorganic NPs have their own advantages and disadvantages, because combining the two into one hybrid drug delivery system provides a multifunctional carrier with better biological properties that can improve therapeutic efficacy and reduce drug resistance. The spontaneous arrangement of molecules in ordered geometric structures is called molecular self-assembly, and it gives inorganic and organic structures with unique properties through non-covalent interactions (Yadav et al. 2020). The spontaneity, versatility, cheapness, and simplicity of self-assembly means that it is widely used in the design of nanoscale biomaterials and has a range of biomedical applications, including drug delivery, tissue engineering, and regenerative medicine. Self-assembled NPs such as micelles, polymeric NPs, liposomes and CNTs have been shown to enhance drug delivery, prolong circulation, regulate drug release kinetics, mediate molecular targeting, enhance tumor accumulation and bioavailability, enable proper encapsulation, and protect drug from the external environment (Wang et al. 2020). These self-assembled or hybrid NPs help to overcome physiological barriers during in vivo drug delivery.

10.4 Targeted Drug Delivery

Targeting of cancer cells specifically is a vital characteristic of nano-carriers for drug delivery, as it enhances the therapeutic efficacy while protecting normal cells from cytotoxicity. The tumor targeting and the nano-carrier system is design to first understand tumor biology and the interaction between nano-carriers and tumor cells (Attia et al. 2019). The targeting mechanisms can be broadly divided into two categories, passive targeting and active targeting (Fig. 10.2) (Holder et al. 2023).

10.4.1 Passive Targeting

Intravenous injection is the most used method of administering anticancer medications based on nanomaterials. This method avoids the stage of intestinal

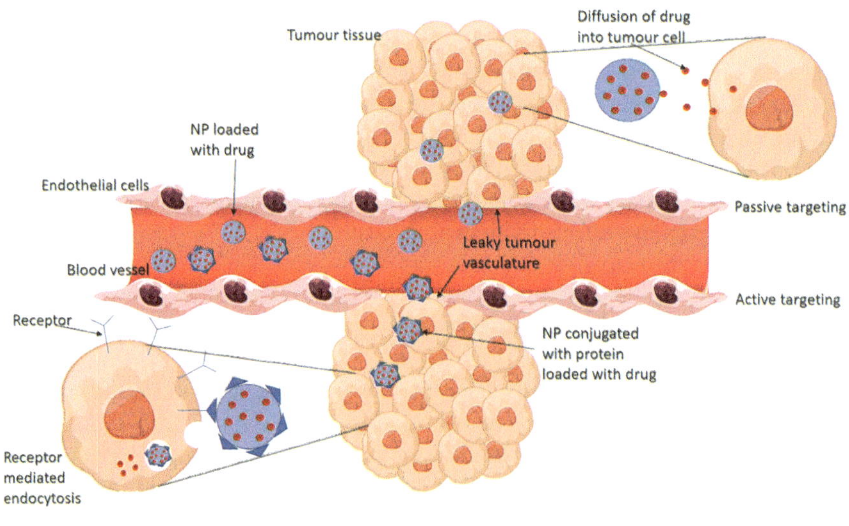

Fig. 10.2 Schematic representation of receptor-mediated active targeting and passive targeting through the EPR effect (Holder et al. 2023)

epithelial absorption that is necessary following oral delivery (Zhao and Castranova 2011). Drugs are successfully transported to the target site by passive targeting in order to perform a therapeutic function. High cancer cell proliferation causes neovascularization, and big holes in the vascular wall make tumor arteries more permeable than healthy vessels. Macromolecules, such as NPs, might leak from blood arteries supplying the tumor and accumulate within tumor tissue as a result of the fast and deficient angiogenesis. Depending on the nature, location, and environment of the tumor, the distances between the endothelial cells in the tumor vasculature might range from 200 to 2000 nm (Rajendra and Stylianopoulos 2010). Additionally, the nanoparticles are not quickly removed and build up in the tumor interstitium because of the poor lymphatic function. Passive targeting is based on a phenomenon known as enhanced permeability and retention (EPR) effect. This passive process of drug buildup at the tumor sites necessitates prolonged drug circulation for effective delivery. Nanocarrier accumulation is largely influenced by physicochemical characteristics such size, shape (morphology), surface charge, and surface chemistry (Bertrand et al. 2014). The first FDA (the Food and Drug Administration, national agency of the United States Department of Health and Human Services) approved nano-drug contains PEGylated liposome-entrapped doxorubicin (DOX), which is intended to treat ovarian cancer and Kaposi sarcoma tumors linked to HIV. The drug's cellular and systemic toxicity was significantly decreased by encasing it inside the lipid substance, which also enhanced the drug's pharmacokinetics and allowed for more controlled biodistribution and release (Navya et al. 2019).

10.4.1.1 Limitations in Passive Targeting

The variety of EPR effects reported inside and between different cancers limits the therapeutic efficacy of passive targeted methods. It may lead to non-uniform extravasation of nanoparticles into the target location because of varying endothelial gaps brought on by aggressive tumorous cell proliferation. Furthermore, extravasation has been shown to depend not only on permeability but also on the blood flow rate around the tumor site. Vascularization exhibits spatial and temporal heterogeneity within and between tumor cells adding another level of challenges to passive targeting (Chauhan and Jain 2013). The nanoparticles permeability is typically higher in the hypoxic core of the tumor area rather than the periphery. The key to enhancing penetration into the tumor matrix is the nanomaterial's size and size-dependent characteristics. The NPs' diameter; research has shown that smaller NPs have better penetration but do not spill into normal channels, whereas micromolecules are selectively eliminated by immune defense mechanisms. Another barrier to the efficient deposition of drug-loaded nanocarriers in solid tumors is interstitial fluid pressure (IFP) (Waite and Roth 2012).

10.4.2 Active Targeting

Active targeting directly engages ligands and receptors to target cancer cells specifically. In order to discriminate between targeted cells and healthy cells, the ligands on the surface of NPs are chosen to target the molecules that are overexpressed on the surfaces of cancer cells. Active targeting is essential for the medication and gene delivery to the desired location, which increases therapeutic effectiveness and limits adverse effects on healthy tissues. Compared to passively targeted NCs, active targeting can increase the amount of medication given to the target tumor cell. Chemical affinities for active targeting are based on many unique molecular interactions, such as interactions between receptors and ligands, contacts based on charges, and facilitated interactions based on motif with substrate molecules (Byrne et al. 2008). A ligand can be made up of a variety of biomolecules, including as antibodies, proteins, nucleic acids, peptides, carbohydrates, and tiny chemical compounds like vitamins. Target substrates include surface molecules expressed by diseased cells, proteins, carbohydrates, or lipids found in organs, chemicals found in the diseased cells' microenvironment (or secreted there by tumor cells), and even the local physicochemical environment (Weissleder et al. 2005; Verma and Stellacci 2010; Mout et al. 2012). Internalized NPs can successfully release therapeutic medications because of receptor-mediated endocytosis, which is brought on by the contact between ligands on NPs and the receptors on the surface of cancer cells. The avidity of the nanoparticles for their appropriate targets increases overall when numerous ligand molecules are added to the nanosystems. Additionally, the cooperativity effects that result from the binding of one ligand molecule typically make the binding of succeeding molecules easier, improving the overall efficiency of binding and the actions that follow. The internalization of nanoparticles by target cells is increased by an active targeting strategy, which also increases the effectiveness of drug delivery. The ligands specifically bind to receptors on targeted cells, which include the transferrin receptor, folate receptor, glycoproteins, and the epidermal growth factor receptor (EGFR).

10.4.3 Targeting to Cancer Cells

10.4.3.1 Epidermal Growth Factor Receptor (EGFRs)-Based Active Targeting

The ErbB family of tyrosine kinase receptors includes the epidermal growth factor receptor. It has already been used as a target for cancer treatment. EGFR is overexpressed in a number of malignancies, including lung, pancreatic, colorectal, and breast cancers. It is also involved in a number of processes that contribute to tumor growth and progression. Lysosomes arise as a result of the nanoparticles being absorbed into the cells by an EGFR-mediated endocytosis process, releasing the enclosed medicines for cancer treatment. Targeting the human epidermal receptor-2 (HER-2) is a frequent treatment for breast and stomach cancer that is HER-2 positive. Numerous monoclonal antibodies (Abs) and their fragments, along with other peptide medicines, have recently been included as targeting units on nanoparticles against HER2 overexpressing tumors. Nanoparticles now have good therapeutic efficacy for the treatment of breast cancer thanks to the development of trastuzumab (TZ), a humanized anti-HER2 Ab. By inhibiting downstream HER2 signaling, it can prevent cell cycle and decrease angiogenesis (Elias et al. 2021). EGFR-overexpressed cancer cells can be targeted using NPs that have been designed to incorporate modified ligands that bind to EGFR. These NPs have improved bioavailability, prolonged blood circulation, enhanced tumor accumulation, and reduced off-target side effects, which significantly increases therapeutic efficacy.

10.4.3.2 Transferrin (Tf) Receptor-Mediated Active Targeting

Transferrin (tf) is an example of a serum glycoprotein, which is involved in iron (Fe) delivery and metabolism. Transferrin receptors (TfR), which have a high affinity for transferrin (Tf), are frequently overexpressed on the surface of a variety of tumors, including pancreatic, breast, prostate, colon, and lung cancer (Wu et al. 2018a, b). In a recent article (Qi et al. 2022) developed a novel transferrin protein corona (Tpc)-modified CuGd nanoplatform (Tpc-CuGd) for tumor-targeting photothermal and chemodynamic synergistic therapy. These Tf-modified nanoparticles have been developed for the active targeted delivery of therapeutic agents to tumor sites, and they can preferentially deliver therapeutics into TfR-overexpressing tumors (Arumov et al. 2021).

10.4.3.3 Estrogen Receptor-Mediated Active Targeting

Estrogen is a steroid hormone that is essential for regulating cardiovascular remodeling, bone homeostasis, brain development, and reproductive system health. E2 is essential for the development of breast, endometrial, and ovarian malignancies among the three types of estradiol (estrone (E1), estradiol (E2), and estriol (E3) (Scabia et al. 2022; Hamilton et al. 2017). The main mechanism by which estrogens function is the binding and subsequent activation of two estrogen receptors (ERα and ERβ), which differ structurally. Intracellular ERs can transport estrogen-modified nanoparticles toward the nucleus for nuclear targeting after intracellular uptake via receptor-mediated endocytosis (Thomas and Gustafsson 2011). On numerous tumor cell surfaces, these receptors have been discovered to be overexpressed. In a recent application, anti- ERα Ab-modified AuNPs were internalized by MCF-7 cells. It was discovered that the ER receptor-mediated endocytosis pathway for increased tumor treatment allowed these modified nanoparticles to be efficiently absorbed by tumor cells (Kapara et al. 2020).

10.4.3.4 Cluster of Differentiation (CD) Receptor-Mediated

Surface receptors from the CD receptor family, such as CD14, CD22, CD36, CD44, and CD133, which are primarily found on cancer stem cells (CSCs), can be exploited as promising delivery targets to prevent tumor spread. A transmembrane adhesion glycoprotein called CD44 has been frequently used to target receptors for targeted tumor treatment among them (Tian et al. 2022). Widespread usage of the biocompatible ligand hyaluronic acid (HA) in CD44 receptor-mediated active targeted delivery systems. Due to its availability as a natural polymer as opposed to polymers that require a multi-step chemical synthesis, it can be easily obtained (Through CD44 receptor-mediated active targeting, HA-functionalized nanoplatforms can efficiently deliver therapeutic drugs to tumor cells with a superior cytotoxic profile and tumor killing (Poudel et al. 2020). Compared to groups without the HA modification, HA-enveloped nanoparticles substantially inhibited tumor growth. HA is typically modified on the surface of nanoparticles to preferentially bind to CD44 receptors that are overexpressed in tumor cells, leading to tumor endocytosis. Additionally, hyaluronidase, which is widely distributed in cancer cells, has the ability to break down HA into smaller fragments. The adaptable properties of HA as a targeted and enzyme-responsive ligand make it an attractive option for use in particular drug delivery systems (Lei et al. 2021).

10.4.3.5 Folate Receptor (FR)-Mediated Active Targeting

Three subtypes of FRs, a group of glycoproteins, have been identified: FRα, FRβ, and FRγ. It should be emphasized that, in contrast to FRγ, which has only been identified in hematopoietic cells, FRα and FRβ can bind tightly to the tumor cell membrane via a glycosylphosphatidylinositol anchor (Holm and Hansen 2020). The most often created FR subtype, FRα, is overexpressed in a variety of tumor cells, particularly those that are cancerous of the breast, lung, kidney, cervical, and ovarian types (Scaranti et al. 2020). Through receptor-mediated endocytosis, FR can provide folate to tumor cells. For improved uptake of therapeutic drugs by tumor cells, a number of FA-based nanoplatforms have been developed. A well-known example of an organic/inorganic hybrid nanoplatform modified by FA-chitosan conjugates to load upconverting NaYF4nanoparticles and daunorubicin for tumor therapy (Fernández et al. 2017). The FA modification dramatically increased the cellular absorption of the nanoparticles. Another example is the preparation of a palladium nanoplatform for FA-mediated targeted treatment using an FA-conjugated chitosan loaded with rutin. Through a process known as FR-mediated endocytosis, FRs can recognize and bind to extracellular FA-modified nanoparticles before delivering them into the tumor cells (Sathiyaseelan et al. 2022). The FA (folic acid) component serves as a tumor-targeting ligand in these nanoparticles. The cell membrane can invade and pinch off to produce endosomes, which then travel to lysosomes or other organelles, upon binding to the FR on tumor cells. The drug-encapsulated nanoparticles have the ability to separate from the FR and successfully release the medication at the tumor environment for the purpose of treating the tumor. The endocytosis efficiency of the nanoparticles in breast cancer cells was dramatically increased by the addition of FA to the developed nanoplatform. A cell viability study shown that the developed nanoplatform significantly reduced cell proliferation (Barz et al. 2010).

10.4.3.6 Glycoproteins-Mediated Active Targeting Systems

The many types of glycoproteins, including lectins, which are non-immunological proteins that particularly detect and bind to particular carbohydrates, are typically expressed by cancer cells. The direct lectin targeting pathway targets the carbohydrates on the surface of cancer cells with lectins attached to NPs, whereas the reverse lectin targeting mechanism targets the lectins on cancer cells with carbohydrates moieties integrated into NPs. Another strategy for active targeting is the conjugation of two cancer-specific ligands onto a single NP, which can increase target specificity.

10.4.3.7 Other Receptor-Mediated Active Targeting Systems

Targeted anticancer nanoparticles have also been created using the other receptors, such as chemokine, biotin, and luteinizing hormone-releasing hormone (LHRH) receptors. A G-protein-coupled receptor family known as chemokine receptor type 4 (CXCR4), for instance, plays a crucial role in tumor dissemination by attracting tumor cells along chemokine gradients (Zheng et al. 2019; Xiao et al. 2021). The development of circular peptide T22-functionalized mesoporous silica for the efficient delivery of chemotherapeutic agents to tumor cells (De La Torre et al. 2015), the preparation of epirubicin-encapsulated polymeric nanoparticles that significantly increased therapeutic efficacy in hepatocellular carcinoma by conjugating the LFC131 peptide to increase the affinity, recently cellulose nanoparticles with the LFC131 peptide for targeted tumor treatment was designed (Di-Wen et al. 2016; Chittasupho et al. 2017). A brand-new nanoplatform to successfully stimulate p53 expression in hepatocellular cancer models by targeting CXCR-4. In established hepatocellular carcinoma models, the CXCR4-targeted p53 mRNA nanoplatform and anti-PD-1 therapy efficiently triggered cellular reprogramming and immunological components of the tumor microenvironment (Xiao et al. 2022).

10.4.4 Targeting to Endothelium

Some NPs have an impact on angiogenesis, another strategy for treating cancer, rather than directly targeting cancer cells. Vascularization depends on the interaction between vascular endothelial growth factor (VEGF) and VEGF receptors (VEGFRs), with liposomes simultaneously targeting the two main VEGF receptors, VEGFR-2 and VEGFR-3, to improve therapeutic efficacy (Apte et al. 2019; Orleth et al. 2016). Integrins are extracellular matrix protein-binding cell surface receptors that are crucial for the invasion and motility of malignancy cells (Desgrosellier and Cheresh 2010). The calcium-dependent route that promotes endothelial cell migration depends on the αβv3 integrin, which is significantly expressed in tumor neovascular endothelial cells but not in resting endothelial cells or normal cells. the effective medication delivery into tumor cells using cationic NPs in conjunction with ligands that target the αβv3 integrin (Nisato et al. 2003). The glycoprotein known as Vascular Cell Adhesion Molecule-1 (VCAM-1) interacts with vascular endothelial cells to play a role in angiogenesis. It is also found on the surface of tumor endothelium. VCAM-1 overexpression has been seen in a number of malignancies, suggesting that it may play a part in the active targeting of NPs for drug delivery (Dienst et al. 2005). The extracellular matrix remodeling and tumor neovascularization processes are carried out by the matrix metalloproteinase (MMP), a component of the tumor microenvironment. Breast cancer, pancreatic cancer, and melanoma are among the tumors for which MMP-sensitive NPs have been suggested to have possible antitumor effects (Cun et al. 2019).

10.5 Stimuli-Responsive Targeting Strategies

An acidic environment, elevated levels of GSH and ROS, and enhanced production of certain enzymes (MMP-2/cathepsin B) are only a few of the TME's (tumor microenvironment) peculiar properties. As a result, tumor treatment and targeted medication delivery are made possible by nanoparticles that incorporate TME responsive components (Banstola et al. 2021). These endogenous cues can cause changes in molecular function, dispersion behavior, morphology, and degradation kinetics. This makes it easier for pharmaceutical active components to enter cells or to escape endosome/lysosomal breakdown and be released. Exogenous stimuli-responsive nanoparticles can also respond to endogenous stimuli like pH, ATP, H_2O_2, enzyme, redox-potential, and hypoxia in addition to exogenous stimuli like temperature, magnetic field, light, ultrasound, and lasers. although the stimuli might be emerged in TME or inside cancer cells (Fig. 10.3) (Tian et al. 2022).

10.5.1 Endogenous Stimuli

Also known as an intrinsic stimulus, endogenous stimuli include changes in the internal pH levels, enzyme activity, redox activity, hypoxic conditions of the body.

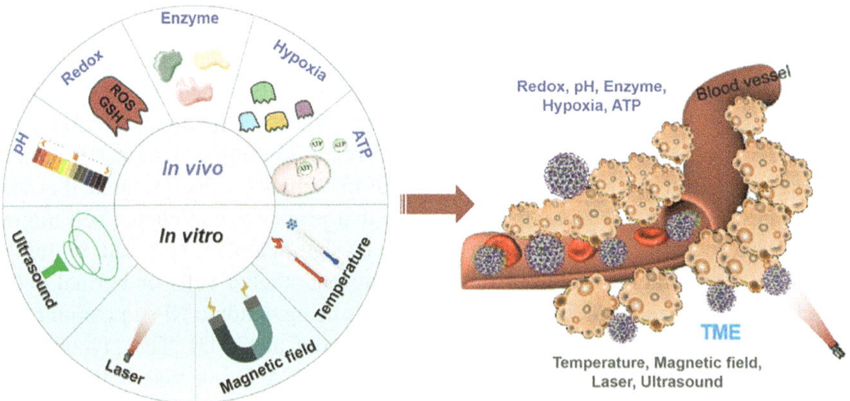

Fig. 10.3 Schematic representation of exogenous and endogenous stimuli-responsive nanoplatforms for tumor therapy (Tian et al. 2022)

10.5.1.1 pH-Responsive Targeting Strategies

pH-sensitive nanoparticles have been extensively investigated to design versatile nanoplatforms for targeted drug delivery. The TME typically has a lower extracellular pH (pHex) with a mean value of ~6.5 compared to healthy tissue. In general, compared to healthy cells, tumor cells rapidly consume glucose for glycolysis with rapid lactate production to obtain the energy needed to sustain their proliferation regardless of oxygen content; consequently, the higher metabolic rate of tumor cells was recognized as the main cause of acidic TME (Stubbs et al. 2000). The tumor cell, lysosomes, and endosomes also have a lower pH (endosomal pH(pHen)) compared to pHex (Lee et al. 2007; Yang et al. 2016). Therefore, pH-responsive nanoplatforms have been developed as an effective tumor treatment tool, significantly increasing the accumulation of charged therapeutic agents in the tumor and facilitating cargo release in the acidic tumor microenvironment, encapsulation of chemotherapeutics inside pH-responsive nanoparticles is an effective approach to prolong the blood circulation time of encapsulated agents and their retention inside the nanoparticles in the physiological environment (Xu et al. 2022). In addition, pH-responsive nanoparticles are also able to improve the pharmacokinetics and biodistribution of the encapsulated payload. This is necessary to delay the metabolism and subsequent release of drugs. The main feature of pH-triggered NPs is their transition from hydrophobic to hydrophilic in response to acidic pH levels, which can change their size or behavior. Reported pH-sensitive NPs, a system based on reversibly ionizable carboxyl groups that undergo protonation and deprotonation due to pH changes, leading to swelling, bursting, and drug release (Iversen et al. 2011).

10.5.1.2 Redox-Responsive Targeting Strategies

Redox species in tumor cells form a complex antioxidant defense system for modulating redox homeostasis, which plays an important role in the cell's life cycle (Diaz-Vivancos et al. 2015). Common reactive oxygen species (ROS) include hydroxyl radicals (·OH), singlet oxygen ($^1O_2$), and hydrogen peroxide (H_2O_2). It should be noted that H_2O_2 is a stable and non-toxic ROS, while others have a short half-life and can be efficiently converted into potent toxic agents (Wu et al. 2019). On the one hand, H_2O_2 can be used as a substrate for O_2 production by a specific enzyme to alleviate the hypoxia of the tumor microenvironment in some O_2-demanding therapeutic strategies (Yang et al. 2020). H_2O_2 can also be converted into other highly active ROS, increased ROS can result in oxidative stress such as lipid peroxidation (LPO) and protein and DNA damage (Dong et al. 2019). Glutathione (GSH), as an antioxidant, is normally distributed in tumor cells in concentrations up to 2–10 mM, plays a significant role in the consumption of ROS and the modulation of redox homeostasis (Fu et al. 2021). A high concentration of GSH can make tumor cells resistant to various treatments. Therefore, it is advantageous to develop redox-sensitive nanoparticles to deliver therapeutic agents to trigger treatments such as chemodynamic therapy (CDT). Recently and with an improved therapeutic profile, ROS generation combined with GSH depletion can effectively disrupt redox homeostasis to increase oxidative stress, leading to tumor cell apoptosis (Hu and Liu 2020). To produce redox-responsive NPs, GSH cleaves the disulfide bond of NPs in cancer cells, leading to redox-controlled drug release. In recent years, various redox-responsive drug delivery systems have been described, including liposomes, micelles, nanogels, and prodrug-based systems. Disulfide bonds are a major component in the design of redox-responsive NPs, diselenide bonds can be used to prepare mesoporous silica-based NPs, controlled drug release can be achieved using albumin or myoglobin by conjugating them to mesoporous silica-based NPs through diselenide bonds, providing NPs, which is highly sensitive to GSH and H_2O_2 (Gisbert-garzar 2021; Yan et al. 2021). In addition, they developed a new intrinsic redox-responsive metal–organic framework support using iron, aluminum, or zirconium as metal nodes and 4,4-dithiobisbenzoic acid (4,4-DTBA) as an organic ligand (Lei et al. 2021).

10.5.1.3 Enzyme-Responsive Targeting Strategies

Enzymes, being a sizeable component of the nanobiotechnology toolbox, have tremendous biorecognition abilities in addition to outstanding catalytic houses. Generally, bizarre enzyme expression found in cancer provides many possibilities for designing focused nanoparticles modified with enzyme-responsive linkages. Currently, many smart nanoparticles were organized for intracellular as well as extracellular tumor-precise drug delivery based totally upon enzyme expression on the goal web site. MMP-responsive nanoplatforms. Matrix metalloproteinases (MMPs), which can be overexpressed in numerous sorts of tumors, are intently associated with cancer pathophysiology. MMP-2 and MMP-nine mainly had been explored for getting ready enzyme-responsive nanoplatforms (Jhaveri et al. 2014; Eatemadi et al. 2017). The gelatin-functionalized DOX-loaded mesoporous silica nanoparticles have been implemented for the shipping of healing sellers toward MMP-9 overgenerated in most cancers. As expected, the organized nanoformulation confirmed genuinely superior DOX launch below simulated tumor microenvironment situations and in tumor mobile models (Xu et al. 2013). The synthesized camptothecin-encapsulated mesoporous silica nanoparticles surface functionalized with targeting cRGD and MMP-2 responsive fluorescence imaging companies, performing as a diagnostic platform as well as for tumor location. These advanced nanoplatforms have been capable of correctly going through enzymatic hydrolysis in overexpressed MMP-2 environments to improve tumor remedy via the discharge of their cargoes (Zeng et al. 2021). The superior MMP stimuli-responsive nanoplatform encapsulated with the chemotherapeutic agent curcumin using a block copolymer with surface-adsorbed peptides that would improve endocytosis. The prepared nanoparticles showed a sustained curcumin launch behavior underneath physiological conditions, while release will be elevated below conditions that mimic the tumor microenvironment. these designed MMP-responsive nanoparticles present great tumor specificity and therapeutic efficacy in most cancers models with few side effects (Guo et al. 2020).

10.5.1.4 Hypoxia-Responsive Targeting Strategies

Poor vascularization in cancer leads to low oxygen (hypoxia), which plays an important role in cancer, including distant metastasis. Therefore, to treat hypoxic tumors, mainly oxygen-enhancing and hypoxia-activating drugs, etc. Many methods are used, including so far, many types of nanoparticles have been prepared for the delivery of hypoxic drugs, such as silica nanoparticles, liposomes, layered nanoparticles, polymer micelles, nanovesicles, polymersomes, and albumin nanoparticles (Liu et al. 2018; Im et al. 2019). Cancer hypoxia is often considered an adverse indicator in cancer therapy, but certain biological agents may make it a unique target for cancer therapy. In fact, due to the Warburg effect, tumor cells prefer aerobic glycolysis rather than the oxidative phosphorylation pathway for energy. Therefore, many enzymes involved in electron donation or reduction reactions, such as azoreductase, nitroreductase, methionine synthase reductase, inducible nitrate synthase, and DT-diaphorase (DTD), are overproduced in Oxygen tumor cells. Therefore, great effort has been devoted to the development of hypoxia-sensitive nanoplatforms that can be activated by enzymes for better cancer therapy. Hypoxia-sensitive chemical linkages, including nitro groups, azo groups, quinones, and N-oxide compounds, have also been used in the design of hypoxia-sensitive nanoparticles. By gaining or losing electrons, they can change their conformation and physical properties such as hydrophobicity and electron affinity. As expected, this hypoxia-sensitive nanoparticle exhibits interesting drug activity. They have great potential in cancer therapy, including hypoxia-responsive product release, prolonged circulation time, and improved tumor penetration and accumulation (Wang et al. 2019a, b).

10.5.1.5 ATP-Responsive Targeting Strategies

ATP, which has been called the energy currency of the cell, ATP is fundamental to various cellular signal cascades. ATP concentrations can reach up to 10 mM in tumor cells, while it is only approximately 5 mM in the extracellular fluid. Therefore, a concentration gradient of ATP levels between extracellular and intracellular levels has been used to develop ATP-responsive nanoplatforms for tumor treatment. The nanoparticles/nanocarrier was prepared using various barrel-shaped chaperonin groups assembled via coordination with Mg^{2+} into tubular structures that protected loaded therapeutic agents from biological metabolism and degradation (Biswas et al. 2013). Upon internalization by tumor cells, hydrolysis of ATP to form ADP can trigger protein conformational changes and collapse of the nanoparticles, resulting in the selective release of the contents. The prepared nanoparticles using a gradient of ATP concentrations showed rapid cargo release in an ATP-responsive manner. An enhanced anti-proliferative effect, possibly due to enhanced cell apoptosis in mitochondria-mediated pathways and cell cycle arrest at the G2 phase (Zhang et al. 2017a, b). In another example, designed ATP-binding aptamer DNA functionalized with polymeric nanocarriers encapsulated with chemotherapeutics for targeted delivery to ATP overgenerated environments (Mo et al. 2014). ATP can be considered an efficient stimulus to promote release of preloaded drugs from nanoparticles for specific cancer treatments and for diagnostic purposes.

10.5.1.6 Tumor-Metabolite Responsive Targeting Strategies

Metabolic reprogramming in tumor cells alters metabolic processes to meet unmet nutrient and energy needs. In this case, tumor-associated metabolites are often valuable for timely prediction of tumor progression and progression. Malnutrition affects cancer cells, this lactate (an additional metabolite) lowers the pH of the tumor microenvironment. Oxalate functionalized NPs can activate lactate dehydrogenase (LDH), which forms OxNP-protein complexes. This study validated the idea of changing the sensitivity of nanoparticles by conjugating them with important proteins to target specific metabolites in cancer cells (Fruehauf et al. 2019).

10.5.2 Exogenous Stimuli

When trigger stimuli are caused by external factors, such as a change in temperature, ultrasound, or light responsiveness, they are known as exogenous stimuli.

10.5.2.1 Temperature Stimuli-Responsive Targeting Strategies

Heat-sensitive nanoparticles use temperature changes as a trigger to release their charge because they can change their hydrophilic and hydrophobic balance, solubility or structural properties in response to temperature and cold. These nanoparticles have a low critical solution temperature (LCST), below which the mixture can become miscible (Pasparakis and Tsitsilianis 2020) The best thermal materials with low solubility include poly(2-oxazoline)s (POxs), poly-N-isopropylacrylamide (PNIPAAm), poly (methyl vinyl ether) (PMVE), and poly(vinylcaprolactam) (PNVCL) (Tian et al. 2022). It easily enters the phase transition from solid to liquid depending on external conditions. Among these materials, PNIPAAm is mainly used to prepare nanoparticles due to its low temperature of about 30 °C, disulfiram to cancer cells as a long-term cancer treatment, and the drug-loaded hydrogel has been found to allow cellular uptake.

Graphene oxide (GO) has a high concentration of functional groups on the surface, allows the transfer of bioactive molecules and is highly reactive in aqueous conditions; However, their application is limited by incorrect dosing and lack of control. To overcome these limitations, GO has been modified using biodegradable and hydrophilic polymers such as poly(N vinylcaprolactam) (PNVCL) and poly (glycolic acid), which confer pH and temperature sensitivity to allow the loading and release of oxaliplatin in MCF-7 cells (Kazempour et al. 2019). Similarly, smart temperature and pH sensitive nanoparticles were fabricated from poly (N-isopropylacrylamide-acrylamide) random copolymer. The presence of a hydrophilic group in acrylamide raises the LCST to 37 °C (closer to body temperature), eventually making this sample suitable for drug delivery (Farjadian et al. 2020). This structure is further modified with hydrazine and binds to DOX via a Schiff base bond (an acid-fissile bond), making the nanoparticles heat and pH sensitive. Considering that hyperthermia raises the temperature of human tissue to 113 °C and increases the sensitivity of cancer cells to chemotherapy, these two effects are combined because thermochemotherapy can improve cancer treatment.

10.5.2.2 Magnetic Stimuli-Responsive Targeting Strategies

Magnetic nanoparticles have good physical, chemical, and biological effects and have been recognized as an ideal platform for many cancer treatments and diagnoses (Jiao et al. 2022). Compared with other magnetic stimulus sensitive nanoplatforms, the encapsulated superparamagnetic Fe_2O_3 nanoparticles (SPION) have good magnetic moments and satisfactory biocompatibility. In addition, magnetic molybdenum disulfide ($mMoS_2$) can be modified by liposomes with phospholipid bilayer membrane structure to form a magnetic nanoplatform that is not easy to collect in physiological solutions and has good biocompatibility, so it shows great promise in large-scale application. nanomedicine. Sensitive nanoparticles involve encapsulating/immobilizing therapeutic agents into magnetically responsive nanoparticles, injecting magnetically responsive nanoparticles into the body, and using an external magnetic field to locate and activate the magnetically responsive nanoparticles of interest (Li et al. 2022a, b). Recent article reported, GSH-responsive metal-organic framework (MOF) that can be loaded with an indoleamine 2,3-dioxygenase (IDO) inhibitor and free s-nitrosothiol donor group to improve antitumor immunotherapy (Du et al. 2022). In this nanoplatform, high relaxation Time (T1) provides magnetic resonance (MR) imaging capability to control the in vivo biodistribution of nanomedicines. Multifunctional nanodiagnostics based on DOX-encapsulated tannic acid-Fe network (TAF) coupled with fibronectin function for combinatorial cancer therapy guided by MR imaging. Developing nanoparticles sensitive to magnetic stimulus by imaging will help determine when a tumor is growing (Xu et al. 2022).

10.5.2.3 Light Stimuli-Responsive Targeting Strategies

Various nanoparticles can respond to light and have the ability to adjust the irradiation wavelength, power, and domain, and photosensitive nanoparticles have recently been developed for controlled drug release. Nanoparticles can be sensitive to any type of light, including ultraviolet, visible or near infrared, which can affect biological diseases such as cancer or tumors. By controlling various irradiation, photosensitive tumor treatment can be done precisely to reduce the negative effects of the body (Yan and Li 2016). Polymer complexes, polymer micelles, liposomes, nanogels, and nanorods have all been developed as photosensitive NCs using materials such as gold nanocomposites, carbon nanotubes, graphene, and organic molecules. Nanoparticles can change shape in response to light by changing the shape of the molecules. For instance, photo-switching nanoparticles that absorb UV-visible light have been shown to be able to transport drugs for cancer treatment, including DOX, docetaxel, and paclitaxel (Tong et al. 2013). The short wavelength of UV-vis light limits the use of nanoparticles. Therefore, near-infrared photosensitive nanoparticles would be suitable for controlled drug delivery (Timko et al. 2014).

10.5.2.4 Ultrasound Stimuli-Responsive Targeting Strategies

Ultrasound has emerged as an external stimulus that can stimulate the destruction of nanoparticles and release their respective charge. The solution is released by the heat generated by the cavitation. Ultrasound sensitive nanocarriers/particles can be safely and cost-effectively used for cancer diagnosis and are widely used in hospitals. Gases and perfluoropentans combine nanoparticles to form gases (e.g., CO_2) presents cancer in a special place in the biological atmosphere (Wang et al. 2015). In addition, the ultrasound device can be used to facilitate siRNA or DNA delivery in the tumor (Carson et al. 2012). However, the large size of ultrasound-sensitive nanoparticles limits their penetration into tissues due to their poor penetration. In general, drug-loaded ultrasound-sensitive nanoparticles can be used for cancer therapy and theranostics. Recently, cisplatin-encapsulated solid lipid nanoparticles have been prepared under ultrasound guidance.

In ultrasound therapy, engineered nanoparticles exhibit satisfactory drug release, cellular uptake, and other pharmacokinetic properties as well as excellent antitumor effect in glioma models. Another way to prepare DOX-loaded mesoporous silica nanoparticles sensitive to ultrasound stimuli characterized by ultrasound release of cargo for glioma treatment (Wu et al. 2018a, b). In a mouse model of glioma, they found that tumor growth and development was significantly inhibited, as well as improved survival. Using HIFU, nanoparticles can be fragmented, allowing drug delivery in dogs with idiopathic chondrosarcoma, osteosarcoma, hepatocellular carcinoma, and prostate cancer (Horise et al. 2019). Biocompatible piezoelectric nanoparticles were also encapsulated with DSPEG and modified with anti-HER2 Ab to target cancer therapy. As expected, nanoplatforms sensitive to engineered ultrasound stimuli could release effective drugs in a controlled manner, interfere with cell division, and inhibit tumors.

10.6 Stimuli-Responsive Gene Delivery Nanocarriers for Cancer Therapy

Stimulus-sensitive gene carrier nanocarriers (GDNCs) have great potential in gene therapy due to their effective protection, prolonged blood circulation, selective delivery targets, and controlled nucleic acid release. Demonstrate nanocarriers that respond to endogenous stimuli such as pH, reactive oxygen species, glutathione, and enzymes, as well as exogenous stimuli such as light, heat, ultrasound, and magnetic fields, and demonstrate the biomedical potential of stimulus-response gene delivery. for the treatment of cancer. Over the last few years, more than 4600 gene therapy trials have been completed worldwide, of which 29% have entered phase II/III/IV clinical trials for cancer treatment (Zhang et al. 2023).

Major nucleic acid drugs used to treat cancer include small interfering RNA (siRNA), antisense oligonucleotides, microRNA, messenger RNA, short hairpin RNA, and gene therapy (such as the CRISPR/Cas9 system) (Wu et al. 2022; Lopez-Cantu et al. 2022; Zhang et al. 2021). Each nucleic acid type can act on its specific target cell at the genetic level. Due to non-biodegradation, lack of cellular

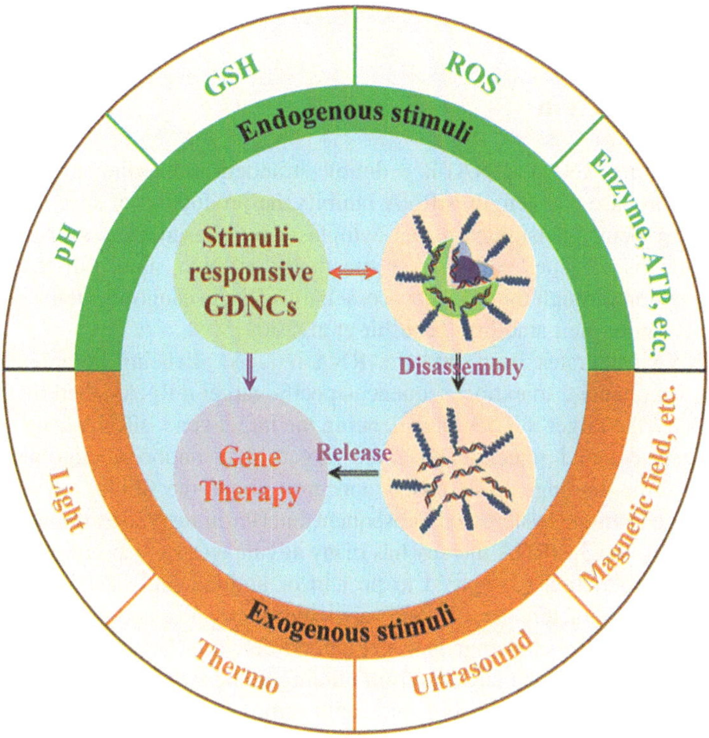

Fig. 10.4 Schematic diagram of stimuli-responsive nanocarriers for gene delivery and cancer therapy (Zhang et al. 2023)

uptake, rapid clearance, and easy degradation by enzymes, bare nucleic acid drugs cannot achieve the desired results with direct administration.

Low toxicity, low immunogenicity, high packaging capacity, nanoparticle formulations not limited by the type of gene load, and non-viral vectors (such as cationic polymers, cationic liposomes, inorganic or organic nanoparticles) have become research topics. In recent years, stimuli-sensitive GDNCs, which can effectively deliver nucleic acid to the target and respond to specific stimuli, thereby increasing the effectiveness of gene therapy, have attracted many attentions. (Mohammadinejad et al. 2020; Liu et al. 2021a, b). These GDNCs often contain nanocarriers that can be activated by endogenous stimuli (pH, reactive oxygen species (ROS), glutathione (GSH), enzymes, etc.) and exogenous stimuli (light, heat, ultrasound (US), magnetic field (MF), etc.) for gene delivery in cancer treatment (Fig. 10.4).

10.7 Application of siRNA (Nanoparticle-Based RNA) Cancer Therapy

10.7.1 Introduction

Small interfering RNA (siRNA) is a double-stranded non-coding RNA of 20–25 nucleotides in base pair length. siRNA inhibits the production of specific genes by binding after being transmitted to the cytoplasm by transcription. siRNAs bind to specific regions of genes with greater specificity. siRNAs silence disease-specific gene expression through the RNAi process. bare siRNAs cannot be delivered due to uncertainty about their stability (Elbashir et al. 2001).

The RNA integrates itself into the RNA-induced silencing complex (RISC), directing Argonaute-2 to extract sequence-specific target mRNAs, thereby blocking translation of the target mRNA and silencing the target gene. siRNA gene silencing is more efficient and selective for more targets than antibody oligonucleotides. Although some findings have raised concerns about the ability of siRNAs to stimulate the immune system and subsequent interferon responses through interaction with Toll-like 3, siRNA therapy has many advantages such as ease of synthesis, low cost of production (compared to protein or antibodies), practical goals, high specificity and characteristics of good medicine. Designing and producing siRNA drugs is not easy. The challenge is to get a good delivery (Draz et al. 2014).

Many problems prevent siRNAs from binding to the right place, such as the ease of degradation by blood nucleases and rapid removal by the immune system, the difficulty of targeting the drug, the lack of cellular uptake, and the lack of escape from endosomes. Delivery efficiency is considered the most important issue of RNAi technology and clinical application. Nanoparticles have better specificity than other carriers and can affect the efficient delivery of siRNA in the bloodstream. In 2014, Atu027, a lipid-based RNAi therapy targeting protein kinase N3 (PKN3) mRNA in the vascular endothelium for the treatment of pancreatic ductal adenocarcinoma (PDAC), achieved promising results in the dose escalation phase. More than 90% of Atu027 adverse events are grade 1 or 2, indicating a high level of safety (Schultheis et al. 2020).

In addition, another lipid-based siRNA nanoparticle (called DCR-MYC) has been developed to suppress MYC, an oncoprotein that is dysregulated in most malignancies. In a phase I dose escalation trial, five doses of DCRMYC therapy were given to patients with multiple types of cancer, including neuroendocrine tumors (NET), metastatic breast cancer (MBC), colorectal cancer (CRC), or other tumors. After treatment, tumor shrinkage was observed in many patients (Tolcher et al. 2015). All these results suggest that the nanoparticle-based RNAi approach is a promising method for cancer therapy. In Table 10.3 details of the RNAi-based cancer therapies in clinical trials (Dolly et al. 2023).

RNAi starts with mature siRNA in the cytoplasmic perinuclear region. Exogenous double-stranded RNA (dsRNA) is usually split into 21–24 nucleotides by an endoribonuclease or an RNase motif helicase, Dicer, and is referred to as mature siRNA. The guide helix then binds to RISC to restore its gene silencing activity,

Table 10.3 Clinical trials on siRNA for cancer therapy (Dolly et al. 2023)

Clinical trial ID	Disease	Drug	Delivery system	Target	Delivery route	Phase	Stage	Company
NCT00882180	Liver cancer	ALN-VSP02	Lipid NPs	VEGF and KSP	IV	Phase 1	Completed	Alnylam Pharmaceuticals
NCT01262235	Advanced solid tumors	TKM 080,301	Lipid NPs	PLK-1	IV	Phase 1 and 2	Completed	Arbutus Biopharma Corporation
NCT01437007	Primary or secondary liver cancer	TKM 080,301	Lipid NPs	PLK-1	IV	Phase 1	Completed	National Cancer Institute (NCI)
NCT01120288	Advanced solid tumors with liver metastases	EZN-2968	Locked nucleic acid	HIF-1	IV	Phase 1	Completed	National Cancer Institute (NCI)
NCT00938574	Advanced solid tumors	Atu027	Lipoplex-siRNA	PKN3	IV	Phase 1	Completed	Silence Therapeutics GmbH
NCT01188785	Pancreatic Cancer	siG12D LODER	Biodegradable polymeric matrix	KRAS	Intratumoral administration	Phase 1	Completed	Silenseed Ltd
NCT01591356	Advanced Malignant Solid Neoplasm	siRNA-EphA2-DOPC	Liposome	EphA2	IV	Phase 1	Not recruiting	M.D. Anderson Cancer Center
NCT00689065	Solid tumor	CALAA-01	Cyclodextrin NPs	RRM2	IV	Phase 1	Terminated	Calando Pharmaceuticals
NCT03608631	Pancreatic cancer	KRAS G12D siRNA	Exosomes	KrasG12D	IV	Phase 1	Recruiting	M.D. Anderson Cancer Center

(continued)

Table 10.3 (continued)

Clinical trial ID	Disease	Drug	Delivery system	Target	Delivery route	Phase	Stage	Company
NCT00672542	Metastatic melanoma	Proteasome siRNA and tumor antigen RNA-transfected dendritic cells	Dendritic Cell	LMP2, LMP7, and MECL1	Intradermal injection	Phase 1	Completed	Scott Pruitt
NCT002647	Myeloid leukemia	SV40 vectors carrying siRNA	Viral vector			Phase I	Completed	Hadassah Medical Organization
NCT02191878	Hepatocellular carcinoma	TKM 080,301	Lipid NPs	Polo-kinase-1	IV	Phase I and II	Completed	Arbutus Biopharma Corporation

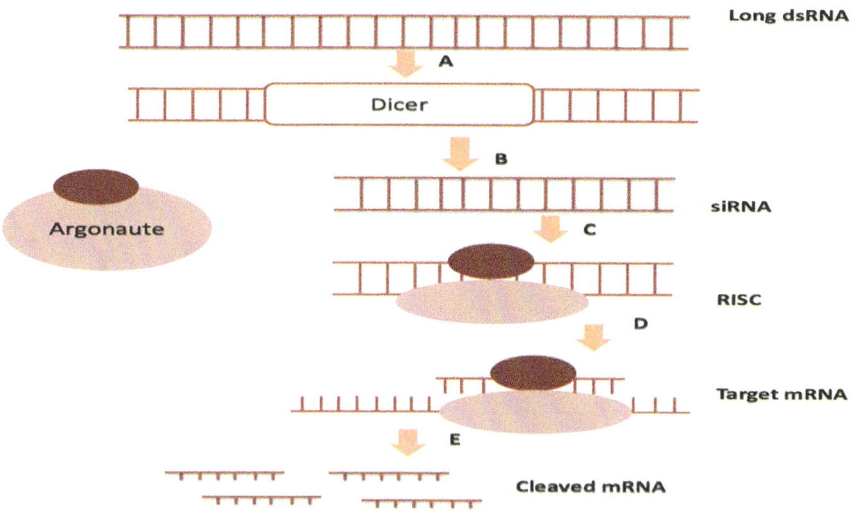

Fig. 10.5 mRNA silencing by RNA interference. (**a**, **b**) Cleavage of long dsRNA precursors by RNase III endonuclease Dicer to form smaller fragments, referred as siRNA. (**b**, **c**) These short dsRNAs are subsequently unwound and the guide strand is assembled into RNA-induced silencing complex (RISC), that contains Argonaute protein (catalytic component) to cleave the passenger strand. (**c**, **d**) RISC containing the guide strand will form complementary base sequence. (**d**, **e**) The sequential will result in mRNA cleavage and degradation (Hattab et al. 2021)

while the non-guiding helix of the old siRNA is degraded by Argonaute 2 (Ago2). Next, the guide strand complexed with RISC searches for target mRNA based on the Watson-Crick base in the assembly. Once localized, the Ago2 protein in RISC cleaves the target mRNA for degradation, at which point the target's ability to express the protein is inhibited. Unlike antisense oligonucleotides, RISC can support multiple replications of target mRNAs, allowing siRNAs to exert their silencing effects more effectively (Fig. 10.5) (Hattab et al. 2021).

10.7.2 Difficulties in siRNA Delivery

10.7.2.1 Physiological Barriers

When siRNA is administered and injected into the bloodstream, many structural and anatomical barriers block the pathway of siRNA to its domain of action. The main stage of coordination is elimination by the reticuloendothelial system (RES) in the kidneys and other organs (e.g., bones, liver, lungs, and spleen). Particles larger than 100 nm are captured by it. Therefore, the recommended RNA molecule diameter should be between 20 and 100 nm for renal passage and elimination of RES (Wang et al. 2010). In addition, therapeutics of this size exhibit enhanced passive uptake of tumor cells from blood vessels with a pore size of 100–800 nm. Penetration of siRNAs from target tissue after crossing the RES is inhibited by local barriers in the

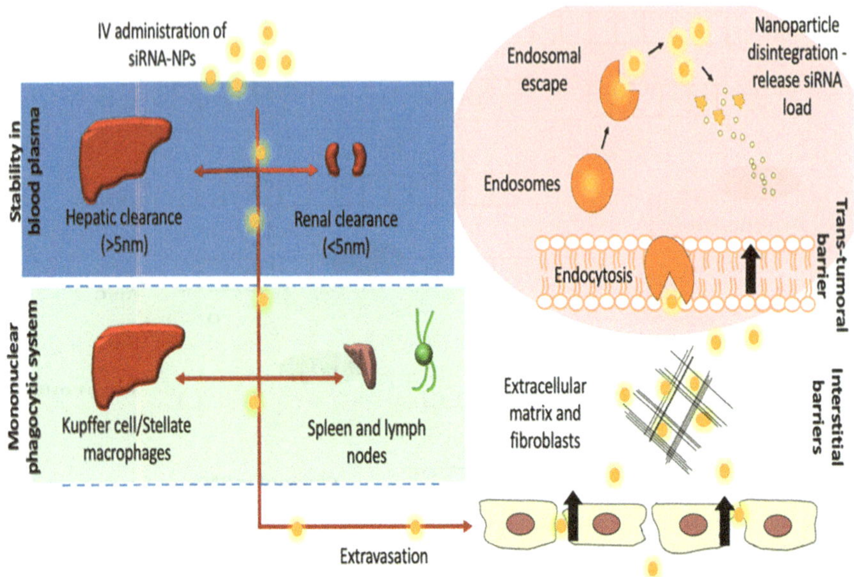

Fig. 10.6 Physiological barriers to in vivo delivery of siRNA-nanoparticles. Extracellular barriers include hepatic and renal clearance, which are dependent on the particle sizes. Opsonization and macrophage engulfment via the mononuclear phagocytic system (MPS) further reduce the bioavailability of siRNA nanoparticles in the plasma. Extracellular matrix (ECM) trapping further decreases the localization of siRNA complexes into the tumor tissues. Intracellular barriers involve effective endosomal escape and siRNA release into the cytosol to initiate RNAi-associated gene knockdown activity (Hattab et al. 2021)

extracellular matrix and endothelial cell layer. The pore size of endothelial tissue is generally 4.5–25 nm, so the largest pore size can pass through particles of 4–5 nm. Only naked siRNA molecules or those associated with small molecule ligands are easily passed through endothelial cells (Gupta et al. 2019). In addition, naked or complex siRNAs must cross the dense extracellular space to reach the target cell surface (Fig. 10.6) (Hattab et al. 2021).

10.7.2.2 Cell Membrane

Negatively charged phospholipid bilayers containing functional proteins form the cell membrane. Because naked siRNA is polyanionic and water-soluble, it cannot enter cells passively through hydrophobic cell membranes. On the other hand, transmembrane transport via endocytosis overcomes this limitation (Shukla et al. 2016). Although siRNA must cross the endosomal membrane and enter the cytoplasm, if not, it is degraded by ribonucleases or lost by exocytosis. Pinocytosis and phagocytosis are two types of endocytosis. The former is characterized by endosomes facing the surface of the cell membrane, thereby suppressing the endosome and pulling it into the cytoplasm. The membrane of the latter spreads outward, absorbs the particle, and attaches to the lysosome. To improve cell recruitment, endocytosis is promoted by receptor-mediated endocytosis using cell peptides and/or ligands. Anionic proteoglycans on the cell surface allow them to interact with the cation-carrying siRNA complex/transporter and enter the cell via endocytosis.

10.7.2.3 Stability

The in vivo stability of siRNA is an important issue. Transfected siRNAs are readily hydrolyzed in the extracellular environment due to diverse enzyme activities in blood and target organs. It is known that the half-life of naked siRNA in blood ranges from a few minutes to an hour. The backbone of the siRNA molecule consists of ribose sugar. Studies have shown that modification of ribosomal loops and nucleobases improves siRNA stability and reduces immune response and resistance. RES contains phagocytes that degrade siRNA. Uptake of polyethylene glycol (PEG) in the nanocomplex reduces interaction with phagocytes, thereby reducing phagocytosis (Tokatlian and Segura 2010). Serum nuclease targets the hydroxyl group at the 2′ position of the pentose ring. These nucleases destroy RNA by disrupting the phosphodiester structure of nucleic acids. Therefore, the delivery of negative siRNA into the blood and maintenance of the therapeutic effect is difficult (Czech et al. 2011). The siRNA must remain in the blood to achieve the desired target in a particular tissue, allowing for proper endocytosis and desired RNAi activity. If unmodified siRNA has high intracellular stability, nuclease stability may not be a limiting factor when siRNA enters the cell. One study found that while nuclease stability is important to prevent degradation of siRNAs in the extracellular environment (such as blood vessels after systemic administration), sexual dominance of cells does not control the persistence of siRNAs internalized into the cytoplasmic compartment.

10.7.2.4 Off-Target Effects

Off-target effects can be divided into three categories: (1) In the cell, the RNAi process becomes saturated, resulting in upregulation of miRNA-regulated genes. To solve this problem, siRNA can be mutated or a stronger siRNA can be used. (2) Passenger silencing leads to the reduction of target genes that are partially or fully homologous to the passenger sequence. To solve this problem, blunt end, asi-RNA, si-siRNA, phosphorylated double target blocking siRNA, ss-siRNA or aiRNA at the 5′ end of the passenger strand were used. (3) Seed region-induced silencing leads to some downregulation of complementary genes (Naik et al. 2021). This can be avoided by using seeds enriched with AU and the addition of seeds can be avoided.

10.7.3 Applications of Nanocarriers for siRNA Delivery

NPs are highly effective in changing the fate of siRNAs in vivo. Because of their size, quality, ability to protect siRNA from degradation, and biological behavior, various NPs are currently being investigated for siRNA delivery in various biological systems, cancer therapy (Table 10.4) (Dolly et al. 2023). Treatment of

Table 10.4 Advantages and disadvantages of various NPs to siRNA and drug delivery

Types of NPs	Advantages	Disadvantages
• Metal-based NPs	• Metallic NPs have a large surface area, small particle size, and simple synthesis • It has strong plasma absorption with a minimum boiling point • Improves stability and biocompatibility • It is potent for multimodal uses (e.g., targeting, diagnostic, and therapy)	• Metallic NPs production on large scale is time-consuming • Limited in vivo stability • Non-biodegradability • Unstable at high temperature
• Solid lipid NPs	• Enhanced stability • pH-responsive release of the payload • High surface functionality	• Chemical degradation reactions • Poorly absorbed bioactive compounds
• Polymeric NPs	• Polymeric NPs production is easy and cheap • Polymers have a fine and clear structure and properties • It has good retention of the drug at the active site • Ease of loading and complexions with nucleic acids by electrostatic interactions • NPs have better bioavailability • It can be modified for a large range of molecular weights • Polymers are biodegradable, biologically acceptable, biocompatible, and non-immunogenic • NPs are soluble in water	• It has limited stability • Cellular toxicity may cause by synthetic polymers • Necrosis and apoptosis • Polymeric NPs may cause autonomic dysfunction • It can directly affect the heart and vascular function
• Magnetic NPs	• The surface area is larger • It provides long circulation and improves tissue penetration • It has high magnetization for fast delivery • Clustering can be controlled • Multifunctionality (e.g., diagnostic, targeting, and therapy)	• Colloidal instability • Poor biological activity, low biodegradability and cytotoxicity
• Ceramic NPs	• Easy method of preparation • It does not affect swelling or porosity • It is highly stable and has a high loading capacity • It can be easily integrated into both hydrophobic and hydrophilic systems	• High cost • High density • Low biodegradability
• Mesoporous silica NPs	• It has a larger surface area, smaller in size, and has high surface reactivity • It is stable, biocompatible, and biodegradable • It controlled porosity and allowed multifunctional sequential delivery • Its functionalization is easy	• High toxicity • In vivo toxicity • Interaction with red blood cells results in hemolysis

(continued)

Table 10.4 (continued)

Types of NPs	Advantages	Disadvantages
• Dendrimers	• It has an accurate size and shape • It is highly biocompatible and water-soluble • The immune response is negligible • Small electrostatic interaction with nucleic acid • Protect the nuclease activity	• Non-specific cytotoxicity, and rapid clearance • Limited release of the associated drug • Chemical degradation reactions

naked siRNA in the host presents several challenges. siRNAs do not reach the target at best because of the short half-life, rapid renal clearance, and rapid blood loss. siRNA can be encapsulated in NPs or attached to the surface by covalent or non-covalent bonds.

Nanoparticles are designed and used as drug carriers because they provide important benefits in biotechnology and enable the spread of nanomedicine. Due to the addition of anionic charge, free siRNA alone cannot infect cells and is degraded in biological fluids. The ideal tool should protect siRNA from RNase degradation, increase cellular uptake, facilitate gene silencing, and be non-toxic. Polycationic formulations that have been used to deliver siRNA include PEI, polyarginine, polylysine, histoylated polylysine peptides, chitosan, and polyamide dendrimers (Zhang et al. 2022a, b) (Table 10.4). Conjugation of nucleic acids to nanoparticles has many benefits, including preventing cellular damage, increasing circulation time, improving cell membrane permeability, and avoiding endosomal digestion. NPs can easily bind to various targets such as RGD peptide or transferrin for target-specific siRNA delivery. Features of NPs for siRNA-level delivery include: (1) better stability and quieter activity (2) improved pharmacokinetics, (3) targeting of EPR effect, and (4) more than half of siRNA in long serum. Most biodegradable formulations that release siRNA use polymeric materials that incorporate the siRNA into a polymeric core.

10.7.4 Nanocarriers for Co-delivery of siRNA with Anticancer Drug

The use of single drugs in cancer treatment is not very successful, so there is increasing interest as the combination of anticancer drugs and siRNA can be beneficial in this way. Apoptosis has been defined as an aberrant immune-mediated cell death that can occur by two distinct but interconnected processes involving the proteolysis of many cellular proteins by caspases. Mitochondrial or intrinsic pathway (direct DNA damage, hypoxia, depletion of oncogenes and survival factors) and death receptor or extrinsic pathway (extracellular). p53 is an important activator of the intrinsic pathway. When combined with anti-inflammatory drugs, siRNAs should reduce anti-inflammatory drugs, reduce anti-inflammatory drugs, and promote apoptotic pathways in cells (Khelghati et al. 2021). After siRNA release, the concentration of anticancer drugs in cancer cells will increase, thereby increasing the anticancer effect. Multidrug resistance develops during long-term therapy and describes environmental resistance to drugs that do not interact with different drug models. The negative apoptotic pathway and drug proliferation are important mechanisms for the progression of multidrug resistance (MDR) (Charbe et al. 2020; Zhu et al. 2021). In studies to improve doxorubicin (DOX) sensitivity, siRNA can downregulate pathways such as STAT3, Notch1, Mcl-1 and Nrf2, and siRNA can be used to inhibit the activity of P-gp to act as a cell membrane drug transporter. results in increased DOX accumulation. Both DOX and siRNA were incorporated into NPs, which promoted accumulation in cancer cells while protecting siRNA from enzymatic degradation (Paskeh et al. 2022). Others have developed cationic liposomes (CTL) for the delivery of paclitaxel (PTX) and crizotinib (CRI) with siRNA (CRI-PTX-CTL-siRNA). CRI-PTX-CTL-siRNA was shown to increase cellular uptake in MCF-7 cells and strongly suppress Bcl-xL expression in cells (Li et al. 2022a, b). Many studies have been conducted to increase the efficacy of chemotherapy by blocking or avoiding drug efflux mechanisms through the combination of antibodies and siRNA. The reduction of proteins by siRNAs involved in the drug efflux increases the immune response in the tumor, resulting in a stronger immune response.

10.7.5 Clinical Applications of siRNA-Based Nanotherapies

Ten siRNA treatments have been studied in the clinic to date (Table 10.3). Phase I trial of all siRNA anticancer nanotherapeutics begins: The Phase I trial of CALAA-01 and DCR-MYC in patients with solid tumors, multiple myeloma, and myeloma has been terminated due to dose limitations, toxicities, and decision support. ALNVSP02, Atu027, TKM-PLK1 and siG12D LODER completed the first stage. A phase I trial of siRNA-EphA2-DOPC has begun collecting samples but is not yet complete, a trial of mesenchymal stem cell-derived exosomes using KRAS G12D has yet to be discovered. Four siRNAs entered nanotherapeutic phase II; Atu027 and TKM-PLK1 complete Phase II, while siG12D LODER is currently recruiting. A phase 2 trial of DCR-MYC was performed in patients with hepatocellular carcinoma, but the promoter was subsequently truncated. As of May 2021, there are no siRNA anticancer nanotherapeutics complete Phase III trials (Hattab et al. 2021). Currently, five siRNA drugs have been approved (patisiran, givosiran, inclisiran, lumasiran, vutrisiran) and several others are in late stages of phase III clinical trials. Notably, there is a clear focus on hepatocyte delivery, with GalNAc being the most popular platform for siRNA bioconjugation and delivery, GalNAc bioconjugates are comparatively straightforward to synthesize and administer they show a good biocompatibility/toxicity profile, and very high efficacy. Among the approved siRNA drugs, nanoparticles (LNPs) are only used in the case of patisiran (Friedrich and Aigner 2022) (For more information, visit the Cortellis Drug Discovery Intelligence Database: https://www.cortellis.com/drugdiscovery/). There are 630 patents for these four drugs. All clinical trials at various stages are the first of their kind in open-label human trials. These trials recruited participants with various types of cancer that were advanced and/or metastatic and did not respond to standard of care (SOC) chemotherapy.

10.7.6 Current Challenges and Opportunities

The development of nanoformulated siRNA-based cancer therapy is one of the most important and potentially revolutionary research areas in biomedicine. However, there are many problems that need to be resolved before these treatments can be used effectively in the clinic. siRNA molecules are small and poorly structured, allowing them to be easily degraded and removed by the immune system. One of the biggest problems is the lack of siRNA delivery systems that can reach and enter the target site. Currently, most siRNA delivery systems are based on cationic lipids or polymers that do not cross cell membranes well. Another challenge is to create siRNA molecules that are resistant to degradation by nucleases, as well as other molecules that can affect the activity of siRNA molecules.

In addition, methods are needed to monitor the in vivo delivery, uptake, and activity of siRNA molecules. Finally, creating siRNA molecules that target cancer cells and developing ways to mitigate the effects of the target is an ongoing challenge. These challenges need to be addressed for nanoformulation-based siRNA therapeutics to be effective for cancer therapy.

10.7.7 Design Concerns for Future Development of RNAi-Mediated Anticancer Nanotherapeutics

The development of new nanoparticle delivery systems and the optimization of siRNA formulations are important not only to improve cellular uptake of anticancer nanotherapeutics, but also to improve their specificity and enable therapy. Several problems with the development of new models for human trials should be considered: (1) Production of large nanoparticle particles (compared to small particles in animal research) can affect the behavior and activity, aggregation, agglomeration, off-target problem of nanoparticle and prevention; therefore safety limits should be taken into account and the production of nanoparticles should be continuously improved. (2) The diameter of nanoparticles was optimized and narrowed in the range of 20–30 nm; many things like this show better reproduction and recreation in humans. Simultaneous delivery of two or more siRNAs at delivery can improve antitumor effect without increasing toxicity and off-target increase. (4) Replace standard nanoparticles with biodegradable nanoplatforms as they have stable profiles. (5) Nanoparticles (but not siRNA) can be functionalized with target ligands (Hattab et al. 2021). While targeting may not improve tumor cell recruitment, targeting ligands may address other biological concerns, including intracellular trafficking and intracellular siRNA escape. Identification of novel nanoformulations and delivery as well as specific oncogenes for siRNA targeting could lead to effective treatment of these therapies (Cuellar et al. 2015).

10.8 Ultrasound Linked Nano-cancer Therapeutics

10.8.1 Introduction

Ultrasound is often used in cancer treatment, such as using ultrasound-guided biopsy to collect tumors from the right place, diagnosing tumors with cheaper and easier equipment such as MRI and CT, or removing tumors using high intensity. focused ultrasound (HIFU) preparation. The efficacy of this technique can be further improved by combining it with various nanosystems to achieve: (1) better resolution in ultrasound examinations, e.g., allowing visualization of angiogenic vessels, (2) specific anti-inflammatory drugs. Drugs or gases that bind or encapsulate into nanosystems and are released into the tumor under a control as in ultrasound application, (3) cancer in the tumor uses less heat than HIFU (Alphandéry 2022). Also, some nanosystems show tunable dimensions, i.e., to nanobubbles can turn into microbubbles. This dual dimension is good because it leads to target aggregation of nanobubbles through the EPR effect and the positive effect of ultrasound on microbubbles in the same cell. Interestingly, the way the nanosystems protect the tumor in principle can be adjusted by the correct selection of a wide range of nanosystems and by adjusting the importance of ultrasound parameters, some of which will come true, such as special effects, as they are all products. Electromagnetic pollution is observed as cavitation is usually absent, but may cause tumor damage (Fig. 10.7) (Alphandéry 2022). The combination of the nanometer system with ultrasound is of particular interest. In fact, it combines a number of advantages. First, by treating tumor-specific nanosystems through force, magnetic, or active use, it could enable the treatment of cancer in a specific region of the tumor. Second, it can also increase the sensitivity of cancer diagnosis by increasing the resolution of ultrasound, such as ultrasound. The nano/microbubbles can help visualize the blood vessels perfusing the tumor or provide a standard US assessment by other methods such as PA, MRI or CT (Zhang et al. 2017a, b). In some cases, ultrasound can help complete this plan by penetrating certain substances, such as the brain or skin, that the nanosystem needs to diffuse before reaching the tumor (Tran et al. 2020). They can also be excited to activate nanosystems in tumors in different ways, such as cellular internalization of nanosystems in tumor cells or thermal or immunological or controlled drug release (Timin et al. 2017; Canavese et al. 2018; Han et al. 2019).

10.8.2 Ultrasound Parameters Used for Cancer Therapy

Ultrasound is a mechanical wave that periodically oscillates at a frequency f greater than the sound frequency, such as $f > 20$ kHz. These waves are usually produced by sensors that convert electrical signals into transducers. Two types of ultrasound can be distinguished from High Intensity Focused Ultrasound (HIFU), which operates at high intensity $i \sim 0.1$–10 kW/cm^2, using low-intensity ultrasound (LIU), which is usually less powerful than HIFU and produces low-intensity ultrasound, for example, $i < 1$–5 W/cm^2 (Elhelf et al. 2018). On the other hand, HIFU can specifically

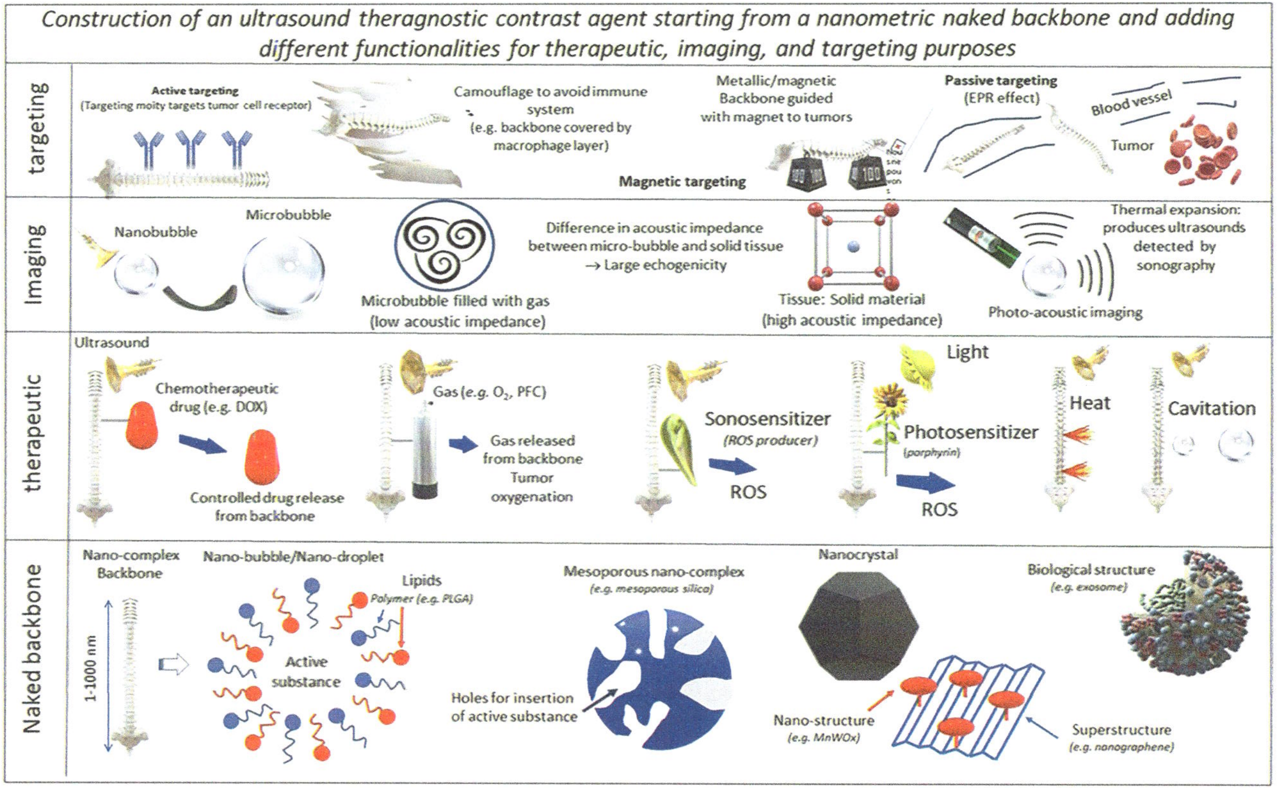

Fig. 10.7 A schematic diagram representing the various types of nano-systems (NS) that have been used for cancer treatment/diagnosis. NS are presented in a hierarchical order, starting from a backbone consisting of nano-bubbles/nano-droplets, mesoporous nano-complexes, nanocrystals, biological nanostructures, to

Fig. 10.7 (continued) which one or several functional element(s) is/are added. Such functional elements provide the backbone with a therapeutic, imaging, or targeting activity. In some cases, another radiation than ultrasound can be used, such as laser for PDT/PTT treatment or PA imaging, or a magnetic field to yield magnetic targeting when the backbone is made of metallic elements. A unique feature of these nano-systems, which is not often encountered in nano-systems not excited by ultrasounds, lies in their gaseous content, which can either be used for cancer treatment, e.g., through O_2 release to fight against tumor hypoxia, or for tumor imaging, e.g., by using microbubbles with a different acoustic impedance from that of the tissular environment (Alphandéry 2022)

target the tumor site and reach a thermal ablation temperature, usually 80 °C, in this region, causing coagulation and death of tumor cells (Van den Bijgaart et al. 2017). While HIFU is good at treating small tumors, it faces some problems in treating large or vascularized tumors. For safety reasons, the intensity of ultrasound should be kept below 0.05–0.50 W/cm^2 to prevent tissue heating. Ultrasonic frequency f can be divided into high frequency usually 3–60 MHz, used in diagnosis (Sharma and Bhargava 2013), medium frequency 0.7–3.0 MHz, used in medicine (Mason 2011) and low frequency 20–200 kHz (Schoellhammer and Traverso 2016) improve drug delivery very deep depths. In order to achieve the desired results, ultrasound frequency and intensity should be adjusted in cancer treatment.

10.8.3 Ultrasound Interactions with Nanoparticles

The effect of ultrasound associated with nanoparticles and cells facilitates the distribution of drugs in cancer and increases the effectiveness of treatment. In addition, nanoparticle ultrasonic drug delivery could overcome the limitations of current cancer therapy. Because ultrasound can interact with nanoparticles, it is possible to design drug-bearing nanoparticles that disintegrate in response to ultrasound and release the therapeutic payload. Therefore, ultrasound can be used to make nanoparticles release drugs on demand, called ultrasound-mediated drug release. In addition to hitting nanoparticles, causing drug release, ultrasound can be used to change the body's pressure and also increase noise, cavitation, and local hyperthermia (Tharkar et al. 2019). Ultrasound can improve clinical outcomes in several ways, particularly:

- Triggering the release of the drug from nanoparticles,
- Promoting uptake and accumulation of nanoparticle in cells,
- Enhancing the penetration of nanoparticles in tumors.

10.8.4 Types of Ultrasound-Sensitive Materials and Nanoparticles

Ultrasound exposure of nanosystems can lead to the generation of ROS and other processes such as cavitation, heating or mechanical changes. Different nanosystems (NS) considered for the treatment of cancer in the presence of a backbone of ultrasound application are formed from the following sources: (1) nanobubbles or nanodroplets, i.e., hollow nanospheres containing oil or liquid, with a polymerized layer. (2) Silica NPs with different porosity or mesoporosity levels, (3) such as PMMA, PLGA or lipids or phospholipids to stabilize inorganic NPs in the form of nanocrystalline or porous structures such as gold NPs, iron oxide NPs, TiO_2 substances natural NPs such as NPs, Au-TiO2 NPs, MnOx NPs, bismuth NPs, ZnO NPs, those containing the heme pigment biliverdin or albumin, (4) exosomes, membrane vesicles, bacterial protein certain organisms such as vesicles, polymersomes, pith—assembly of peptides isolated from their original biological environment or amplified by chemical synthesis from living materials. (Alphandéry 2022). In general, functional content is added by integrating into the outer or inner core of the bone, with theoretically greater content than is accepted in macroscopic solutions with a small volume-to-volume ratio. The basically aims to prevent bacteria from malfunctioning or degraded before they reach the tumor. Premium active ingredients improve clinical outcomes include: (1) Chemotherapeutic drugs such as DOX, cisplatin or docetaxel (Kim et al. 2020) combined with nanosystems to increase the amount of drug in tumors, chemotherapy and ultrasound antitumor Effect Synergy, (2) two types of gasoline, direct radiation to cancer, e.g., carbon monoxide (CO), nitric oxide (NO), hydrogen sulfide (H2S), hydrogen (H2), sulfur dioxide (SO2), carbon dioxide (CO2), and oxygen (O2) produce tumor oxygen and help chemotherapeutic drugs overcome resistance in hypoxic tumor environment (Chen et al. 2019) (3) ROS enhancers that can generate ROS under ultrasound action, such as protoporphyrin (Pellow et al. 2020), Tetra(4-aminophenyl)porphyrin (TAPP) (Liang et al. 2019). The second class of active materials gives the nanosystem different ultrasound capabilities and includes nanomaterials with different acoustic impedances than the surrounding tissues; it is generally made from gases such as sulfur hexafluoride (SF6) or perfluorocarbons (PFCs), which are found in nanobubbles and can often expand into more echogenic microbubbles under certain conditions (Abenojar et al. 2020). Finally, the spine can also be combined with workers who like their bones. The agent may be a targeting moiety that recognizes tumor cell receptors such as anti-HER2 antibody, rabies glycoprotein peptide targeting neuroblastoma cells, anti-EGFR antibody, folic acid, cycloarginine-glycine-aspartate Pentapeptide, Epiregulin targeting antibody. It is also incorporated into compounds such as PEG in the membrane of macrophages, allowing nanosystems to prevent capture by the immune system. Sometimes, the scaffold itself has a function, such as in the application of ultrasound, as with TiO2 NPs that produce ROS, magnetic NPs guided by magnets, or scaffolds with a composition, geometry, or dimensions that allow passive diffusion into the tumor. It inhibits the tumor with the effect of EPR (Alphandery 2020).

10.9 Exosomes as Anticancer Drug Delivery Vehicles

10.9.1 Introduction

Exosomes are small bilayers between 50 and 150 nm in diameter, secreted by almost all cells. They also play an important role in physiological or pathological processes such as regulation of immunity, development of metabolic and neurodegenerative diseases, growth and development of animals. Exosomes also play a role in tumor growth and development by modulating extracellular factors in the tumor microenvironment, metastasis, and antitumor or antitumor effects. Exosomes are found in all body fluids, including cerebrospinal fluid, blood, saliva, urine, amniotic fluid, ascites, and bronchoalveolar lavage fluid (Kalluri and LeBleu 2020). Exosomes carry a variety of cargo, including proteins, lipids, various RNAs, and DNA (Isaac et al. 2021). Because of the diversity and amount of burden in different diseases, exosomes are widely used in the diagnosis and monitoring of diseases. Exosomes with organelles, according to their biological functions and properties, are a versatile drug for the treatment of cancer and other diseases. For example, curcumin (cur), dopamine, paclitaxel (PTX and doxorubicin), and other compounds that are stable in exosomes (Liao et al. 2019). As natural nanoparticle biocarriers, exosomes are stable, membrane permeable, cell specific, and can even cross the blood-brain barrier (BBB) (Ha et al. 2016). Compared with other biological products, exosome-based therapeutic delivery has better efficacy and fewer side effects. However, some disadvantages of drug delivery vehicles are low efficacy, strong heterogeneity, short half-life, and low exosome loading efficiency, which limit their clinical application. The development of new exosome preparation and drug delivery has spurred the use of exosomes in cancer therapy.

10.9.2 Biogenesis and Uptake of Exosomes

Exosomes originate from endosomal structures and form early sorting endosomes (ESEs) via plasma membrane invaginations. ESEs mature into late-separating endosomes, exchange cargo with the Golgi apparatus, and bud inside to form multivesicular bodies (MVBs) that encapsulate intraluminal vesicles (ILVs) containing exosomal contents at this stage. MVBs can fuse with lysosomes or autophagosomes to degrade or integrate with the plasma membrane to release ILVs as exosomes. It contains many bioactive molecules, including exosomes, proteins and nucleic acids. Tetraspanins CD9, CD63, and CD81 on the exosome membrane are often used as exosome marker proteins (Yubo et al. 2022). Cells recruit exosomes via caveolin-dependent endocytosis, clathrin-dependent endocytosis, macropinocytosis, lipid rafts, phagocytosis, and direct fusion. In addition, exosomes can enter cells in more than one way.

10.9.3 The Sources of Exosomes

Exosomes are found in all fluids of the human body. However, the results of exosomes isolated from body fluids or culture medium are insufficient to meet clinical needs. Quality control is also a problem that limits its clinical application. Exosomes from non-human sources such as bacteria, milk and plants are considered a promising tool for cancer therapy. Due to their low toxicity and easy availability, exosomes can also be used as a personal oral delivery vehicle. Milk-derived exosomes are also considered a good delivery vehicle for hydrophilic and lipophilic drugs, as well as chemotherapeutic drugs, and show no toxicity and immunogenicity in mice.

10.9.4 Isolation of Exosomes

For effective and safe drug delivery, exosomes must be isolated using efficient and reproducible isolation techniques. Exosome isolation technologies have been developed based on speed, size, and antibody properties (Sidhom et al. 2020). The purity, amount and physicochemical properties of exosomes vary depending on the separation method, and the following methods are generally used for separation.

10.9.4.1 Ultracentrifugation

Ultracentrifugation is the most popular and widely accepted gold standard for isolating exosomes. It includes differential ultracentrifugation and density gradient ultracentrifugation. Differential ultracentrifugation separates exosomes from other products by using different centrifugal force and duration during centrifugation, mainly because different molecules have different particle sizes.

10.9.4.2 Ultrafiltration

Ultrafiltration uses membranes with different pore sizes to separate particles into smaller sizes. Ultrafiltration alone or in combination with ultracentrifugation and gel filtration chromatography effectively separates exosomes.

10.9.4.3 Size Exclusion Chromatography (SEC)

SEC can be used to separate exosomes depending on the pore size that separates macromolecules and particles by size. Components with small hydrodynamic radii enter these pores, resulting in a longer retention time, while exosomes and other products with large hydrodynamic radii are removed from the pores, resulting in a shorter retention time.

10.9.4.4 Flow Field-Flow Fractionation (F4)

F4 is a hydrodynamic process that separates exosomes from different diameters. Fractionation takes place in a rectangle in which the migrating flow moves along the axis of the flow field while the sample is held at the secondary flow rate (cross flow at F4). The channel side acts as an outer surface which causes the structure to move along the channel wall. Due to the diffusion effect, the sample is distributed at different locations on the wall of the water flow, which can be separated according to the size of the sample. The delivery system of small particles has a higher mean wall volume height than larger particles, resulting in earlier, exosome separation.

10.9.4.5 Hydrostatic Filtration Dialysis (HFD)

HFD is often used to isolate urinary extracellular vesicles by diffusing solutes through cellulose ester dialysis membranes with defined molecular weight cut-offs. The solvent is also passed through membrane filtration under hydrostatic pressure to obtain the filtration content. The good thing about the HFD is that it can be fixed and focused on quality standards and does not require the operation of special equipment, making this process easy. HFD is significantly better for a wide range of sample volumes than differential centrifugation.

10.9.4.6 Polymer-Based Precipitation

Polymer precipitation uses interactions between polymers such as polyethylene glycol and water molecules and water molecules or salt solutions such as sodium acetate to neutralize the negative charge of phosphatidylserine on the surface of exosomes, thereby reducing the solubility and dispersibility of exosomes and the formation of physical aggregates. Finally, the exosome pellet can be collected by centrifugation.

10.9.4.7 Immunoaffinity Capture-Based Technology

There are many proteins and receptors in the exosome membrane. Immunoaffinity capture uses the specific binding of receptors to their ligands or the immunoaffinity of proteins and their specific antibodies to isolate exosomes. Surface receptors such as CD9, CD63 and CD81 are considered biomarkers of exosomes and are often used to remove exosomes through immune affinity capture. The main technologies developed from this principle are ELISA in microtiter plate format and affinity capture based on immunomagnetic beads.

10.9.4.8 Microfluidic-Based Exosome Isolation

Microfluidic-based exosome isolation is a promising method that uses microfluidic devices to isolate exosomes based on various properties such as size, density, and immunogenicity. In addition, nanowire trapping, acoustics, lateral displacement and viscoelastic flow etc. in microfluidic platforms. There are many new isolation methods such as Compared with other existing exosome extraction methods, the microfluidic method is faster, cheaper, requires less samples and reagents, can isolate exosomes from the source and allow most of the exosomes to be separated by microfluidics into the body's natural form.

10.9.5 Drug Loading on Exosomes

The drug transport mechanism of exosomes can be divided into pre-drug and post-drug release, depending on whether the drug is transported directly by exosomes (Table 10.5). As the name suggests, pre-secretory drug loading refers to drug loading during exosome biogenesis before vesicle separation. By adding to the culture medium, the cell's endogenous machinery is used to transport drugs to the exosomes (Zhang et al. 2021). This creates the biggest problem, which is the inability to control the drug encapsulated in exosomes. The process involves incubation and transfection of the drug with the brain. Post-secretory drug loading transports drug directly to individual exosomes, which is more prominent and the encapsulation efficiency of exosomes and drug loading are easier to control. Electroporation, sonication, freeze-thaw cycles, extrusion, drug-exosome co-incubation, surfactant treatment and dialysis are all in this mechanism.

Table 10.5 Comparisons of exosome drug-loading technology (Yubo et al. 2022)

Loading mechanism	Loading technique	Advantages	Disadvantages
Presecretory drug loading	Co-incubation (drug and cells)	Easy operation	Low loading efficiency, difficulty to control, and the effect of drug toxicity
	Transfection	Overexpression of specific molecules	Low loading efficiency, the toxicity of transfection reagents, and micelles used for transfections that are not easily separated from exosomes
Post secretory drug loading	Electroporation	The gold standard	Low loading efficiency, aggregation of exosomes, and reduced activity caused by the destruction of the protein drug structure
	Sonication	High loading efficiency, not easy to form aggregates, and retention of membrane surface adhesion proteins	Altered immune privilege status and disruption of exosome integrity
	Freeze-thaw cycle	Easy operation	Low loading efficiency, exosome aggregation, and inactivation of protein drugs
	Extrusion	High loading efficiency, the most efficient water-soluble cargo loading technology, and unaffected exosome internalization properties	Altered immune privilege status and disruption of exosome integrity
	Co-incubation (drug and exosomes)	Easy operation	Low loading efficiency and susceptibility to drug hydrophobicity
	Surfactant treatment	High loading efficiency and little impact on the physicochemical properties of exosomes	Hemolytic activity of saponins, requiring additional purification steps
	Dialysis	High loading efficiency	Degradation of protein or peptide drugs and changes in the size distribution of exosomes

10.9.5.1 Presecretory Drug Loading

10.9.5.1.1 Co-Incubation (Drug and Cells)

In the coincubation method, the drug is incubated with the brain so that the drug enters the cell and is then secreted into the exosomal cargo. It is characterized by simple operation, but the drug-loading efficiency is low, and the production and release process of drug-loaded exosomes cannot be controlled. Some drugs can be toxic and affect the state of the cell and the release of exosomes.

10.9.5.1.2 Transfection

Transfection is the best way to package peptides, proteins, nucleic acids, and other active molecules into exosomes. The target gene is introduced into cells by the process of transfection to express the product of interest, which is then encapsulated into exosomes by the cellular machinery. The loading efficiency is very low as the loading process cannot be controlled. Direct drug transfer always leads to destruction or infection of exosomes. HEK293 cells were transfected with plasmids and miRNA and purified in culture supernatant to obtain miRNA-loaded exosomes transfected with epidermal growth factor or GE11 peptide, which can target miRNA delivery to epidermal growth factor receptor expressing cancer tissues.

10.9.5.2 Post Secretory Drug Loading

10.9.5.2.1 Electroporation

Electroporation uses short high voltage pulses to create pores in the exosome membrane, allowing the drug to enter the exosome through the pores. This approach is often used to improve the uptake of siRNA, metal oxide nanoparticles, and small drugs. In particular, electroporation is limited to some RNAs, whereas miRNAs, short hairpin RNAs (shRNAs), and mRNAs cannot be transported into exosomes using this method.

10.9.5.2.2 Sonication

Exosomes purified from host cells and related substances are mixed and then homogenized using an ultrasonic probe. The integrity of the exosome membrane is disrupted by the mechanical shear force, allowing drugs to enter the exosome during membrane deformation. After sonication, the tension of the exosome membrane was reduced, which allowed the incorporation of the drug into the lipid bilayer, which improved its loading capacity. In addition, the fact that the drug will adhere to the surface of the exosomes may cause the drug to be broken down in the early period and sustained release in the later stage.

10.9.5.2.3 Freeze-Thaw Cycle

The solution containing the exosomes and drug in a freeze-thaw cycle was incubated at 37 °C, snap-crozene at −80 °C and thawed at room temperature; This process is repeated for at least three cycles. This method is easy to implement, but the drug loading efficiency is low. The drug-loaded exosome particles are larger and contain multiple smaller exosomes that may result from aggregation.

10.9.5.2.4 Extrusion

Extrusion is a physical process. When the drug and exosome mixture was extruded through a porous polycarbonate membrane with a certain pore size, the exosome membrane was damaged and ruptured, and the drug in the mixture was evenly loaded into the exosomes. This method has high loading efficiency and is considered the most efficient method for transporting water-soluble cargoes.

10.9.5.2.5 Co-Incubation (Drugs and Exosomes)

During co-incubation of drugs and exosomes, they are only incubated at a certain temperature, and the drug concentration gradient inside and outside the exosomes pushes the drug into the exosomes by diffusion. This method only depends on the hydrophobic contact and diffusion between the exosome membrane and the drug, so the hydrophobicity of the drug is one of the determinants of loading. This method is easy to implement, but the disadvantage is its low carrying capacity and the inability of hydrophilic drugs to pass through the lipid layer of exosomes by passive diffusion.

10.9.5.2.6 Surfactant Treatment

Surfactants such as saponins can dissolve exosome membrane molecules, thereby forming pores and increasing membrane permeability. This approach improved the drug loading capacity of exosomes.

10.9.5.2.7 Dialysis

In hypoosmotic dialysis, exosomes and capsules are transferred to a dialysis membrane or dialysis tube and located in a near-impartial buffer, and then obtain drug loaded exosomes are received by means of stirring dialysis. furthermore, the exosomes within the dialysis machine may be pre-dehydrated in ethanol after which rehydrated in an acidic buffer to lessen the pH inside the exosomes and promote the formation of a pH gradient inside and outside the exosome membrane, thereby substantially enhancing the loading efficiency of exosomes.

10.9.6 Exosomes Applications in Cancer Treatment

The excellent physicochemical and biological properties of exosomes make them great potential drug delivery vehicles for the treatment of cancer. Exosomes serve as excellent carriers for different therapeutics, including nucleic acids, proteins, small molecules, and other nanomaterials. The involvement of exosomes as drug delivery systems in cancer therapy for various drugs is summarized in Fig. 10.8.

10.9.6.1 Small Molecule Chemotherapy Drugs

Cur, an anti-inflammatory phytochemical, inhibits transcription of miR-21, which induces PI3K/protein kinase B/phosphatase and tensin homolog (PTEN), nuclear factor kappa B (NF-κB), and programmed cell death protein 4 pathway in influencing apoptosis. It has an inhibitory effect on migration, cell proliferation, resistance, and rooting, and finally on cancer cells. Cur's poor water solubility, lack of stability, and low bioavailability limit its application as a drug (Chen et al. 2015a, b). Stability and enzyme therapy, these properties make exosome-encapsulated Cur an oral anti-inflammatory drug. Taxanes are a class of cytotoxic diterpenoids commonly used to treat malignancies (Münz et al. 2018). The representative drug is PTX, which is used to treat prostate cancer, breast cancer, breast cancer, ovarian cancer, lung cancer, and breast cancer. Similar to cur, PTX is also

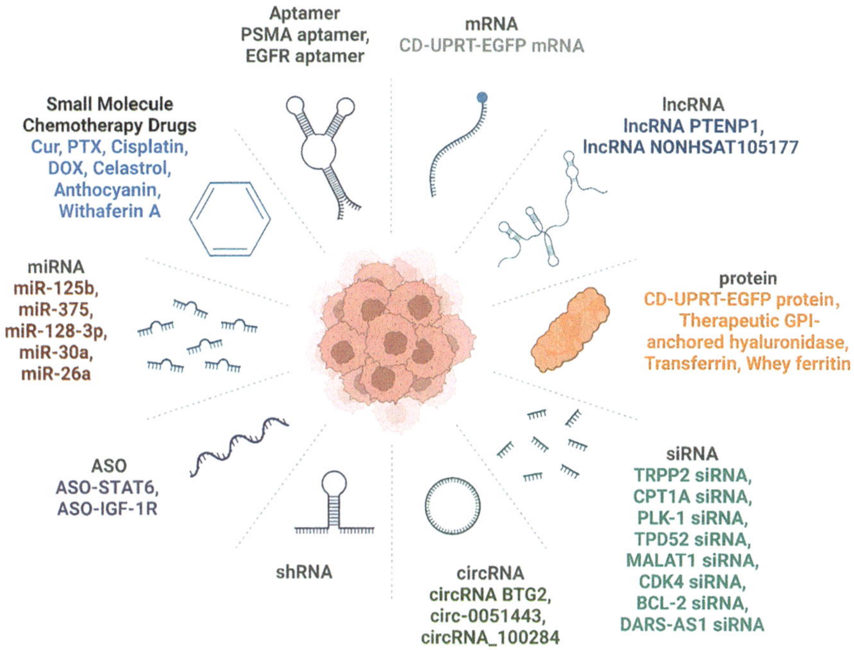

Fig. 10.8 Application of exosome-based cancer therapy. Different therapeutic agents are delivered by exosomes to treat cancer. We summarize the application of small molecule chemotherapeutics, miRNA, antisense oligonucleotides, shRNA, circRNA, siRNA, protein, lncRNA, mRNA, and aptamer in cancer therapy when exosomes are used as delivery vehicles (Yubo et al. 2022)

sparingly water soluble which limits its clinical applications. Exosomes have also been used to transport PTX. PTX is delivered to tumor cells via exosomes secreted by M1 macrophages. The study found that exosomes take advantage of transporters to deliver more PTX to the tumor site while activating the NF-κB pathway, thereby enhancing its antitumor effect (Wang et al. 2019a, b). By electroporation of a mixture of doxorubicin (DOX) and exosomes, DOX-loaded exosomes were successfully processed, improving the therapeutic index of the drug and avoiding cardiovascular complications (Peng et al. 2019). Recently, incorporation of docetaxel into exosomes by electroporation resulted in significant tumor suppression in non-small cell lung cancer models.

10.9.6.2 Therapeutic Nucleic Acid

The microRNA (miRNA) is a form of small, single-stranded RNA, 18–25 nucleotides long, found in all eukaryotic cells that can bind complementaryly to the 3' untranslated region of mRNA, thereby negatively regulating the transcription of the target gene. miRNA-based cancer therapy has many advantages, such as broad regulation, which can effectively target the target while simultaneously regulating other genes of interest to produce clinical results. In addition, miRNAs also have low toxicity and low immunogenicity properties. However, miRNA-based drugs face many problems, such as poor blood quality and integrity, difficulty to enter the tumor, and off-target effects. Exosomes can be used as a delivery vehicle to ameliorate these problems. siRNAs are approximately 21 nucleotides long derived from double-stranded RNA and induce gene silencing by sequence-specific cleavage of all complementary mRNAs. As a natural drug delivery carrier, exosomes can deliver siRNA to tissues or cells in vivo, regulate gene expression, and inhibit tumor growth. It has the advantage of avoiding phagocytosis, cytotoxicity and low immunogenicity (Chen et al. 2015a, b).

10.9.6.3 Other Therapeutic Compounds

Exosomes can also provide other types of therapeutics. For example, HEK-293T cells were used to stably express the fusion protein of uracil phosphoribosyltransferase and cytosine deaminase by transfection, and exosomes loaded with the fusion protein were isolated and exosomes converted to 5-fluoro-exchangeable Cytosine prodrugs active forms exert anticancer effects (Mizrak et al. 2013). In addition to proteins and mRNAs, antisense oligonucleotides, long non-coding RNA (lncRNA), shRNA, circular RNA (circRNA), and aptamers have also entered cancer exosome therapy (Zhang et al. 2022a, b). Other nanomaterials are also included in exosomes e.g., Superparamagnetic iron oxide nanoparticles (SPION) loaded with melanoma-derived exosomes. Recently, exosomes from cancer patients were functionalized by amphiphilic polymer-modified gold nanoparticles and bound to bovine serum albumin containing chlorine 6 (PMA/Au-BSA@Ce6) via electroporation, enabling deep penetration and higher retention behavior in tumors (Eftekhari et al. 2023).

10.9.7 Targeted Delivery of Exosomes

Cell-specific exosomes exhibit a natural response for the cancer microenvironment, which is important for the development of targets for nanomedicine (Basoglu et al. 2018). For example, macrophage-derived exosomes can target cancer cells via LFA-1 protein expressed on their surface and ICAM-1 overexpressed in most cancers. Similarly, exosomes from T cell lines are taken up by myeloid cells, while mature DC-derived exosomes are highly enriched by T cell utilization (György et al. 2015). Artificial modification of exosome surface ligands may also facilitate the development of receptor-mediated tissue targeting.

10.10 Conclusion

The chapter discuss the roles of nanoparticles and hybrid nanoparticles for drug delivery in chemotherapy, targeted therapy, and immunotherapy and describe the targeting mechanism of nanoparticle-based drug delivery as well as its function on reversing drug resistance, compared to conventional drugs, nanoparticle-based drug delivery has specific advantages, such as improved stability and biocompatibility, enhanced permeability and retention effect, and precise targeting. The application and development of nano-system, which incorporates the combined properties of different nanoparticles, has led this type of drug-carrier system to the next level. The Chapter also focuses on combinations of siRNA with chemotherapeutic drug delivery systems for the treatment of cancer and gives an overview of several nanocarrier formulations in both research and clinical applications. The use of various combinations of ultrasounds and nano-systems for cancer treatment has been reviewed, the current chapter introduces the biogenesis and uptake of exosomes and compares different approaches for isolation and drug loading, focusing on the application and current challenges of exosomes as drug delivery vehicles in cancer therapy.

References

Abenojar EC, Bederman I, de Leon AC, Zhu J, Hadley J, Kolios MC, Exner AA (2020) Theoretical and experimental gas volume quantification of micro- and nanobubble ultrasound contrast agents. Pharmaceutics 12:208

Ali HR, Selim SA, Aili D (2021) Effects of macrophage polarization on gold nanoparticle-assisted plasmonic photothermal therapy. RSC Adv 11:25047

Alphandery E (2020) Iron oxide nanoparticles for therapeutic applications. Drug Discov Today 25: 141–149

Alphandéry E (2022) Ultrasound and nanomaterial: an efficient pair to fight cancer. J Nanobiotechnol 20:139

Apte RS, Chen DS, Ferrara N (2019) VEGF in signaling and disease: beyond discovery and development. Cell 176:1248–1264. https://doi.org/10.1016/j.cell.2019.01.021

Arumov A, Liyanage PY, Trabolsi A, Roberts ER, Li L, Ferreira BCLB, Gao Z, Ban Y, Newsam AD, Taggart MW, Vega F, Bilbao D, Leblanc RM, Schatz JH (2021) Optimized doxorubicin chemotherapy for diffuse large B-cell lymphoma exploits nanocarrier delivery to transferrin receptors. Cancer Res 81(3):763–775

Attia MF, Anton N, Wallyn J, Omran Z, Vandamme TF (2019) An overview of active and passive targeting strategies to improve the nanocarriers efficiency to tumour sites. J Pharm Pharmacol 71:1185–1198

Bahrami B, Hojjat-Farsangi M, Mohammadi H, Anvari E, Ghalamfarsa G, Yousefi M et al (2017) Nanoparticles and targeted drug delivery in cancer therapy. Immunol Lett 190:64–83. https://doi.org/10.1016/j.imlet.2017.07.015

Banstola A, Poudel K, Kim JO, Jeong JH, Yook S (2021) Recent progress in stimuli-responsive nanosystems for inducing immunogenic cell death. J Control Release 337:505–520

Barz M, Canal F, Koynov K, Zentel R, Vicent MJ (2010) Synthesis and in vitro evaluation of defined HPMA folate conjugates: influence of aggregation on folate receptor (FR) mediated cellular uptake. Biomacromolecules 11(9):2274–2282

Basoglu H, Goncu B, Akbas F (2018) Magnetic nanoparticle-mediated gene therapy to induce Fas apoptosis pathway in breast cancer. Cancer Gene Ther 25:141–147. https://doi.org/10.1038/s41417-018-0017-2

Bertrand N et al (2014) Cancer nanotechnology: the impact of passive and active targeting in the era of modern cancer biology. Adv Drug Deliv Rev 66:2–25

Biswas S, Kinbara K, Niwa T, Taguchi H, Ishii N, Watanabe S, Miyata K et al (2013) Biomolecular robotics for chemomechanically driven guest delivery fuelled by intracellular ATP. Nat Chem 5(7):613–620

Byrne JD, Betancourt T, Brannon-Peppas L (2008) Active targeting schemes for nanoparticle systems in cancer therapeutics. Adv Drug Deliv Rev 60(15):1615–1626

Canavese G, Ancona A, Racca L, Canta M, Dumontel B, Barbaresco F, Limongi T, Cauda V (2018) Nanoparticle-assisted ultrasound: a special focus on sonodynamic therapy against cancer. Chem Eng J 340:155–172

Carson AR, McTiernan CF, Lavery L, Grata M, Leng X, Wang J et al (2012) Ultrasound-targeted microbubble destruction to deliver siRNA cancer therapy. Cancer Res 72:6191–6199

Charbe NB, Amnerkar ND, Ramesh B, Tambuwala MM, Bakshi HA, Alaa AA, Aljabali, Khadse SC et al (2020) Small interfering RNA for cancer treatment: overcoming hurdles in delivery. Acta Pharm Sin B 10(11):2075–2109

Chauhan VP, Jain RK (2013) Strategies for advancing cancer nanomedicine. Nat Mater 12(11):958

Chen Y, Gao D, Huang L (2015a) In vivo delivery of miRNAs for cancer therapy: challenges and strategies. Adv Drug Deliv Rev 81:128–141

Chen J, Tiefeng X, Chen C (2015b) The critical roles of miR-21 in anti-cancer effects of curcumin. Ann Transl Med 3:21

Chen L, Zhou SF, Su L, Song J (2019) Gas-mediated cancer bioimaging and therapy. ACS Nano 13:10887–10917

Cheng CA, Deng T, Lin FC, Cai Y, Zink JI (2019) Supramolecular nanomachines as stimuli-responsive gatekeepers on mesoporous silica nanoparticles for antibiotic and cancer drug delivery. Theranostics 9:3341–3364. https://doi.org/10.7150/thno.34576

Chenglong W, Shuhan X, Jiayi Y, Wencai G, Guoxiong X, Hongjing D (2020) Dextran-based coacervate nanodroplets as potential gene carriers for efficient cancer therapy. Carbohydr Polym 231:115687

Chittasupho C, Kewsuwan P, Murakami T (2017) CXCR4-targeted nanoparticles reduce cell viability, induce apoptosis and inhibit SDF-1α induced BT-549-Luc cell migration in vitro. Curr Drug Deliv 14(8):1060–1070

Chou MJ, Yu HY, Hsia JC et al (2018) Highly efficient intracellular protein delivery by cationic polyethyleneimine-modified gelatin nanoparticles. Materials 11(2):301

Cuellar TL, Barnes D, Nelson C, Tanguay J, Yu SF, Wen X, Scales SJ, Gesch J, Davis D, van Brabant SA et al (2015) Systematic evaluation of antibody-mediated siRNA delivery using an industrial platform of THIOMAB–siRNA conjugates. Nucleic Acids Res 43:1189–1203. https://doi.org/10.1093/nar/gku1362

Cun X, Chen J, Li M, He X, Tang X, Guo R et al (2019) Tumor-associated fibroblast-targeted regulation and deep tumor delivery of chemotherapeutic drugs with a multifunctional size-switchable nanoparticle. ACS Appl Mater Interfaces 11(43):39545–39559

Czech MP, Aouadi M, Tesz GJ (2011) RNAi-based therapeutic strategies for metabolic disease. Nat Rev Endocrinol 7(8):473–484

De La Torre C, Casanova I, Acosta G, Coll C, Moreno MJ, Albericio F, Aznar E et al (2015) Gated mesoporous silica nanoparticles using a double-role circular peptide for the controlled and target-preferential release of doxorubicin in CXCR4-expresing lymphoma cells. Adv Funct Mater 25(5):687–695

Debnath SK, Srivastava R (2021) Drug delivery with carbon-based nanomaterials as versatile nanocarriers: progress and prospects. Front Nanotechnol 3:1–22

Decuzzi P, Pasqualini R, Arap W, Ferrari M (2009) Intravascular delivery of particulate systems: does geometry really matter? Pharm Res 26:235–243. https://doi.org/10.1007/s11095-008-9697-x

Desgrosellier JS, Cheresh DA (2010) Integrins in cancer: biological implications and therapeutic opportunities. Nat Rev Cancer 10:9–22. https://doi.org/10.1038/nrc2748

Diaz-Vivancos P, de Simone A, Kiddle G, Foyer CH (2015) Glutathione-linking cell proliferation to oxidative stress. Free Radic Biol Med 89:1154–1164

Dienst A, Grunow A, Unruh M, Rabausch B, Nör JE, Fries JW et al (2005) Specific occlusion of murine and human tumor vasculature by VCAM-1-targeted recombinant fusion proteins. J Natl Cancer Inst 97:733–747. https://doi.org/10.1093/jnci/dji130

Di-Wen S, Pan G, Hao L, Zhang J, Xue Q, Wang P, Yuan Q (2016) Improved antitumor activity of epirubicin-loaded CXCR4-targeted polymeric nanoparticles in liver cancers. Int J Pharm 500(1-2):54–61

Dolly J, Prajapati SK, Jain A, Singhal R (2023) Nano-formulated siRNA-based therapeutic approaches for cancer therapy. Nano Trends 1:100006. https://doi.org/10.1016/j.nwnano.2023.100006

Dong Z, Feng L, Chao Y, Hao Y, Chen M, Gong F, Han X et al (2019) Amplification of tumor oxidative stresses with liposomal fenton catalyst and glutathione inhibitor for enhanced cancer chemotherapy and radiotherapy. Nano Lett 19(2):805–815

Draz MS, Fang BA, Zhang P, Hu Z, Gu S, Weng KC, Gray JW, Chen FF (2014) Nanoparticle-mediated systemic delivery of siRNA for treatment of cancers and viral infections. Theranostics 4(9):872–892. https://doi.org/10.7150/thno.9404

Du L, He H, Xiao Z, Xiao H, An Y, Zhong H, Lin M et al (2022) GSH-responsive metal-organic framework for intratumoral release of NO and IDO inhibitor to enhance antitumor immunotherapy. Small 18(15):e2107732

Eatemadi A, Aiyelabegan HT, Negahdari B, Mazlomi MA, Daraee H, Daraee N, Eatemadi R et al (2017) Role of protease and protease inhibitors in cancer pathogenesis and treatment. Biomed Pharmacother 86:221–231

Eftekhari A, Kryschi C, Pamies D, Gulec S, Ahmadian E, Janas D, Davaran S, Khalilov R (2023) Natural and synthetic nanovectors for cancer therapy. Nanotheranostics 7(3):236–257

Elbashir SM et al (2001) Duplexes of 21-nucleotide RNAs mediate RNA interference in cultured mammalian cells. Nature 411(6836):494–498

Elhelf IAS, Albahar H, Shah U, Otoe A, Cressman E, Almekkawy M (2018) High intensity focused ultrasound: the fundamentals, clinical applications and research trends. Diagn Interv Imaging 99:349–359

Elias SS, Nogueira KAB, Fernandes LCC, Martins JRP, Reis AVF, de Brito J, Neto V, José I, da Silva C, Pessoa RP, Eloy JO (2021) EGFR targeting for cancer therapy: pharmacology and immunoconjugates with drugs and nanoparticles. Int J Pharm 592:120082. https://doi.org/10.1016/j.ijpharm.2020.120082

Fan Z, Chang Y, Cui C et al (2018) Near infrared fluorescent peptide nanoparticles for enhancing esophageal cancer therapeutic efficacy. Nat Commun 9(1):2605

Farjadian F, Ghasemi S, Andami Z, Tamami B (2020) Thermo-responsive nanocarrier based on poly(N-isopropylacrylamide) serving as a smart doxorubicin delivery system. Iran Polym J 29: 197–207

Feltrin FDS, Agner T, Sayer C, Lona LMF (2022) Curcumin encapsulation in functional PLGA nanoparticles: a promising strategy for cancer therapies. Adv Colloid Interf Sci 300:102582

Fernández M, Javaid F, Chudasama V (2017) Advances in targeting the folate receptor in the treatment/imaging of cancers. Chem Sci 9(4):790–810

Fontana F, Shahbazi MA, Liu D, Zhang H, Mäkilä E, Salonen J et al (2017) Multistaged nanovaccines based on porous silicon@acetalated dextran@cancer cell membrane for cancer immunotherapy. Adv Mater 29:1603239. https://doi.org/10.1002/adma.201603239

Friedrich M, Aigner A (2022) Therapeutic siRNA: state-of-the-art and future perspectives. BioDrugs 36(5):549–571. https://doi.org/10.1007/s40259-022-00549-3. Epub 2022 Aug 23. PMID: 35997897; PMCID: PMC9396607.

Fruehauf KR, Kim TI, Nelson EL, Patterson JP, Wang S-W, Shea KJ (2019) Metabolite responsive nanoparticle-protein complex. Biomacromolecules 20:2703–2712

Fu L, Wan Y, Qi C, He J, Li C, Yang C, Xu H et al (2021) Nanocatalytic theranostics with glutathione depletion and enhanced reactive oxygen species generation for efficient cancer therapy. Adv Mater 33(7):e2006892

Gao Y, Wang K, Zhang J, Duan X, Sun Q, Men K (2023) Multifunctional nanoparticle for cancer therapy. Media Commun 4(1):e187. https://doi.org/10.1002/mco2.187. PMID: 36654533; PMCID: PMC9834710.

Ghezzi M, Pescina S, Padula C, Santi P, Del Favero E, Cantù L et al (2021) Polymeric micelles in drug delivery: an insight of the techniques for their characterization and assessment in biorelevant conditions. J Control Release 332:312–336

Gisbert-garzar M (2021) Redox-responsive mesoporous silica nanoparticles for cancer treatment: recent updates. Nanomaterials 11:2222

Guo F, Fu Q, Zhou K, Jin C, Wu W, Ji X, Yan Q, Yang Q et al (2020) Matrix metalloprotein-triggered, cell penetrating peptide-modified star-shaped nanoparticles for tumor targeting and cancer therapy. J Nanobiotechnol 18(1):48

Gupta N, Rai DB, Jangid AK, Pooja D, Kulhari H (2019) Nanomaterials-based siRNA delivery: routes of administration, hurdles and role of nanocarriers. Nanotechnol Mod Anim Biotechnol 31:67–114

György B, Hung ME, Breakefield XO, Leonard JN (2015) Therapeutic applications of extracellular vesicles: clinical promise and open questions. Annu Rev Pharmacol Toxicol 55:439–464

Ha D, Yang N, Nadithe V (2016) Exosomes as therapeutic drug carriers and delivery vehicles across biological membranes: current perspectives and future challenges. Acta Pharm Sin B 6(4):287–296

Hailing Y, Xiufang L, Lili W et al (2020) Doxorubicin-loaded fluorescent carbon dots with PEI passivation as a drug delivery system for cancer therapy. Nanoscale 12(33):17222–17237

Hamilton KJ, Hewitt SC, Arao Y, Korach KS (2017) Estrogen hormone biology. Curr Top Dev Biol 125:109–146

Han X, Wang R, Xu J, Chen Q, Liang C, Chen J, Zhao J, Chu J, Fan Q, Archibong E, Jiang L, Wang C, Liu Z (2019) In situ thermal ablation of tumors in combination with nano-adjuvant and immune checkpoint blockade to inhibit cancer metastasis and recurrence. Biomaterials 224: 119490

Hattab D, Gazzali AM, Bakhtiar A (2021) Clinical advances of siRNA-based nanotherapeutics for cancer treatment. Pharmaceutics 13:1009. https://doi.org/10.3390/pharmaceutics13071009

Holder JE, Christopher F, Elisabete O, Carlos L, Trim CM, Byrne LJ, Emilia B, Wilson CM (2023) The use of nanoparticles for targeted drug delivery in non-small cell lung cancer. Front Oncol 13:1154318

Holm J, Hansen SI (2020) Characterization of soluble folate receptors (folate binding proteins) in humans. Biological roles and clinical potentials in infection and malignancy. Biochim Biophys Acta, Proteins Proteomics 1868(10):140466

Horise Y, Maeda M, Konishi Y, Okamoto J, Ikuta S, Okamoto Y, Ishii H et al (2019) Sonodynamic therapy with anticancer micelles and high-intensity focused ultrasound in treatment of canine cancer. Front Pharmacol 10:545

Hu J, Liu S (2020) Modulating intracellular oxidative stress via engineered nanotherapeutics. J Control Release 319:333–343

Huang G, Huang H (2018) Hyaluronic acid-based biopharmaceutical delivery and tumor-targeted drug delivery system. J Control Release 278:122–126

Im S, Lee J, Park D, Park A, Kim Y-M, Kim WJ (2019) Hypoxia-triggered transforming immunomodulator for cancer immunotherapy via photodynamically enhanced antigen presentation of dendritic cell. ACS Nano 13:476–488

Isaac R, Reis FCG, Ying W, Olefsky JM (2021) Exosomes as mediators of intercellular crosstalk in metabolism. Cell Metab 33(9):1744–1762

Iversen TG, Skotland T, Sandvig K (2011) Endocytosis and intracellular transport of nanoparticles: present knowledge and need for future studies. Nano Today 6(2):176–185

Jhaveri A, Deshpande P, Torchilin V (2014) Stimuli-sensitive nanopreparations for combination cancer therapy. J Control Release 190:352–370. Return to ref 231 in article.

Jiao W, Zhang T, Peng M, Yi J, He Y, Fan H (2022) Design of magnetic nanoplatforms for cancer theranostics. Biosensors 12(1):38

Kalluri R, LeBleu VS (2020) The biology, function, and biomedical applications of exosomes. Science 367(6478):6977

Kapara A, Brunton V, Graham D, Faulds K (2020) Investigation of cellular uptake mechanism of functionalised gold nanoparticles into breast cancer using SERS. Chem Sci 11(22):5819–5829
Kaushik N, Borkar SB, Nandanwar SK et al (2022) Nanocarrier cancer therapeutics with functional stimuli-responsive mechanisms. J Nanobiotechnol 20:152
Kazempour M, Edjlali L, Akbarzadeh A, Davaran S, Farid SS (2019) Synthesis and characterization of dual pH-and thermo-responsive graphene-based nanocarrier for effective anticancer drug delivery. J Drug Deliv Sci Technol 54:101158
Khelghati N, Mokhtarvand JS, Mir M, Alemi F, Asemi Z, Sadeghpour A, Maleki M et al (2021) The importance of co-delivery of nanoparticle-siRNA and anticancer agents in cancer therapy. Chem Biol Drug Des 97(4):997–1015
Kim TY, Kim DW, Chung JY, Shin SG, Kim SC, Heo DS et al (2004) Phase I and pharmacokinetic study of Genexol-PM, a Cremophor-free, polymeric micelle-formulated paclitaxel, in patients with advanced malignancies. Clin Cancer Res 10:3708–3716
Kim D, Lee SS, Yoo WY, Moon H, Cho A, Park SY, Kim YS, Kim HR, Lee HJ (2020) Combination therapy with doxorubicin-loaded reduced albumin nanoparticles and focused ultrasound in mouse breast cancer xenografts. Pharmaceuticals 13:235
Lee ES, Oh KT, Kim D, Youn YS, Bae YH (2007) Tumor pH-responsive flower-like micelles of poly(L-lactic acid)-b-poly(ethylene glycol)-b-poly(L-histidine). J Control Release 123(1): 19–26
Lei W, Sun C, Jiang T, Gao Y, Yang Y, Zhao Q et al (2019) Polydopamine-coated mesoporous silica nanoparticles for multi-responsive drug delivery and combined chemo-photothermal therapy. Mater Sci Eng C Mater Biol Appl 105:110103. https://doi.org/10.1016/j.msec.2019.110103
Lei C, Liu XR, Chen QB, Li Y, Zhou JL, Zhou LY, Zou T (2021) Hyaluronic acid and albumin based nanoparticles for drug delivery. J Control Release 331:416–433
Li J, Yang N, Yang M, Lu C, Xie M (2022a) Development of a magnetic MoS system camouflaged by lipid for chemo/phototherapy of cancer. Colloids Surf B Biointerfaces 213:112389
Li M, Li S, Li Y, Li X, Yang G, Li M, Xie Y et al (2022b) Cationic liposomes co-deliver chemotherapeutics and siRNA for the treatment of breast cancer. Eur J Med Chem 233:114198
Liang S, Deng X, Chang Y, Sun C, Shao S, Xie Z, Xiao X, Ma P, Zhang H, Cheng Z, Lin J (2019) Intelligent hollow Pt-CuS janus architecture for synergistic catalysis-enhanced sonodynamic and photothermal cancer therapy. Nano Lett 19:4134–4145
Liao W, Yu D, Zhang C, Pan F, Yao Y, Zhang T, Peng Q (2019) Exosomes: the next generation of endogenous nanomaterials for advanced drug delivery and therapy. Acta Biomater 86:1–14
Liu Y, Jiang Y, Zhang M, Tang Z, He M, Bu W (2018) Modulating hypoxia via nanomaterials chemistry for efficient treatment of solid tumors. Acc Chem Res 51:2502–2511
Liu J, Tian B, Liu Y, Wan JB (2021a) Cyclodextrin-containing hydrogels: a review of preparation method, drug delivery, and degradation behavior. Int J Mol Sci 22(24):13516
Liu Y, Wu W, Wang Y, Han S, Yuan Y et al (2021b) Recent development of gene therapy for pancreatic cancer using non-viral nanovectors. Biomater Sci 9(20):6673–6690. https://doi.org/10.1039/D1BM90082J
Lopez-Cantu DO, Wang X, Carrasco-Magallanes H, Afewerki S, Zhang X et al (2022) From bench to the clinic: the path to translation of nanotechnology-enabled mRNA SARS-CoV-2 vaccines. Nano-Micro Lett 14(1):41. https://doi.org/10.1007/s40820-021-00771-8
Mandriota G, Di Corato R, Benedetti M, De Castro F, Fanizzi FP, Rinaldi R (2019) Design and application of cisplatin-loaded magnetic nanoparticle clusters for smart chemotherapy. ACS Appl Mater Interfaces 11:1864–1875. https://doi.org/10.1021/acsami.8b18717
Mason TJ (2011) Therapeutic ultrasound an overview. Ultrason Sonochem 18:847–848
Mizrak A, Bolukbasi MF, Ozdener GB, Brenner GJ, Madlener S, Erkan EP, Ströbel T, Breakefield XO, Saydam O (2013) Genetically engineered microvesicles carrying suicide mRNA/protein inhibit schwannoma tumor growth. Mol Ther 21(1):101–108
Mo R, Jiang T, DiSanto R, Tai W, Gu Z (2014) ATP-triggered anticancer drug delivery. Nat Commun 5:3364

Mohammadinejad R, Dehshahri A, Madamsetty VS, Zahmatkeshan M, Tavakol S et al (2020) In vivo gene delivery mediated by non-viral vectors for cancer therapy. J Control Release 325: 249–275. https://doi.org/10.1016/j.jconrel.2020.06.038
Mout R et al (2012) Surface functionalization of nanoparticles for nanomedicine. Chem Soc Rev 41(7):2539–2544
Münz F, Perez RL, Trinh T, Sisombath S, Weber K-J, Wuchter P, Debus J, Saffrich R, Huber PE, Nicolay NH (2018) Human mesenchymal stem cells lose their functional properties after paclitaxel treatment. Sci Rep 8(1):312
Naik S, Shreya AB, Raychaudhuri R, Pandey A, Lewis SA, Hazarika M, Bhandary SV, Rao BSS, Mutalik S (2021) Small interfering RNAs (siRNAs) based gene silencing strategies for the treatment of glaucoma: Recent advancements and future perspectives. Life Sci 264:118712
Navya PN, Kaphle A, Srinivas SP et al (2019) Current trends and challenges in cancer management and therapy using designer nanomaterials. Nano Convergence 6:23
Nisato RE, Tille JC, Jonczyk A, Goodman SL, Pepper MS (2003) Alpha beta 3 and alpha beta 5 integrin antagonists inhibit angiogenesis in vitro. Angiogenesis 6:105–119. https://doi.org/10.1023/B:AGEN.0000011801.98187.f2
Nordstrom R, Zhu L, Harmark J et al (2021) Quantitative Cryo-TEM reveals new structural details of doxil-like PEGylated liposomal doxorubicin formulation. Pharmaceutics 13(1):123
Orleth A, Mamot C, Rochlitz C, Ritschard R, Alitalo K, Christofori G et al (2016) Simultaneous targeting of VEGF-receptors 2 and 3 with immunoliposomes enhances therapeutic efficacy. J Drug Target 24:80–89. https://doi.org/10.3109/1061186x.2015.1056189
Paskeh MD, Abad HS, Mahabady MK, Orouei S, Hushmandi K, Entezari M, Hashemi M et al (2022) Overcoming doxorubicin resistance in cancer: siRNA-loaded nanoarchitectures for cancer gene therapy. Life Sci 298:120463
Pasparakis G, Tsitsilianis C (2020) LCST polymers: thermoresponsive nanostructured assemblies towards bioapplications. Polymer 211:123146
Pellow C, Abenojar EC, Exner AA, Zheng G, Goertz DE (2020) Concurrent visual and acoustic tracking of passive and active delivery of nanobubbles to tumors. Theranostics 10:11690–11706
Peng P-C, Hong R-L, Tsai T, Chen C-T (2019) Co-encapsulation of chlorin e6 and chemotherapeutic drugs in a pegylated liposome enhance the efficacy of tumor treatment: pharmacokinetics and therapeutic efficacy. Pharmaceutics 11(11):617
Perrault SD, Walkey C, Jennings T, Fischer HC, Chan WC (2009) Mediating tumor targeting efficiency of nanoparticles through design. Nano Lett 9:1909–1915. https://doi.org/10.1021/nl900031y
Poudel K, Banstola A, Tran TH, Thapa RK, Gautam M, Ou W, Pham LM et al (2020) Hyaluronic acid wreathed, trio-stimuli receptive and on-demand triggerable nanoconstruct for anchored combinatorial cancer therapy. Carbohydr Polym 249:116815
Qi X, Wang G, Wang P, Pei Y, Zhang C, Yan M, Wei P et al (2022) Transferrin protein corona-modified CuGd core-shell nanoplatform for tumor-targeting photothermal and chemodynamic synergistic therapies. ACS Appl Mater Interfaces 14(6):7659–7670
Rajendra KJ, Stylianopoulos T (2010) Delivering nanomedicine to solid tumors. Nat Rev Clin Oncol 7:653
Rong L, Qin S-Y, Zhang C et al (2018) Biomedical applications of functional peptides in nano-systems. Mater Today Chem 9:91–102
Sathiyaseelan A, Saravanakumar K, Manivasagan P, Jeong MS, Jang E-S, Wang M-H (2022) Folic acid conjugated chitosan encapsulated palladium nanoclusters for NIR triggered photothermal breast cancer treatment. Carbohydr Polym 280:119021
Scabia V, Ayyanan A, De Martino F, Agnoletto A, Battista L, Laszlo C, Treboux A, Zaman K et al (2022) Estrogen receptor positive breast cancers have patient specific hormone sensitivities and rely on progesterone receptor. Nat Commun 13(1):3127
Scaranti M, Cojocaru E, Banerjee S, Banerji U (2020) Exploiting the folate receptor α in oncology. Nat Rev Clin Oncol 17(6):349–359

Schoellhammer CM, Traverso G (2016) Low-frequency ultrasound for drug delivery in the gastrointestinal tract. Expert Opin Drug Deliv 13(8):1045–1048
Schultheis B, Strumberg D, Kuhlmann J, Wolf M, Link K, Seufferlein T, Kaufmann J, Feist M, Gebhardt F, Khan M et al (2020) Safety, efficacy and pharcacokinetics of targeted therapy with the liposomal RNA interference therapeutic Atu027 combined with gemcitabine in patients with pancreatic adenocarcinoma. A randomized phase Ib/IIa study. Cancer 12:3130
Sharma P, Bhargava M (2013) Applications and characteristics of nanomaterials in industrial environment. Res Dev 3(4):63–72
Shukla RS, Jain A, Zhao Z, Cheng K (2016) Intracellular trafficking and exocytosis of a multi-component siRNA nanocomplex. Nanomedicine 12(5):1323–1334
Sidhom K, Obi PO, Saleem A (2020) A review of exosomal isolation methods: is size exclusion chromatography the best option? Int J Mol Sci 21(18):6466
Stubbs M, McSheehy PM, Griffiths JR, Bashford CL (2000) Causes and consequences of tumour acidity and implications for treatment. Mol Med Today 6(1):15–19
Tharkar P, Varanasi R, Wong WSF, Jin CT, Chrzanowski W (2019) Nano-enhanced drug delivery and therapeutic ultrasound for cancer treatment and beyond. Front Bioeng Biotechnol 7:324. https://doi.org/10.3389/fbioe.2019.00324. PMID: 31824930; PMCID: PMC6883936.
Thomas C, Gustafsson JÅ (2011) The different roles of ER subtypes in cancer biology and therapy. Nat Rev Cancer 11(8):597–608
Tian H, Zhang T, Qin S et al (2022) Enhancing the therapeutic efficacy of nanoparticles for cancer treatment using versatile targeted strategies. J Hematol Oncol 15:132
Timin AS, Muslimov AR, Lepik KV, Okilova MV, Tcvetkov NY, Shakirova AI, Afanasyev BV, Gorin DA, Sukhorukov GB (2017) Intracellular breakable and ultrasound-responsive hybrid microsized containers for selective drug release into cancerous cells. Part Syst Charact 34: 1600417
Timko BP, Arruebo M, Shankarappa SA, McAlvin JB, Okonkwo OS, Mizrahi B et al (2014) Near-infrared-actuated devices for remotely controlled drug delivery. Proc Natl Acad Sci U S A 111: 1349–1354
Tokatlian T, Segura T (2010) siRNA applications in nanomedicine. Wiley Interdiscip Rev 2(3): 305–315
Tolcher AW, Papadopoulos KP, Patnaik A, Rasco DW, Martinez D, Wood DL, Fielman B, Sharma M, Janisch LA, Brown BD et al (2015) Safety and activity of DCR-MYC, a first-in-class Dicer-substrate small interfering RNA (DsiRNA) targeting MYC, in a phase I study in patients with advanced solid tumors. J Clin Oncol 33:11006
Tong R, Chiang HH, Kohane DS (2013) Photoswitchable nanoparticles for in vivo cancer chemo-therapy. Proc Natl Acad Sci U S A 110:19048–19053
Tran VL, Novell A, Tournier N, Gerstenmayer M, Schweitzer-Chaputa A, Mateos C, Jego B, Bouleau A, Nozach H, Winkeler A, Kuhnast B, Larrat B, Truillet C (2020) Impact of blood-brain barrier permeabilization induced by ultrasound associated to microbubbles on the brain delivery and kinetics of cetuximab: an immunoPET study using 89Zr-cetuximab. J Control Release 328:304–312
Van den Bijgaart RJE, Eikelenboom DC, Hoogenboom M, Fütterer JJ, Den Brok MH, Adema GJ (2017) Thermal and mechanical high-intensity focused ultrasound: perspectives on tumor ablation, immune effects and combination strategies. Cancer Immunol Immunother 66:247–258
Venturoli D, Rippe B (2005) Ficoll and dextran vs. globular proteins as probes for testing glomerular permselectivity: effects of molecular size, shape, charge, and deformability. Am J Physiol Renal Physiol 288:F605–F613. https://doi.org/10.1152/ajprenal.00171.2004
Verma V, Stellacci F (2010) Effect of surface properties on nanoparticle–cell interactions. Small 6(1):12–21
Waite CL, Roth CM (2012) Nanoscale drug delivery systems for enhanced drug penetration into solid tumors: current progress and opportunities. Crit Rev Biomed Eng 40(1):21–41. https://doi.org/10.1615/critrevbiomedeng.v40.i1.20. PMID: 22428797; PMCID: PMC3639314.
Wang et al (2010) Delivery of siRNA therapeutics: barriers and carriers. AAPS J 12(4):492–503

Wang X, Niu D, Li P, Wu Q, Bo X, Liu B et al (2015) Dual-enzyme-loaded multifunctional hybrid nanogel system for pathological responsive ultrasound imaging and T2-weighted magnetic resonance imaging. ACS Nano 9:5646–5656

Wang Y, Shang W, Niu M, Tian J, Xu K (2019a) Hypoxia-active nanoparticles used in tumor theranostic. Int J Nanomedicine 14:3705–3722. https://doi.org/10.2147/IJN.S196959. PMID: 31190820; PMCID: PMC6535445.

Wang P, Wang H, Huang Q, Peng C, Yao L, Chen H, Qiu Z, Yifan W, Wang L, Chen W (2019b) Exosomes from M1-polarized macrophages enhance paclitaxel antitumor activity by activating macrophages-mediated inflammation. Theranostics 9(6):1714

Wang Z, Chen J, Little N, Lu J (2020) Self-assembling prodrug nanotherapeutics for synergistic tumor targeted drug delivery. Acta Biomater 111:20–28

Wang F, Li J, Chen C, Qi H, Huang K, Hu S (2022) Preparation and synergistic chemo-photothermal therapy of redox-responsive carboxymethyl cellulose/chitosan complex nanoparticles. Carbohydr Polym 275:118714

Weissleder R et al (2005) Cell-specific targeting of nanoparticles by multivalent attachment of small molecules. Nat Biotechnol 23(11):1418–1423

Wong C, Stylianopoulos T, Cui J, Martin J, Chauhan VP, Jiang W et al (2011) Multistage nanoparticle delivery system for deep penetration into tumor tissue. Proc Natl Acad Sci U S A 108:2426–2431. https://doi.org/10.1073/pnas.1018382108

Wu M, Chen W, Chen Y, Zhang H, Liu C, Deng Z, Sheng Z et al (2018a) Focused ultrasound-augmented delivery of biodegradable multifunctional nanoplatforms for imaging-guided brain tumor treatment. Adv Sci 5(4):1700474

Wu Y, Xu J, Chen J, Zou M, Rusidanmu A, Yang R (2018b) Blocking transferrin receptor inhibits the growth of lung adenocarcinoma cells in vitro. Thorac Cancer 9(2):253–261

Wu M, Ding Y, Li L (2019) Recent progress in the augmentation of reactive species with nanoplatforms for cancer therapy. Nanoscale 11(42):19658–19683

Wu L, Zhou W, Lin L, Chen A, Feng J et al (2022) Delivery of therapeutic oligonucleotides in nanoscale. Bioact Mater 7:292–323. https://doi.org/10.1016/j.bioactmat.2021.05.038

Xiao K, Liu Q, Suby N, Xiao W, Agrawal R, Vu M, Zhang H et al (2021) LHRH-targeted redox-responsive crosslinked micelles impart selective drug delivery and effective chemotherapy in triple-negative breast cancer. Adv Healthc Mater 10(3):e2001196

Xiao Y, Chen J, Zhou H, Zeng X, Ruan Z, Pu Z, Jiang X et al (2022) Combining p53 mRNA nanotherapy with immune checkpoint blockade reprograms the immune microenvironment for effective cancer therapy. Nat Commun 13(1):758

Xu J, Gao F, Li L, Ma H, Fan Y, Liu W, Guo S et al (2013) Gelatin-mesoporous silica nanoparticles as matrix metalloproteinases-degradable drug delivery systems in vivo. Microporous Mesoporous Mater 182:165–172

Xu Y, Guo Y, Zhang C, Zhan M, Jia L, Song S, Jiang C et al (2022) Fibronectin-coated metal-phenolic networks for cooperative tumor chemo-/chemodynamic/immune therapy via enhanced ferroptosis-mediated immunogenic cell death. ACS Nano 16(1):984–996

Yadav S, Sharma AK, Kumar P (2020) Nanoscale self-assembly for therapeutic delivery. Front Bioeng Biotechnol 8:1–24

Yan L, Li X (2016) Biodegradable stimuli-responsive polymeric micelles for treatment of malignancy. Curr Pharm Biotechnol 17:227–236

Yan H, Dong J, Huang X, Du X (2021) Protein-gated upconversion nanoparticle-embedded mesoporous silica nanovehicles via diselenide linkages for drug release tracking in real time and tumor chemotherapy. ACS Appl Mater Interfaces 13:29070–29082

Yang Q, Jones SW, Parker CL, Zamboni WC, Bear JE, Lai SK (2014) Evading immune cell uptake and clearance requires PEG grafting at densities substantially exceeding the minimum for brush conformation. Mol Pharm 11:1250–1258. https://doi.org/10.1021/mp400703d

Yang Q, He C, Zhang Z, Tan L, Liu B, Zhu Z, Shao Z et al (2016) Redox-responsive flower-like micelles of poly (L-lactic acid)-b-poly (ethylene glycol)-b-poly (L-lactic acid) for intracellular drug delivery. Polymer 90:351–362

Yang N, Xiao W, Song X, Wang W, Dong X (2020) Recent advances in tumor microenvironment hydrogen peroxide-responsive materials for cancer photodynamic therapy. Nano Lett 12(1):15

Yao Y, Zhou Y, Liu L, Xu Y, Chen Q, Wang Y, Wu S, Deng Y, Zhang J, Shao A (2020) Nanoparticle-based drug delivery in cancer therapy and its role in overcoming drug resistance. Front Mol Biosci 7:193. https://doi.org/10.3389/fmolb.2020.00193

Yubo Z, Li J, Gao W, Xie N (2022) Exosomes as anticancer drug delivery vehicles: prospects and challenges. Front Biosci 27(10):293

Zeng W, Wu L, Sun Y, Wang Y, Wang J, Ye D (2021) Ratiometric imaging of MMP-2 activity facilitates tumor detection using activatable near-infrared fluorescent semiconducting polymer nanoparticles. Small 17(36):e2101924

Zhang J, Wang Y, Chen J, Liang X, Han H, Yang Y, Li Q et al (2017a) Inhibition of cell proliferation through an ATP-responsive co-delivery system of doxorubicin and Bcl-2 siRNA. Int J Nanomedicine 12:4721–4732

Zhang N, Li J, Hou R, Zhang J, Wang P, Liu X, Zhang Z (2017b) Bubble-generating nano-lipid carriers for ultrasound/CT imaging-guided efficient tumor therapy. Int J Pharm 534:251–262

Zhang D, Cai L, Wei X, Wang Y, Shang L et al (2021) Multiplexed CRISPR/Cas9 quantifications based on bioinspired photonic barcodes. Nano Today 40:101268. https://doi.org/10.1016/j.nantod.2021.101268

Zhang Y, Li J, Gao W, Xie N (2022a) Exosomes as anticancer drug delivery vehicles: prospects and challenges. Front Biosci 27(10):293. https://doi.org/10.31083/j.fbl2710293. PMID: 36336863.

Zhang Y, Liu Q, Zhang X, Huang H, Tang S, Chai Y et al (2022b) Recent advances in exosome-mediated nucleic acid delivery for cancer therapy. J Nanobiotechnol 20(1):279

Zhang Q, Kuang G, Li W, Wang J, Ren H, Zhao Y (2023) Stimuli-responsive gene delivery nanocarriers for cancer therapy. Nano Lett 15(1):44. https://doi.org/10.1007/s40820-023-01018-4

Zhao J, Castranova V (2011) Toxicology of nanomaterials used in nanomedicine. J Toxicol Environ Health B 14(8):593–632

Zheng N, Liu W, Li B, Nie H, Liu J, Cheng Y, Wang J et al (2019) Co-delivery of sorafenib and metapristone encapsulated by CXCR4-targeted PLGA-PEG nanoparticles overcomes hepatocellular carcinoma resistance to sorafenib. J Exp Clin Cancer Res 38(1):232

Zhu J, Li H, Xiong Z et al (2018) Polyethyleneimine-coated manganese oxide nanoparticles for targeted tumor PET/MR imaging. ACS Appl Mater Interfaces 10(41):34954–34964

Zhu Y-X, Jia H-R, Duan Q-Y, Fu-Gen W (2021) Nanomedicines for combating multidrug resistance of cancer. Wiley Interdiscip Rev Nanomed Nanobiotechnol 13(5):e1715

Zylberberg C, Matosevic S (2016) Pharmaceutical liposomal drug delivery: a review of new delivery systems and a look at the regulatory landscape. Drug Deliv 23:3319–3329

SPRINGER NATURE

GPSR Compliance

The European Union's (EU) General Product Safety Regulation (GPSR) is a set of rules that requires consumer products to be safe and our obligations to ensure this.

If you have any concerns about our products, you can contact us on ProductSafety@springernature.com

In case Publisher is established outside the EU, the EU authorized representative is:

Springer Nature Customer Service Center GmbH
Europaplatz 3
69115 Heidelberg, Germany

The manufacturer's authorised representative in the EU is Springer
Nature Customer Service Centre GmbH, Europaplatz 3, 69115 Heidelberg,
Germany. If you have any concerns regarding our products, please
contact ProductSafety@springernature.com

Printed and bound by CPI Group (UK) Ltd, Croydon, CR0 4YY
30/09/2026
02225322-0004